MECHANICAL
CATALYSIS

MECHANICAL CATALYSIS

Methods of Enzymatic, Homogeneous, and Heterogeneous Catalysis

GERHARD F. SWIEGERS

A JOHN WILEY & SONS, INC., PUBLICATION

Copyright © 2008 by John Wiley & Sons, Inc. All rights reserved

Published by John Wiley & Sons, Inc., Hoboken, New Jersey
Published simultaneously in Canada

No part of this publication may be reproduced, stored in a retrieval system, or transmitted in any form or by any means, electronic, mechanical, photocopying, recording, scanning, or otherwise, except as permitted under Sections 107 or 108 of the 1976 United States Copyright Act, without either the prior written permission of the Publisher, or authorization through payment of the appropriate per-copy fee to the Copyright Clearance Center, Inc., 222 Rosewood Drive, Danvers, MA 01923, (978) 750-8400, fax (978) 750-4470, or on the web at www.copyright.com. Requests to the Publisher for permission should be addressed to the Permissions Department, John Wiley & Sons, Inc., 111 River Street, Hoboken, NJ 07030, (201) 748-6011, fax (201) 748-6008, or online at http://www.wiley.com/go/permission.

Limit of Liability/Disclaimer of Warranty: While the publisher and author have used their best efforts in preparing this book, they make no representations or warranties with respect to the accuracy or completeness of the contents of this book and specifically disclaim any implied warranties of merchantability or fitness for a particular purpose. No warranty may be created or extended by sales representatives or written sales materials. The advice and strategies contained herein may not be suitable for your situation. You should consult with a professional where appropriate. Neither the publisher nor author shall be liable for any loss of profit or any other commercial damages, including but not limited to special, incidental, consequential, or other damages.

For general information on our other products and services or for technical support, please contact our Customer Care Department within the United States at (800) 762-2974, outside the United States at (317) 572-3993 or fax (317) 572-4002.

Wiley also publishes its books in variety of electronic formats. Some content that appears in print may not be available in electronic formats. For more information about Wiley products, visit our web site at www.wiley.com.

Library of Congress Cataloging-in-Publication Data:

Mechanical catalysis : methods of heterogeneous, homogeneous, and enzymatic catalysis / editor, Gerhard F. Swiegers.
    p. cm.
  Includes index.
  ISBN 978-0-470-26202-3 (cloth)
 1. Heterogeneous catalysis.   2. Catalysis.   3. Enzymes–Biotechnology.
 I. Swiegers, Gerhard F.
  QD505.M39 2008
  541'.395–dc22
                                                   2008007592

Printed in the United States of America
10 9 8 7 6 5 4 3 2 1

To Jane, Matthew, and Daniel

CONTENTS

PREFACE xxi

CONTRIBUTORS xxv

GLOSSARY xxvii

1 Introduction to Thermodynamic (Energy-Dependent) and Mechanical (Time-Dependent) Processes: What Are They and How Are They Manifested in Chemistry and Catalysis?
Gerhard F. Swiegers

- 1.1 Thermodynamic (*Energy-Dependent*) and Mechanical (*Time-Dependent*) Processes   1
- 1.2 What Is a Thermodynamic Process?   5
- 1.3 What Is a Mechanical Process?   7
- 1.4 The Difference between *Energy-Dependent* (Thermodynamic) and *Time-Dependent* (Mechanical) Processes   9
  - 1.4.1 Time-Dependent (Mechanical) Processes Are Path-Reliant and Spatiotemporal in Character   9
  - 1.4.2 Time-Dependent (Mechanical) Processes Have a Flat Underlying Energy Landscape (or Are Unaffected by the Energy Landscape)   10

	1.4.3	Time-Dependent (Mechanical) Processes Display Deterministic Chaos; This Causes Them to be Stochastic and Complex	11
	1.4.4	Time-Dependent (Mechanical) Processes Often Involve Synergies of Action	14
	1.4.5	Time-Dependent (Mechanical) Processes Characterize Numerous Aspects of Human Experience	15
1.5	*Time*- and *Energy-Dependence* in Chemistry and Catalysis		17
	1.5.1	The Origin of Time- and Energy-Dependent Processes in Chemistry	17
	1.5.2	Examples of Time-Dependent Processes in Chemistry	19
	1.5.3	Time- and Energy-Dependent Processes in Catalysis	21
	1.5.4	Is There Such a Thing as a Time-Dependent Process in Catalysis?	23
1.6	The Aims, Structure, and Major Findings of this Series		24
	1.6.1	Summary of the Key Finding: Many Enzymes Seem to be Time-Dependent Catalysts	25
	1.6.2	The Aims and Structure of this Series. Summary: Other Major Findings of this Series	28
References			34

2 Heterogeneous, Homogeneous, and Enzymatic Catalysis. A Shared Terminology and Conceptual Platform. The Alternative of Time-Dependence in Catalysis 37

Gerhard F. Swiegers

2.1	Introduction: The Problem of Conceptually Unifying Heterogeneous, Homogeneous, and Enzymatic Catalysis? Trends in Catalysis Science		37
2.2	Background: What Is Heterogeneous, Homogeneous, and Enzymatic Catalysis		38
	2.2.1	Homogeneous and Heterogeneous Catalysis	38
	2.2.2	Hybrid Homogeneous–Heterogeneous Catalysts	40
	2.2.3	Enzymatic Catalysis	41
	2.2.4	Theories and Mimicry of Enzymatic Catalysis	42
2.3	Distinctions Within Homogeneous Catalysis: *Single-Centered* and *Multicentered* Homogeneous Catalysis		44
	2.3.1	Single-Centered Homogeneous Catalysts. Most Manmade Homogeneous Catalysts Are Single-Centered Catalysts	44
	2.3.2	Multicentered Homogeneous Catalysts: Most Enzymes Are Multicentered Homogeneous Catalysts	46
2.4	The Distinction between *Single-Site/Multisite* Catalysts and *Single-Centered/MultiCentered* Catalysts in *Heterogeneous Catalysis*: An Important Convention Used in This Series		48

	2.4.1	A Key Convention Used in This Series: A Catalytic Site Is a Collection of Atoms about Which a Reaction Is Catalyzed. A Catalytic Center Is an Atom Within that Site Which Binds and Facilitates the Transformation of a Reactant	48
2.5		The Alternative of *Time-Dependence* in Catalysis	48
References			52

3 A Conceptual Description of Energy-Dependent ("Thermodynamic") and Time-Dependent ("Mechanical") Processes in Chemistry and Catalysis — 55

Gerhard F. Swiegers

3.1	Introduction		55
3.2	Theoretical Considerations: Common Processes in Uncatalyzed Reactions		56
	3.2.1	Reactions as Collisions Between Molecules	56
	3.2.2	The Fundamental Origin of Energy-Dependent and Time-Dependent Reactions	57
	3.2.3	Time-Dependent and Energy-Dependent Domains Were First Observed in Unimolecular Gas-Phase Reactions	58
	3.2.4	The Pathway of the Reaction Is also Controlled by the Least-Likely Step in the Sequence	59
	3.2.5	Transition State Theory (TST) Describes the Pathway and Rate of Energy-Dependent Reactions. Transition State Theory Corresponds to the High-Pressure Limit of Hinshelwood–RRK Theory	61
	3.2.6	Time-Dependent Reactions in the Liquid Phase: Some Examples	63
	3.2.7	The Transition between Energy-Dependence and Time-Dependence as a Function of Temperature. Curvature in Arrhenius Plots	65
	3.2.8	Methods of Creating Time-Dependent Reactions	67
	3.2.9	Summary: The Key Properties of Time-Dependent and Energy-Dependent Reactions	68
3.3	Theoretical Considerations: Common Processes in Catalyzed Reactions		68
	3.3.1	Catalyzed Reactions Are More Likely to be Time-Dependent than Are Uncatalyzed Reactions	68
	3.3.2	Catalysis Changes the Reaction Processes	69
	3.3.3	Physical Manifestation of Time- and Energy-Dependence in Catalysts	72
	3.3.4	The Distinction Between Time-Dependent Catalysis and Diffusion-Controlled Catalysis	72
	3.3.5	Energy-Dependent and Time-Dependent Control of Catalysis	73

	3.3.6	The Influence of the Product Release Step	74
3.4		Conclusions: Energy- and Time-Dependent Catalysis	75
Acknowledgments			75
References			76

4 Time-Dependence in Heterogeneous Catalysis. Sabatier's Principle Describes Two Independent Catalytic Realms: Time-Dependent ("Mechanical") Catalysis and Energy-Dependent ("Thermodynamic") Catalysis — 77

Gerhard F. Swiegers

4.1	Introduction			77
4.2	Sabatier's Principle in Heterogeneous Catalysis			79
	4.2.1	Volcano Plots		79
	4.2.2	Some Important Points about Volcano Plots		82
	4.2.3	Time-Dependent Catalysis in Volcano Plots		82
		4.2.3.1	How Is Time-Dependence Created on the Left-Hand Side of the Volcano Plot?	82
		4.2.3.2	Why Do Volcano Plots Slope Upward on the Left	84
		4.2.3.3	The Rate-Determining Step in a Time-Dependent Catalyst	86
		4.2.3.4	The Physical Manifestation of Time-Dependent Catalysis. "Saturation" of a Time-Dependent Catalyst	87
	4.2.4	Energy-Dependent Catalysis in Volcano Plots		88
		4.2.4.1	How Is Energy-Dependence Created on the Right-Hand Side of the Volcano Plot?	88
		4.2.4.2	Why Do Volcano Plots Slope Downward on the Right?	88
		4.2.4.3	The Rate-Determining Step in an Energy-Dependent Catalyst	89
		4.2.4.4	The Physical Manifestation of Energy-Dependence. Saturation in an Energy-Dependent Catalyst	89
	4.2.5	The Physical Origin of Sabatier's Principle		89
	4.2.6	Other Plots Illustrating Sabatier's Principle		90
	4.2.7	Modeling of Volcano Plots		91
	4.2.8	Reaction Pathway as a Function of the Most-Favored Transition State		92
4.3	Exceptions to Sabatier's Principle			93
4.4	Sabatier's Principle in Homogeneous Catalysis			93
4.5	Conclusions. Sabatier's Principle Describes Two Independent Catalytic Domains: Energy- and Time-Dependent Catalysis			94
Acknowledgments				95
References				95

5 Time-Dependence in Homogeneous Catalysis. 1. Many Enzymes Display the Hallmarks of Time-Dependent ("Mechanical") Catalysis. Nonbiological Homogeneous Catalysts Are Typically Energy-Dependent ("Thermodynamic") Catalysts   97

Robin Brimblecombe, Jun Chen, Junhua Huang, Ulrich T. Mueller-Westerhoff, and Gerhard F. Swiegers

5.1	Introduction	97
5.2	Historical Background: Are Enzymes Generally *Energy-Dependent* or *Time-Dependent* Catalysts?	99
5.3	The Methodology of This Chapter: Identify, Contrast, and Rationalize the Common Processes Present in Biological and Nonbiological Homogeneous Catalysts	100
5.4	Does Michaelis–Menten Kinetics in Enzymes Indicate that They Are *Time-Dependent* Catalysts?	102
	5.4.1 Michaelis–Menten Kinetics	102
	5.4.2 Kinetics in Most Nonbiological Catalysts	103
	5.4.3 The Contradiction of Saturation Kinetics in Enzymes	103
	5.4.4 Saturation in Time- and Energy-Dependent Catalysts. Saturation Kinetics Is Necessarily an Indication of Time-Dependence	104
	5.4.5 Physical Studies of the Rate Processes in Enzymes Are Consistent with a Time-Dependent Action	106
	5.4.6 A Time-Dependent Catalyst Cannot Become an Energy-Dependent Catalyst, or vice versa, Without Changing the Temperature or Chemically Altering the Reactivity of the Reactants	107
	5.4.7 The Current View of Michaelis–Menten Kinetics Is Flawed by an Unwarranted Assumption	107
	5.4.8 Summary: Michaelis–Menten Kinetics Is Characteristic of Time-Dependent Catalysis. Time-Dependent Catalysis Provides an Explanation for Michaelis–Menten Kinetics in Enzymes	109
5.5	Other General Characteristics of Catalysis by Enzymes and Comparable Nonbiological Homogeneous Catalysts	110
	5.5.1 Enzymes Employ Weak and Dynamic Individual Binding Interactions with Their Substrates. Nonbiological Catalysts Do Not	110
	5.5.2 Enzymes Display Transition State Complementarity. Nonbiological Catalysts Do Not	111
	5.5.3 Enzymatic Catalysis Is "Structure-Sensitive." Nonbiological Catalysis Is "Structure-Insensitive"	112
	5.5.4 Enzymes Transform Catalytically Unconventional Groups into Potent Catalysts. Nonbiological Catalysts Use Only Conventional Catalytic Groups	113
	5.5.5 Enzymes Catalyze Forward and Reverse Reactions. Nonbiological Catalysts Do Not	113

		5.5.6	Enzymes Display High Selectivity and Activity. Nonbiological Catalysts Do Not	115
		5.5.7	Enzymes Display Convergent Synergies. Nonbiological Catalysts Display Complementary Synergies	115
		5.5.8	Summary	116
	5.6	Rationalization of the Underlying Processes. The Mechanism of Action in Time-Dependent and Energy-Dependent Catalysts		117
		5.6.1	Common Processes in Multicentered Homogeneous Catalysts	117
		5.6.2	The Influence of the Strength of the Individual Catalyst–Reactant Binding Interactions	119
		5.6.3	The Coexistence of Transition State Complementarity, Structure-Sensitive Catalysis, and Unconventional Catalytic Groups in Enzymes Is Caused by their Weak Individual Binding Interactions	122
		5.6.4	The Origin of the Time-Dependence and the Synergies of Enzymes	123
		5.6.5	The Mechanism of Time-Dependence in Enzymes Resolves the Contradiction of a Kinetically Observed Rapidly Forming and Dissociating Intermediate in the Face of Strong Overall Substrate Binding	125
		5.6.6	Catalysis in Enzymes Involves Synchronization of Enzyme Binding and Enzyme Flexing	125
		5.6.7	Summary: The Origin of the General Properties of Enzymes	127
		5.6.8	Catalysis in Nonbiological Analogues Depends on the Activation Energy E_a	127
		5.6.9	Enzymatic Selectivity and Synergies Derive from Time-Dependence	128
		5.6.10	Enzymatic Activity Is Consistent with Time-Dependence	129
	5.7	All Generalizations Support Time-Dependence in Enzymes		129
	5.8	Time-Dependence in a Nonbiological Catalyst Generates the Distinctive Properties of Enzymes		130
	5.9	Conclusion: Many Enzymes Are Time-Dependent Catalysts		133
	Acknowledgments			134
	References			134
6	**Time-Dependence in Homogeneous Catalysis. 2. The General Actions of Time-Dependent ("Mechanical") and Energy-Dependent ("Thermodynamic") Catalysts**			**137**
	Robin Brimblecombe, Jun Chen, Junhua Huang, Ulrich T. Mueller-Westerhoff, and Gerhard F. Swiegers			
	6.1	Introduction		137
	6.2	*Time-* and *Energy-Dependent*, Multicentered Homogeneous Catalysts		139

6.3	The Action of *Energy-Dependent*, Multicentered Homogeneous Catalysts	141
6.4	The Action of *Time-Dependent*, Multicentered Homogeneous Catalysts	146
	6.4.1 The Activation Energy E_a Does Not Provide a True Measure of the Threshold Energy in Time-Dependent Catalysts	148
	6.4.2 Weak and Dynamic Binding and Activation Is Sufficient to Fulfill the Threshold Energy in Time-Dependent Catalysts	149
	6.4.3 Transition State Formation in a Time-Dependent Catalyst Can Be Thought of as a Coordinated Mechanical Process	150
	6.4.4 Time-Dependent Catalysts Are Machine-Like (Mechanical) in Their Catalytic Action	150
	6.4.5 The Origin of Michaelis–Menten Kinetics in Time-Dependent Catalysts	151
	6.4.6 Time-Dependent Catalysts like Many Enzymes Display All of the Characteristic Hallmarks of Mechanical Processes	153
	6.4.7 Additional Insights into Enzymatic Catalysis: The Bidirectionality of Enzymatic Catalysis Originates from the Mechanical Nature of the Catalytic Action	154
	6.4.8 Additional Insights into Enzymatic Catalysis: Many Enzymes Select the First-Encountered Transition State, Rather than the Lowest Energy Transition State	155
6.5	The Importance of Recognizing *Time-Dependent* Catalysis	155
6.6	Time-Dependent Catalysis Is Very Different to Energy-Dependent Catalysis and Therefore Seems Unfamiliar	156
6.7	Conclusions for Biology	157
6.8	Conclusions for Homogeneous Catalysis	157
6.9	The "Ideal" Homogeneous Catalyst	158
6.10	Conclusions for the Conceptual Unity of the Field of Catalysis	158
Acknowledgments		159
References		159

7 Unifying the Many Theories of Enzymatic Catalysis. Theories of Enzymatic Catalysis Fall into Two Camps: Energy-Dependent ("Thermodynamic") and Time-Dependent ("Mechanical") Catalysis — **161**

Gerhard F. Swiegers

7.1	Introduction	161
7.2	Theories of Enzymatic Catalysis	163
	7.2.1 Adsorption Theory	163
	7.2.2 "Lock-and-Key" Theory	163
	7.2.3 Haldane's Strain Theory	164
	7.2.4 Pauling's Theory of Transition State Complementarity	165

	7.2.5	Koshland's Induced Fit Theory. Fersht's Concept of Stress and Strain	165
	7.2.6	Intramolecularity	165
	7.2.7	Orbital Steering	167
	7.2.8	Entropy Traps	168
	7.2.9	The Proximity (Propinquity) Effect	168
	7.2.10	"Coupled" Protein Motions	168
	7.2.11	The Spatiotemporal Hypothesis	169
7.3	Theories Explaining Enzymatic Catalysis Fall into Two Camps: Energy-Dependent and Time-Dependent Catalysis		169
	7.3.1	Haldane's Strain Theory and Fersht's Concept of Stress and Strain Are Valid Explanations for Rate Accelerations but Do Not Seem to be Responsible for the Rate Accelerations of Many Enzymes	171
	7.3.2	Theories Based on Reaction Entropy Are Valid Explanations for Rate Accelerations but Do Not Seem to be Behind the Rate Accelerations of Many Enzymes	172
	7.3.3	Experiments Studying Intramolecular Reaction Rates Were Probably Often Conceptually Contradictory	172
	7.3.4	Theories of "Coupled" Protein Motions and Machine-Like Catalytic Actions Seem to Be Generally Accurate Descriptions of Enzymatic Catalysis	173
7.4	Studies Verifying Pauling's Theory in Model Systems Are Correct, but Describe Energy-Dependent and not Time-Dependent Catalysis		174
7.5	The Anomaly Described in the Spatiotemporal Hypothesis Originates, in Part, from the Onset of Time-Dependence		176
Acknowledgments			177
References			177

8 Synergy in Heterogeneous, Homogeneous, and Enzymatic Catalysis. The "Ideal" Catalyst — 181

Gerhard F. Swiegers

8.1	Introduction		181
8.2	Synergy in Heterogeneous Catalysts		183
8.3	Single-Centered Nonbiological Homogeneous Catalysts and Their *'Mutually Enhancing'* Synergies		184
	8.3.1	Facial Selectivity in Single-Centered Catalysts	184
	8.3.2	Energy-Dependent, Single-Centered Homogeneous Catalysts Display 'Mutually Enhancing' Synergies	187
	8.3.3	The Synergies in Time-Dependent, Single-Centered Homogeneous Catalysts	188
	8.3.4	The Selectivity of Single-Centered Catalysts	189
8.4	Multicentered, Energy-Dependent Homogeneous Catalysts and Their *Functionally Complementary* Synergies		190

8.5	Enzymes and Their *Functionally Convergent* Synergies	194
8.6	Biomimetic Chemistry and Its *Pseudo-Convergent* Synergies	197
	8.6.1 Cyclodextrin-Appended Epoxidation Catalysts: Pseudo-Convergence in a Nonbiological, Multicentered Catalyst	198
8.7	The Spectrum of Synergistic Action in Homogeneous Catalysis	200
	8.7.1 The Relationship Between Complementary and Convergent Synergies	202
	8.7.2 The Ideal Catalyst	203
8.8	Synergy in Catalysis Is Conceptually Related to Other Synergistic Processes in Human Experience	205
	References	206

9 A Conceptual Unification of Heterogeneous, Homogeneous, and Enzymatic Catalysis — 209

Gerhard F. Swiegers

9.1	Introduction	209
9.2	*Diffusion-Controlled* and *Reaction-Controlled* Catalysis	210
9.3	The Diversity of Catalytic Action in Heterogeneous Catalysts	211
9.4	The Diversity of Catalytic Action in Nonbiological Homogeneous Catalysts	212
9.5	The Diversity of Catalytic Action in Enzymes	214
9.6	Heterogeneous Catalysis and Enzymatic Catalysis Has, Effectively, Involved Combinatorial Experiments that Have Produced Time-Dependent Catalysts. Nonbiological Homogeneous Catalysis Has Not	214
9.7	Homogeneous and Enzymatic Catalysts Are the 3-D Equivalent of 2-D Heterogeneous Catalysts	215
9.8	A Conceptual Unification of Heterogeneous, Homogeneous, and Enzymatic Catalysis	216
	References	218

10 The Rational Design of Time-Dependent ("Mechanical") Homogeneous Catalysts. A Literature Survey of Multicentered Homogeneous Catalysis — 219

Junhua Huang and Gerhard F. Swiegers

10.1	Introduction	219
10.2	The Rational Design of Time-Dependent Homogeneous Catalysts	221
	10.2.1 Design Criteria for a Time-Dependent Homogeneous Catalyst	221
	10.2.2 The Problem of Simultaneously Identifying Suitable Catalytic Groups and Their Active Spatial Arrangement	223

	10.2.3	Time-Dependent Homogeneous Catalysis May Conceivably Be Achieved by Mimicry of a Natural Time-Dependent Catalyst	225
	10.2.4	Time-Dependent Homogeneous Catalysis May Conceivably Be Achieved in the form of a Combinatorial Experiment Involving a "Statistical Proximity" Effect	226
		10.2.4.1 A Time-Dependent Combinatorial Catalyst May Display Unique Kinetics	228
		10.2.4.2 Previous Attempts at Concentration-Based Biomimetic Catalysis Involved Energy-Dependent Systems	229
	10.2.5	Time-Dependent Catalysis May Be Useful in Transformations of Small Gaseous Molecules	230
	10.2.6	Why Do We Need New Time-Dependent Catalysts?	230
10.3	Elements of Rational Design in Multicentered Catalysis	230
	10.3.1	Modes of Binding in Multicentered Catalysts	230
	10.3.2	Optimizing the Spatial Arrangement of Catalytic Groups	231
		10.3.2.1 Intramolecular Catalysts	231
		10.3.2.2 Intermolecular Catalysts	233
		10.3.2.3 Unconventional Approaches to Optimizing the Spatial Organization of Catalytic Groups	233
	10.3.3	Creating Functionally Convergent Catalysts	234
		10.3.3.1 Practical Approaches to Achieving Functionally Convergent Catalysis	234
10.4	A Review of Nonbiological, Multicentered Molecular Catalysts Described in the Chemical Literature	235
	10.4.1	Intramolecular Catalysts	235
		10.4.1.1 Functionally Convergent Catalysis (Class A Type): Cofacial and Capped Metalloporphyrins as Oxygen Reduction Catalysts	235
		10.4.1.2 Functionally Convergent Catalysis (Class B Type): [1.1]Ferrocenophanes and Related Compounds as Hydrogen Generation Catalysts	241
		10.4.1.3 Pseudoconvergent Catalysis: Supramolecular, Bifunctional Catalysts of Organic Reactions	245
		10.4.1.4 Probable Functionally Convergent Catalysis: Rhodium-Phosphine Hydroformylation Catalysts	248
		10.4.1.5 Possible Functionally Convergent Catalysis: Ruthenium-Based Water Oxidation Catalysts	249
		10.4.1.6 Functionally Complementary Catalysis: Intramolecular Epoxidation Catalysts	252
		10.4.1.7 Metal Clusters in Multicentered Molecular Catalysis: Triruthenium Dodecacarbonyl Hydrogenation Catalysts	252

	10.4.1.8	Statistical Approaches to Functionally Convergent Catalysis: Macromolecular Intramolecular Catalysts	254
10.4.2	Intermolecular Catalysts		259
	10.4.2.1	Functionally Complementary Catalysis	259
	10.4.2.2	Statistical Approaches to Functionally Convergent Catalysis: Concentration Effects in Intermolecular Catalysts	260
	10.4.2.3	Statistical Approaches to Functionally Convergent Catalysis: Self-Assembled, Supramolecular Catalysts	261
10.4.3	Footnote: Unexpected Mechanistic Changes in Multicentered Catalysts		262

Acknowledgments 263
References 263

11 Time-Dependent ("Mechanical"), Nonbiological Catalysis. 1. A Fully Functional Mimic of the Water-Oxidizing Center (WOC) in Photosystem II (PSII) 267

Robin Brimblecombe, G. Charles Dismukes, Greg A. Felton, Leone Spiccia, and Gerhard F. Swiegers

11.1	Introduction		267
11.2	The Physical and Chemical Properties of the *Cubanes* **1a-b**		273
	11.2.1	Chemical Structures	273
	11.2.2	Stepwise Hydride Abstraction, Leading to Water Release	275
	11.2.3	Dioxygen Generation	275
	11.2.4	A Possible Catalytic Cycle	277
	11.2.5	Other Reactions	277
	11.2.6	Summary	278
11.3	Nafion Provides a Means of Solubilizing and Immobilizing Hydrophobic Metal Complexes		278
11.4	Photoelectrochemical Cells and Dye-Sensitized Solar Cells for Water-Splitting		279
11.5	Photocatalytic Water Oxidation by *Cubane* **1b** Doped into a Nafion Support		282
	11.5.1	Solution Electrochemistry	282
	11.5.2	Electrochemistry of **1b** Doped into a Nafion Membrane	283
	11.5.3	Electrocatalytic Effects Are Observed Under CV Conditions	283
	11.5.4	A Photo-Electrocatalytic Effect Is Observed at 1.00 V (vs. Ag/AgCl)	284
	11.5.5	If the Photocurrent Is Caused by Water Oxidation Catalysis, This Involves a Decrease in the Overpotential of 0.4 V	285

		11.5.6	The Photocurrent Is Observed only in the Presence of Water. The System Saturates at Low Water Content, Consistent with a Time-Dependent Catalytic Action	286

11.5.6 The Photocurrent Is Observed only in the Presence of Water. The System Saturates at Low Water Content, Consistent with a Time-Dependent Catalytic Action — 286

11.5.7 The pH Dependence of the Photocurrent Is Consistent with Water Oxidation — 287

11.5.8 Bulk Water Is a Reactant and Oxygen Is Generated — 287

11.5.9 The Quantity of Gas Generated Matches the Current Obtained. Notable Turnover Frequencies Are Implied — 287

11.5.10 Photocurrent as a Function of the Illumination Wavelength — 290

11.5.11 The Photoaction Spectrum of the Catalysis Corresponds to the Main LMCT Absorption Peak of **1b** — 290

11.6 The Challenge of Dye-Sensitized Water-Splitting — 291

11.7 The Mechanism of the Catalysis — 292

11.8 Conclusions — 293

References — 294

12 Time-Dependent ("Mechanical"), Nonbiological Catalysis. 2. Highly Efficient, "Biomimetic" Hydrogen-Generating Electrocatalysts — 297

Jun Chen, Junhua Huang, Gerhard F. Swiegers, Chee O. Too, and Gordon G. Wallace

12.1 Introduction — 297

12.2 Monomer and Polymer Preparation — 301

12.3 Catalytic Experiments — 302

12.3.1 PPy-**9** and PPy-**12** Display Anodic Shifts in the Most Positive Potential for Hydrogen Generation — 302

12.3.2 PPy-**9** and PPy-**12** Increase the Rate of Hydrogen Generation on Pt by ca. **7**-Fold after **12** h at -0.44 V — 304

12.3.3 PPy-**9** and PPy-**12** Increase the Rate of Hydrogen Generation on Pt per Unit Area by ca. 3.5-Fold — 307

12.3.4 The Mechanism of Catalysis in PPy-**9**. Is PPy-**9** a Combinatorial ("Statistical Proximity") Catalyst? — 308

12.3.5 Polypyrrole Is Likely Involved in the Catalytic Cycle — 309

12.3.6 Other Evidence for the Involvement of Polypyrrole in the Catalytic Cycle — 311

12.3.7 The Pyrrole in Polypyrrole Is a Powerful, Time-Dependent, Combinatorial, "Statistical Proximity" Catalyst — 313

12.4 Conclusions: A Combinatorial "Statistical Proximity" Catalyst Was Obtained as a Bulk, Hybrid Homogeneous–Heterogeneous Catalyst — 316

Acknowledgments — 317

References — 317

13 Time-Dependent ("Mechanical"), Nonbiological Catalysis. 3. A Readily Prepared, Convergent, Oxygen-Reduction Electrocatalyst   319

Jun Chen, Gerhard F. Swiegers, Gordon G. Wallace, and Weimin Zhang

- 13.1 Introduction   319
- 13.2 Cofacial Diporphyrin Oxygen-Reduction Catalysts   321
- 13.3 Vapor-Phase Polymerization of Pyrrole as a Means of Immobilizing High Concentrations of Monomeric Catalytic Groups at an Electrode Surface   323
- 13.4 Preparation and Catalytic Properties of PPy-3   324
  - 13.4.1 Vapor-Phase Preparation of Polypyrrolle-Co Tetraphenylporphyrin, PPy-3   324
  - 13.4.2 Electrochemistry of, and Oxygen Reduction by, Polypyrrolle-Co Tetraphenylporphyrin, PPy-3   324
  - 13.4.3 Rotating Disk Electrochemistry (RDE) of Polypyrrolle-Co Tetraphenylporphyrin, PPy-3   326
  - 13.4.4 Rotating Ring Disk Electrochemistry (RRDE) of Polypyrrolle-Co Tetraphenylporphyrin, PPy-3   328
  - 13.4.5 The Product Distribution Relative to the Proportion of **3** in the Polypyrrolle-Co Tetraphenylporphyrin, PPy-3   329
- 13.5 PPy-3 as a Fuel Cell Catalyst   330
  - 13.5.1 PPy-3 on Carbon Fiber Paper   330
  - 13.5.2 Electrochemical Characterization of PPy-3 on Carbon Fiber Paper   330
  - 13.5.3 Morphology of the PPy-3 Carbon Fiber Composite Film   330
  - 13.5.4 Oxygen-Reduction Catalysis by the PPy-3 Carbon Fiber Composite Film in Simple Fuel Cell Test Apparatus   331
- 13.6 Conclusions   334
- References   335

Appendix A   Why Is Saturation Not Observed in Catalysts that Display Conventional Kinetics?   337

Appendix B   Graphical Illustration of the Processes Involved in the Saturation of Molecular Catalysts   341

Index   347

PREFACE

For over 200 years chemistry has focused, with an ever-increasing clarity, on the principles, laws, and notions of thermodynamics. From Sadi Carnot (the Carnot heat engine), to Willard Gibbs (Free Energy), to Henry Eyring (Transition State Theory), there has been an almost continuous growth in our understanding of how thermodynamics affects and drives chemical reactions. The intellectual brilliance and the empirical power of these insights have captured the imagination of the chemical scientific community. Thermodynamics is today a well-established and deeply entrenched, core discipline within chemistry.

However, in the shadow of this development has resided another, almost forgotten field of chemistry involving chemical reactions that occur without regard to thermodynamics. Such reactions do not involve a thermodynamic equilibrium and are, consequently, said to be *nonequilibrium* in character. They are today typically termed "diffusion-controlled" reactions and are driven by the simple, mechanical interaction of the reactants. That is, such reactions are subject to the principles, notions, and laws of *mechanics*, not of thermodynamics.

The field of mechanics pre-dates thermodynamics by several hundred years. Nevertheless, mechanics is currently undergoing a veritable renaissance, with its principles finding application in a growing and diverse range of fields, both within physical science and outside of it. The most high profile of these applications has, undoubtedly, been the development of *Chaos Theory* in mathematics. However, applications exist in fields as far removed as economics, traffic management, and weather forecasting.

The new science of mechanics has not as yet significantly influenced chemistry and the chemical sciences. In particular, it has not been seriously considered in the field of catalysis, which involves species that accelerate chemical reactions without themselves being transformed.

In this work, we examine the principles of mechanics as they apply in chemistry and, more particularly, catalysis. We show that many reactions have a purely mechanical character, in that their rates and pathways are determined by the mechanics of reactant encounter, rather than by its thermodynamics. In other words, the reaction is governed by the nature of the *first* encounter between the reactants, rather than by the encounter having the *lowest energy*.

Since catalysts act to diminish the thermodynamic barrier to reaction during a collision between reactant molecules, one may reasonably expect such "mechanical" reactions to be more prevalent in the field of catalysis. We will show that the use of these mechanical reactions is, indeed, widespread and that "mechanical catalysis" is, in fact, well known, being found in many of the enzyme catalysts of biology as well as in certain heterogeneous catalysts. The reaction rates and pathways of such catalyzed reactions are determined by the ability of the catalyst to bring its bound reactants into mechanical contact with each other.

"Mechanical catalysts" of this type display many characteristic hallmarks of mechanical devices. Thus, they typically depend on synchronization in the actions of their components to achieve a catalytic effect. They also require complementarity in the shape and arrangement of their components. Once operational, they act like machines, dynamically taking in reactants, bringing them into a reactive encounter with each other, and then ejecting the resulting products. As with many mechanical devices, they carry out these actions because of mechanical impulses—conformational flexing in the case of molecular catalysts—rather than as a response to an energy gradient or barrier. Like a machine, they may operate in either the forward or the reverse direction.

The action of such catalysts does not mediate a thermodynamic equilibrium. Rather, it is an intrinsically *nonequilibrium* event. It happens because the impulses and the mechanics of the situation are such that *no alternative exists but for it to happen*. In asking why "mechanical catalysis" of this type takes place, one may as well ask: Why does a bottle-making machine make bottles? The answer is that it makes bottles because when it is switched on and raw materials are inserted, the mechanical actions within the machine are such that bottles are generated.

Many present-day chemists will, undoubtedly, find such a conceptualization of chemistry and catalysis difficult to understand. But one has to ask: If we can accept that a bottle-making machine makes bottles whenever it is activated, why should we not accept that the mechanical actions within some catalysts are such that they have no alternative but to repeatedly make products when they are activated? If humans can make machines, why can nature also not do so in the form of many enzymes? Such enzymes are, arguably, true molecular machines of the type that humans have only aspired to create. In the latter chapters of this work, we will examine how one may set about designing and making such molecular machines.

PREFACE

The material in this book is intended for undergraduate, graduate, and professional scientists in the fields of hetereogeneous (*chemical engineers*), homogeneous (*chemists*), and enzymatic catalysis (*biochemists*). It also interfaces with the nascent field of *complex systems science* and, through it, with a diversity of other fields involving *complex* and *chaotic* phenomena. We hope that it will be of interest to all and that it will stimulate wide-ranging intellectual discussions.

In the preparation of this book, I was assisted by several very able and talented individuals. To begin with, I would like to thank the many reviewers who read all or parts of this work and provided comments.

Thanks are due to the external reviewers who considered the entirety of the initial draft. These reviewers were: Bob Williams (Oxford University), Chuck Dismukes (Princeton University), Fred Menger (Emory University), Steve Benkovic (Pennsylvania State University), and John Moss (University of Cape Town, South Africa). Although all made useful and helpful comments, I must particularly single out Bob Williams, with whom I had a long and meaningful interaction. Bob was exceedingly helpful in drawing out and illuminating the key issues of this work. Thanks also to Chuck Dismukes who hosted me for a six-week visit to Princeton University in September–October 2005. Several of the concepts discussed here came out of that visit. Thank you to Bob and Chuck, and thanks to all of the external reviewers.

Several internal reviewers also helped by reading and considering the individual chapters. These reviewers are mentioned in the acknowledgments at the end of each chapter. I must, however, especially thank Wolf Sasse (CSIRO) who spent many long hours working with me through the very first drafts. To Wolf and all of the internal reviewers: thank you. This work could not have been completed without your input.

The coauthors of the individual chapters are, of course, also highly deserving of acknowledgment. Ulli Mueller-Westerhoff was my PhD supervisor many years ago. He instilled in me a curiosity about the role of kinetic dynamism in catalysis that proved to be truly invaluable in the construction of this book. Thanks are also due to Jewel Huang, Jun Chen, Rob Brimblecombe, Greg Felton, and Weimin Zhang, who as postdoctoral fellows and PhD students, helped put several of the individual chapters together. They put up with what can only be described as an extraordinary conceptual ambiguity during the years that we were trying to understand the processes at play. It is to their credit that this ambiguity did not deter them from giving of their best. My long-suffering collaborators on this project should also be noted. Gordon Wallace, Chee Too, Leone Spiccia, and Chuck Dismukes acted as very capable cosupervisors and supported this work throughout, even when its significance was not yet clear.

The final set of acknowledgments goes to friends and family. Thank you to my friends Don and Anne Vandegrift of Doylestown, Pennsylvania, who so very kindly provided me with accommodation during a recent extended trip to the United States. Some chapters in this volume were born in their townhouse.

Last, but certainly not least, my most sincere and heartfelt gratitude goes to my family for its support and patience during my consuming fascination with this topic. Without the love, dedication, and backing of my family, I would never have completed this work. Thank you to my wife, Jane, and my sons Matthew and Daniel. This book is dedicated to them.

GERRY SWIEGERS

Melbourne, Australia, 2007

CONTRIBUTORS

Robin Brimblecombe, School of Chemistry, Monash University, Clayton, Victoria 3800, Australia

Jun Chen, ARC Centre of Excellence for Electromaterials Science, Intelligent Polymer Research Institute, University of Wollongong, New South Wales 2522, Australia

G. Charles Dismukes, Department of Chemistry and the Princeton Environmental Institute, Princeton University, Princeton, New Jersey 08544, U.S.A.

Greg A. Felton, Department of Chemistry and the Princeton Environmental Institute, Princeton University, Princeton, New Jersey 08544, U.S.A.

Junhua Huang, Division of Energy Technology, Commonwealth Scientific and Industrial Research Organisation (CSIRO), Bag 10, Clayton, Melbourne, Victoria 3169, Australia

Ulrich T. Mueller-Westerhoff, Department of Chemistry, University of Connecticut, Storrs, Connecticut 06268, U.S.A.

Leone Spiccia, School of Chemistry, Monash University, Clayton, Victoria 3800, Australia

Gerhard F. Swiegers, Datatrace DNA Pty Ltd and Division of Molecular and Health Technologies, Commonwealth Scientific and Industrial Research Organisation (CSIRO), Bag 10, Clayton, Melbourne, Victoria 3169, Australia

Chee O. Too, ARC Centre of Excellence for Electromaterials Science, Intelligent Polymer Research Institute, University of Wollongong, New South Wales 2522, Australia

Gordon G. Wallace, ARC Centre of Excellence for Electromaterials Science, Intelligent Polymer Research Institute, University of Wollongong, New South Wales 2522, Australia

Weimin Zhang, ARC Centre of Excellence for Electromaterials Science, Intelligent Polymer Research Institute, University of Wollongong, New South Wales 2522, Australia

GLOSSARY

Cascade

Cascade process

Chain

Chain process

"Cascade" is a mathematical term for a chain process in which each step initiates the next step by a tangible and an observable action–reaction event. Such processes are not influenced by underlying energy considerations; what you see is what you get. Examples include billiard balls colliding, stacked dominos falling, evolution, family trees, and laissez-faire economic systems. Chain processes of this type are dependent on well-defined and path-dependent spatial interactions in time. The more perfectly these interactions are set up, the smaller is the impulse needed to initiate them. Cascade or chain processes typically occur when the individual steps of the process are not subject to strong underlying energy drivers, that is, where the underlying energy landscape for the step is relatively flat. (See Chapter 1.)

Activation energy

E_a

The threshold energy barrier that must be crossed in going from the reactants to the products in a collision between reactant molecules. The term *activation energy* (E_a) originates in collision theory where it describes the proportion of molecules in a gas that collide with sufficient energy to react in an uncatalyzed process. In common usage, it is used to describe the energy barrier to reactions in general.

	Activation energy is calculated from the Arrhenius equation: $$k = A \exp\left(\frac{-E_a}{kT}\right)$$ (See Chapter 3.)
Collision frequency **Frequency factor**	The term *collision frequency* originates in collision theory where it describes the frequency with which molecules in a gas collide with each other. In this work, the term is used to describe the rate at which reactants are brought into physical contact with each other by a catalyst (or the rate at which they are pulled apart from each other). For the purposes of clarity and consistency of argument, the term *collision* is used here in preference to the term *encounter* that is normally employed in the liquid phase. *Collision frequency* is formally stated to be the pre-exponential term A in the Arrhenius equation: $$k = A \exp\left(\frac{-E_a}{kT}\right)$$ (See Chapter 3.)
Energy-dependent **"energetic"** **Energetically controlled** **Thermodynamic**	A thermodynamic process whose pathway and speed is determined by an intangible, underlying energy landscape or energy field. Examples include objects reacting to a gravitational or magnetic field (e.g., a ball rolling into a well, water flowing downhill, or a compass needle). In chemistry, this process refers to reactions (catalyzed and uncatalyzed) whose rates and pathways are determined by the energy change (and energy landscape) involved in the reaction and, especially, about the activated complex (the reaction transition state). Such processes are governed by Transition State Theory. (See Chapters 1 and 3.)
Time-dependent **"Temporal"** **Temporally controlled** **Mechanical**	A mechanical process whose pathway and speed is determined by the tangible interaction of individual elements in space and time. In chemistry and catalysis, time-dependence refers to a reaction whose rate and pathway are determined by the frequency with which the reactants collide (the "collision frequency"). Such reactions are independent of the underlying energy landscape of the reaction and are not subject to Transition State Theory. (See Chapters 1 and 3.)

Stochastic **Chaos** **Complex** **Complexity** **Complex systems science**	*Stochastic*, from the Greek *stochos* or "goal," means of, relating to, or characterized by randomness. In the context of this work, stochastic behavior comes from the dominance of time-dependent (mechanical) processes in chemical or catalytic systems containing extremely large numbers of molecules. When a chain process runs in many parallel, randomly intersecting pathways, unpredictable and complicated overall outcomes may be created. This unpredictability is known as *chaos* or *complexity*; it serves as the basis for *complex systems science*. The mathematical study of randomness is termed probability theory and is used in everything from actuarial science to weather forecasting. An example of a stochastic process in chemistry is the pressure of a gas. Although each molecule is moving in a well-defined path, the motion of a collection of molecules is computationally and practically unpredictable. A large enough set of gas molecules will exhibit stochastic characteristics, such as filling the container, exerting equal pressure, diffusing along concentration gradients, and so on. These characteristics are termed *emergent properties* of the system. (See Chapter 1.)
Synergy **Synergies** **Synergistic**	A term that refers to the situation when the whole of a system exceeds the sum of the properties of its individual components. In multifunctional catalysis, synergy describes a catalyst whose properties exceed the sum of the catalytic properties of its individual catalytic groups. (See Chapter 1.)
Complementarity **Functional complementarity**	A form of synergy in which every element of a system carries out a complementary task. The synergy comes from the simple combination of these tasks. (See Chapter 8.)
Convergence **Functional convergence**	A form of synergy in which every element of a system depends on the other elements to carry out its specific task. The tasks are interrelated and must be performed simultaneously. All elements, therefore, succeed or fail together. If they succeed, it is because their actions have "converged"; that is, they have occured simultaneously and in a mutually reinforcing manner. (See Chapter 8.)
"Pseudo-convergence"	A form of synergy that has the superficial appearance of convergence but actually involves complementarity

	that happens to often occur simultaneously. It can be distinguished from true convergence by the fact that simultaneous action is not essential to the overall effect. (See Chapter 8.)
"Mutual enhancement"	A form of synergy in which one element passively assists another without becoming formally involved. Generally, the passive participant acts by deforming the underlying energy landscape so as to favor a particular outcome in the process. (See Chapter 8.)
Cooperativity (positive or negative)	A biological term to describe "mutually enhancing" synergies. (See Chapter 8.)
Combinatorial	In the context of this work, the term "combinatorial" refers to multiple possible reactions or processes, either catalyzed or uncatalyzed, that occur simultaneously, in parallel, and in the same vessel or location. In a combinatorial reaction, multiple species, arrangements, or chemical structures may be present, each of which may be involved in a different reaction or process. In effect, a "library" of different species, arrangements, or structures are available to undertake a plurality of different processes, including ones of interest.

1

INTRODUCTION TO THERMODYNAMIC (ENERGY-DEPENDENT) AND MECHANICAL (TIME-DEPENDENT) PROCESSES: WHAT ARE THEY AND HOW ARE THEY MANIFESTED IN CHEMISTRY AND CATALYSIS?

GERHARD F. SWIEGERS

1.1 THERMODYNAMIC (*ENERGY-DEPENDENT*) AND MECHANICAL (*TIME-DEPENDENT*) PROCESSES

Many processes in human experience involve a drive to the lowest possible energy [1–3]. A ball that is thrown into the air will fall to the ground and roll to the lowest part of the local landscape. Water, similarly, flows downhill, giving us rivers, lakes, and rainfall. The driving force in these systems is an imperative to reduce the *energy* that is stored within the components of the system. Thus, a ball held at a great height has a *gravitational potential energy*, which means that it has the potential to undergo movement under the influence of gravity [4]. When the ball is released, this *potential energy* is converted into motion (known as *kinetic energy*) [4]. As the ball nears the ground during its fall, its potential energy decreases in accordance with its declining potential to fall. Once the ball has struck the ground, it will roll into the deepest hollow available to it (Fig. 1.1). It will then no longer have any gravitational potential energy.

Mechanical Catalysis: Methods of Enzymatic, Homogeneous, and Heterogeneous Catalysis,
Edited by Gerhard F. Swiegers
Copyright © 2008 John Wiley & Sons, Inc.

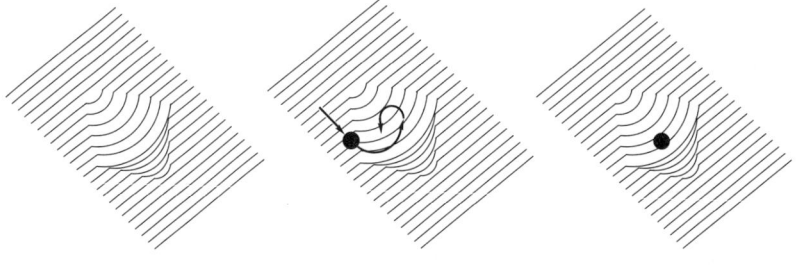

    A surface   A ball rolling onto the surface   The ball at rest on the surface

Figure 1.1. Example of a *thermodynamic (energy-dependent) process*: A ball will roll into the deepest available hollow under the influence of a gravitational energy field.

The imperative for movement and actions driven by a *release of energy* is a well-established principle of physical science that is embodied in the field and laws of *thermodynamics* [1–3].

The term *thermodynamics* derives from the Greek *therme*, which means "heat," and *dynamis*, which means "force." Since *heat* is, effectively, "energy in transit" and *dynamics* relates to "actions" or "processes," *thermodynamics* can be considered to be *the study of how changes in energy create actions or processes*.

Historically, *thermodynamics* developed out of a need to increase the efficiency of early steam engines [5]. Steam engines were the first machines that harnessed an energy differential (in the form of heat) to create motion. That is, heat moved from a hot boiler to a cold condensor and work was extracted.

Thermodynamics is today, however, a far broader discipline that encompasses much more than merely studying the conversion of energy into actions [1–3]. Nevertheless, some utility exists in this original, foundation concept. To capture this principle without any confusion in respect to the modern field of *thermodynamics*, we will, in this work, refer to *all processes that are driven by a release of energy* as being **energy-dependent** in character. That is, any process that originates out of an *external energy gradient* will be said to be to be "energy-dependent" in character. When we use the term *thermodynamic*, we will be referring to a process that is driven by an overall release of energy on the part of the system components.

It is important to make this definition because not all processes or actions are driven by the release of energy. Many processes unfold as a result of physical action-and-reaction sequences [6–8]. For example, a billiard ball struck by another billiard ball on a flat table will be propelled in a certain direction, with a particular momentum, depending on the angle and the force with which it was struck (Fig. 1.2). No overall *release* of energy by the two billiard balls occurs. A physical *transferral* of energy simply occurs from the one ball to the other. That is, the momentum of the one ball is transferred, in some part, to the other ball during the collision. The combined energies (momenta) of the two billiard balls before and after the collision remain unchanged.

Another example in this vein is a series of dominoes stacked in a line. When the first domino falls, it knocks over the next one, which knocks over the next one, and so

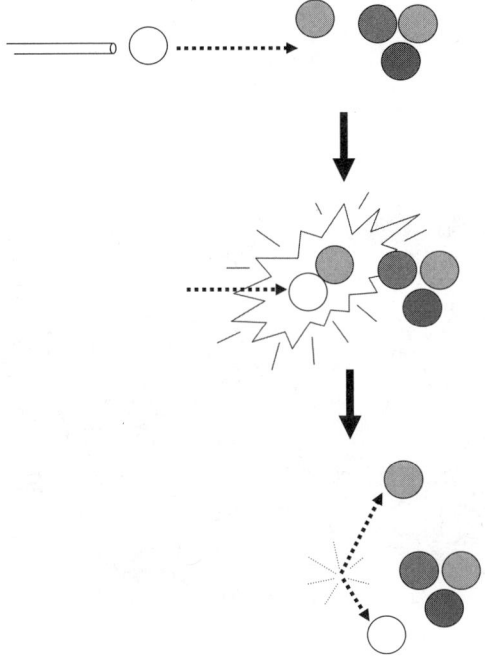

Figure 1.2. Example of a *mechanical (time-dependent) process*: Two billiard balls collide with each other.

on (Fig. 1.3). Thus, the motion occurs as a result of a simple transferral of energy from one domino to the next. Although each domino falls because of gravity, the collapse sequence of the dominoes is not driven by a gravitational energy gradient *per se*. Rather it occurs because each step in this sequence is favored by the physical arrangement of the dominoes; that is, each domino is set up so that it will knock over the next one when it falls.

Such a process need not lead to an overall reduction in the energy of the system components. For example, the dominoes may be stacked up the side of a gently sloping hill, so that the initial impulse which starts the sequence of events is ultimately transmitted to a point of *higher* gravitational potential energy. Moreover, chain sequences of this type may be made to branch into two or more different, parallel progressions, each of which has a different energy outcome. For example, a line of dominoes may be split into two parallel lines, one of which is stacked up the gently sloping hill and the other down the hill. Each of these branches will ultimately produce a different outcome in terms of gravitational potential energy.

Sequential action-and-reaction events of this type are governed by a different set of physical laws, known as the laws of *mechanics* [6,7].

Mechanical processes are dominated by action–reaction sequences in which energy is transferred between the system components but is preserved overall. The system components do not interact in a way that minimizes their overall energy.

Figure 1.3. Example of a *mechanical (time-dependent) process*: (Top) Schematic illustrating a line of dominoes collapsing. (Bottom) Photo showing a curving line of stacked dominoes. [Copyright (C) 2000, 2001, 2002 Free Software Foundation, Inc. 51 Franklin St, Fifth Floor, Boston, MA 02110-1301 USA Everyone is permitted to copy and distribute verbatim copies of this license document, but changing it is not allowed.]

Rather, they interact in a purely physical manner in which their combined energies are redistributed according to the individual circumstances present. This redistribution is purely *statistical*, which means that it is a predictable mathematical function of the way in which the interaction occurs. One type of interaction will *always* give a certain result. Another will *always* give a different result.

Mechanical processes can be distinguished in several ways from processes that are driven by a release of energy. One of the most important is that they are *time-dependent* and not *energy-dependent*. That is, the physical unfolding of the process is dependent on the *time that has elapsed* and the *time required for each step* rather than on the relative energies of the steps [6–8]. In other words, mechanical processes proceed in a perfectly predictable way according to the time that has passed since the initial state and the time taken for each step in the sequence [6–8]. Thus, for example, the movement and spatial position of the cogs in a mechanical engine is entirely dependent on and determined by the time that has elapsed since the start of the last cycle. By contrast, the unfolding of an energy-dependent process can only be predicted by taking into account the energy gradient that exists; it is this gradient that determines how events will proceed.

The field of *mechanics* was first developed by Isaac Newton in the seventeenth century [7]. It has since expanded from the original *classical mechanics* to include,

1.2 WHAT IS A THERMODYNAMIC PROCESS?

inter alia, *relativistic mechanics* (mechanics near to the speed of light) and *quantum mechanics* (mechanics at extremely small distance scales).

Although modern mechanics therefore covers a large diversity of subdisciplines, its original, underlying concept is one of motion or actions created by a physical action–reaction sequence [7]. This most fundamental principle is actually an example of a philosophical concept known as *causality* or *determinism* [8],[1] that predates the formulation of *mechanics* (which was only its first scientific example). To capture the above concept without any confusion regarding the many and various forms of mechanics, we will, in this work, refer to *all processes that are driven by physical action–reaction sequences* as being **time-dependent** in character. Where we use the term *mechanical*, it should be understood that this is used to signify a process or event that is brought about by a physical action–reaction sequence.

1.2 WHAT IS A THERMODYNAMIC PROCESS?

Thermodynamics involves the study of changes in temperature, volume, and pressure, as well as the transfer of energy, on physical systems [1–3]. Of central importance to this study are the laws of thermodynamics, which theorize that energy can be exchanged between a *system* and its *surroundings* [1–3]. The term *system* is defined in this context as a defined set of particles or system components. The *surroundings* are everything else in the universe. Between the system and its surroundings is a *boundary*, which may be the walls of a container in, for example, a steam engine (Fig. 1.4). Alternatively, the boundary may be an imaginary delimitation of a particular volume within which the system is contained. Across the boundary, energy is transferred, to thereby increase or decrease the *internal energy* of the system. That is, the combined, overall internal energies of the system components is altered and changed.

The energy that moves between a system and its surroundings may take several forms [1–3]. The simplest is *heat*, also known as *enthalpy*, which is given the mathematical symbol H. The laws of thermodynamics postulate the existence of another form of energy, known as *entropy*, which is assigned the symbol S. It is beyond the scope of this work to describe entropy in detail [1–3,9] but, effectively, it involves changes in the extent of disorder within the system. The combined effect of enthalpy and entropy is known as the *free energy* of the system or G [1–3,10].

With these tools, thermodynamics describes how systems respond to their surroundings. In other words, if the surroundings favor a transfer of energy over the boundary, the system will respond accordingly. The manner in which it responds will depend on the free energy differential ΔG that exists between the system and its surroundings. ΔG is, formally, a combination of terms that describe the change

---

[1]*Determinism* describes the philosophy that every event is caused by a preceeding event in an action–reaction-type relationship. *Causality* examines the relationship between causes and effects; it is fundamental to all natural science. *Causality* is also studied in philosophy, computer science, and statistics.

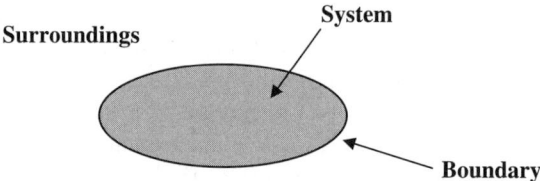

Figure 1.4. *A thermodynamic system*: A boundary exists between the system and its surroundings.

in enthalpy ΔH and in entropy ΔS according to the equation:

$$\Delta G = \Delta H + T\Delta S \tag{1.1}$$

where T is the absolute temperature in Kelvin [1–3].

At the most fundamental level, the field of thermodynamics is therefore concerned with how systems respond under the influence of an external energy imperative. In effect, how do they minimize their internal energy in response to the surroundings?

A *thermodynamic process* can, consequently, be defined as the *energetic* evolution of a system as it proceeds from an initial state to a final state. Once it reaches its final state, it is said to be in *thermodynamic equilibrium*.

It is important to note that the *total* energy of the system and its surroundings remain constant during a thermodynamic process. This is in accord with the first law of thermodynamics, which states that energy is neither created nor destroyed; it can only be transferred. In the case of a thermodynamic system, it is transferred into or out of the system by, for example, heating or cooling, compression or expansion, or some other form of energy transfer. In *mechanics*, by contrast, energy transfer occurs via a physical force, which causes displacement.

The original science of the relationship between heat and power is today termed *classical thermodynamics*. With the understanding of the atomic and molecular nature of chemistry in the late 1800s, thermodynamics has since expanded to encompass a molecular interpretation. This field is called *statistical thermodynamics*, and it extends the laws of thermodynamics to systems that comprise extremely large numbers of individual components, such as are found in typical chemical systems [11]. Statistical thermodynamics describes the macroscopic properties of chemical systems that are subject to energy gradients.

It is important to note that the term *statistical* is used here in a more superficial sense to describe systems that contain extremely large numbers of components. It does not refer to an essential attribute of a thermodynamic system. Thermodynamic systems are not fundamentally statistical in character; that is, the outcome of a thermodynamic process is not dependent on the nature of the interaction between two elements, as it is in mechanics. Rather, it depends on the energy gradient that exists.

The field of thermodynamics is, most generally, concerned with *reversible* systems at equilibrium or moving toward equilibrium. However, a subdiscipline, termed *nonequilibrium thermodynamics*, examines irreversible transformations and open systems [12]. In effect, this discipline studies *time-dependent thermodynamic systems*. The theory and practice of *nonequilibrium thermodynamics* is considerably less developed than that of *equilibrium thermodynamics*, so that it is most sucessful in the study of stationary states which effectively involve no time variation. Methods for handling and predicting the full extent of non-equilibrium systems are notably absent.

1.3 WHAT IS A MECHANICAL PROCESS?

In his original conceptualization of *mechanics*, Newton described three *laws of motion* [6,7]. The most pertinent of these for our purposes, is the *third law*, known as the *law of reciprocal actions*. This law states that *for every action, there is an equal but opposite reaction*. That is, if a body collides with another and thereby changes the motion of the other, that body will undergo an equal but contrary change in its motion.

The billiard ball that initiates a collision with another billiard ball is therefore, itself, diverted in a different and complementary direction. Similarly, when each domino in a line falls, it transfers its momentum entirely to the next domino.

The influence of Newtonian mechanics on scientific thinking in the 1600s was profound and long lasting. For over three centuries, it embodied the principle of *determinism* in science [8]. *Determinism* states that every event or action is the inevitable result of a preceding event or action. In other words, motion and structure in the material world is typically governed by cause-and-effect. Thus, according to deterministic thinking, the universe is like a perfect machine, unfolding over time in a preordained way. This belief is exemplified in the Newtonian model of a collision between billiard balls, which proceeds from the initial conditions over time in a fully predictable sequence, rather like a movie that may be played forward or backward in time.

The deterministic concept of cause-and-effect remains a core philosophy of physical science.

A *mechanical process* can, consequently, be defined as the *physical* progress of a system as it proceeds from an initial state over time.

The key philosophical feature of a mechanical process that is different than a thermodynamic one is the absence of a *boundary* between the system and the surroundings. Such a boundary is not needed nor relevant in a mechanical system because the system does not act to minimize its internal energy in response to its surroundings. In fact, in a mechanical system, the system is not at all influenced by its surroundings. The interaction is purely physical and involves only the components that collide with each other.

The relationship between thermodynamics and mechanics can be thought of as follows. Thermodynamics involves nonphysical, "intangible" influences, whereas mechanics focuses on the tangible, physical effects at play. We use the term *intangible* here to signify something that can be discerned only by its influence on events (e.g., gravity).

In general, when the transfer of energy, including heat or work, between a system and its surroundings is substantial, then the system will typically (but not always) adapt to it in accordance with the laws of thermodynamics. However, when the transfer of energy is small or insignificant, the system will normally evolve according to the laws of mechanics.

Thus, it is important to note that *within* the boundary of a thermodynamic system, the laws of mechanics apply. Mechanics potentially explains what happens within a thermodynamic system under an energy gradient.

Indeed, mechanics may dominate what happens within a thermodynamic system even in the presence of a strong external energy gradient. This dominance will occur when a large barrier exists to the efficient implementation of the energy gradient. Thus, for example, a tennis ball dropped from a certain height will rapidly fall to the ground if it is allowed to fall through the air. However, if it must fall through an equivalent height of water, then this will take much longer. Moreover, if it must fall through Jell-O, then its position at any one time may end up depending more on the viscosity (mechanical properties) of the Jell-O than on the effect of gravity. The tennis ball will, nevertheless, at all times be subject to the gravitational energy gradient. However, the mechanical properties of the system may be such that they dominate the effect of the energy gradient. Systems of this type are said to be *nonequilibrium* in character, which means that a thermodynamic equilibrium does not exist [12].

The field of *statistical mechanics* employs the laws of mechanics to describe the macroscopic properties of reversible thermodynamic systems that comprise large numbers of individual components, such as typical chemical systems [13]. The term *statistical* is here, again, used to denote a probability-based description of an extremely large number of objects. A mechanical interaction is, of course, also fundamentally *statistical* in character in that its outcome is a predictable mathematical function of the interaction.

In effect, then, two possible driving forces exist under which processes may unfold: mechanical or thermodynamic. When a steep energy gradient exists, thermodynamics will typically be the dominant driving force and the process is then *energy-dependent*. However, in the absence of such a gradient, in the presence of only a mild energy gradient, or in a nonequilibrium system, mechanics will typically be the controlling feature and the process is then *time-dependent*.

Although the concepts that underly mechanics are currently considered by many chemists and other scientists to be outdated, they have, in fact, very significant utility. They underpin a diverse range of scientific fields, including, for example, celestial mechanics (the motion of stars and galaxies), astrodynamics (spacecraft navigation) [14], acoustics (the motion of sound) [15], statics (semi-rigid bodies in mechanical equilibrium),[2] fluid mechanics (the motion of liquids) [16], and hydraulics (the transmission and amplification of forces),[3] to name but a few.

---

[2] *Statics* is the branch of applied physics concerned with the analysis of loads (force or torque/moment) on physical systems in static equilibrium.
[3] *Hydraulics* is the science of the mechanical properties of liquids.

In perhaps the best illustration of its relevance, *mechanics* has now, three centuries after Newton, spawned a remarkable new field of research known as *chaos theory* [17]. Chaos theory derives directly from the principles of cause-and-effect that underly mechanics and is central to the topics addressed in this series of works. For this reason, we will, in a following section, discuss and illustrate this field in some detail, particularly in respect to its relationship to *chemistry and catalysis*.

1.4 THE DIFFERENCE BETWEEN *ENERGY-DEPENDENT* (THERMODYNAMIC) AND *TIME-DEPENDENT* (MECHANICAL) PROCESSES

1.4.1 Time-Dependent (Mechanical) Processes Are Path-Reliant and Spatiotemporal in Character

Processes that are fundamentally *mechanical* can be distinguished from their *thermodynamic* counterparts in several ways.

As noted, they are, first, *time-dependent* and not *energy-dependent* [6–8]. That is, the physical unfolding of the process depends on the *time that has elapsed* and on the *time required for each step*, rather than on the relative energies of the steps. By contrast, the course of an energy-dependent process is determined by the energy gradient that exists [1–3].

Second, each action in a mechanical process comes about because of the specific circumstances that exist at that place and at that point in time; it is not energetically favored *per se* [6–8]. For example, each domino in a line of dominoes does not knock over the next one because it is bigger or heavier than the next one. Rather, it knocks it over because its position is such that this is the first one it encounters when it falls. The reason for this particular action-and-reaction sequence is therefore that the position of each element is such that it has a *high likelihood* of initiating the next element.

Mechanical processes are, consequently, associated with *very particular* and often *carefully designed spatial positioning* and *spatial interactions* of the elements at specific points in time. For example, machine parts have to be very carefully designed. They must then be fabricated to tight tolerances of that design. If this process is not performed, they will not function within the machine because the elements of spatial positioning and mutually synchronized pathways will not be present.

In summary, mechanical, time-dependent processes have the properties of being *path-dependent* and *spatiotemporal* (that is, they are fundamentally to do with space and time). They involve specific pathways that are followed at specific times.

This characteristic is not true of energy-dependent processes, which typically arrive at the same, lowest energy state by many different pathways. Thus, there are many different ways in which a ball may roll into a well, but it always ends up with the ball at the bottom of the well. By contrast, every sequence of dominoes falling is different and the way they fall, as well as the speed at which they fall, depends critically on their precise spatial arrangement relative to each other.

1.4.2 Time-Dependent (Mechanical) Processes Have a Flat Underlying Energy Landscape (or Are Unaffected by the Energy Landscape)

The difference between an *energy-dependent* (thermodynamic) and a *time-dependent* (mechanical) process can perhaps be most vividly illustrated by envisioning a collection of billiard balls on a table.

If the table is perfectly flat, then the arrangement of balls at any one instant will depend only on how they have physically interacted with each other over time from an initial starting point. In other words, it will depend on the impulses that were applied to the balls and the collisions that have taken place according to the laws of *mechanics*. The exact pathways that have been followed and the time at which each collision took place will be critical to understanding the arrangement of the table at any one instant and to predicting it from there.

If, however, the billiard table is now tilted, an energy gradient in the form of a gravitational differential would be created. The billiard balls present on the table would then move to the lowest point on the table to minimize their individual as well as their overall gravitational potential energies.

Thus, if the table was not flat, but instead steeply contoured, then the arrangement of the balls would depend mostly on the landscape of the table. Each ball would ultimately end up rolling into and staying in the deepest well that was accessible to it. In other words, the arrangement of the table immediately after the balls have been placed on it, would *not* unfold according to their momenta and interactions over time *per se*. Rather, it would be controlled by the landscape that exists and the initial positions of the balls on it. In effect, it would be governed by the imperative of each the ball to minimize its *gravitational potential energy* according to the landscape of the table. The arrangement of the table would then be understood and predicted accordingly.

In the same way, *time-dependent* (mechanical) processes have flat underlying energy landscapes. That is, their components are not subject to strong energy gradients and are, instead, controlled by the precise spatial positions of the elements and their interactions over time, beginning at a certain starting point. Time-dependent processes, therefore, unfold according to the position and path of their components in time.

By contrast, *energy-dependent* (thermodynamic) processes have steeply contoured energy landscapes that play a decisive role in the unfolding of the process. That is, their components are subject to a powerful imperative driving them to change their internal energy. This imperative derives from the laws of *thermodynamics*. Energy-dependent processes are, consequently, path-*independent*.

In other words, all processes involve a tangible and an intangible component. The tangible component comprises the observable, physical actions that take place, namely, the movements and positions in time and space of the elements involved. The intangible component comprises the underlying energy changes that take place during the process; these include heat (enthalpy), entropy, and changes in free energy.

Many processes are driven and governed by the intangible energy changes that take place. In such cases, the spatial arrangement of the participating elements at any particular time is important only insofar as it affects the thermodynamic properties of the system. The process unfolds according to these thermodynamic characteristics. In other words, the system explores its underlying energy hypersurface, eventually settling in the deepest well, which comprises the state of lowest energy.

In such an *energy-dependent* system, the tangible arrangement of the components in space and time is nothing more than a proxy for an intangible, underlying, controlling dynamic that is energy-based. Processes of this type are subject to the laws of thermodynamics, and one must look at the overall changes in energy in order to understand the way in which they unfold.

In other systems, however, the spatial arrangement of the components in time is such that their tangible, *physical* interaction dominates the process. In effect, the underlying thermodynamic driving force for each step is sufficiently small that it does not govern the overall unfolding of the process. Instead the components interact in a purely mechanical way. The course of the process is then controlled by the manner of their arrangement and interactions in space and time, starting at the initial conditions. Such processes are subject to the laws of mechanics, and one must look at the spatio-temporal arrangement of the components in order to understand the process.

As a final remark in this section, we should note that the overall outcome of every process will always involve a combination of thermodynamics and mechanics. Thus, the path of a billiard ball on a table will be affected by both its mechanical impulses *and* the landscape of the table. However, one of these will usually dominate the other. The more it dominates, the more the system will take on its characteristics overall.

Thus, the relationship between thermodynamics and mechanics is always relative. In other words, the steepness of the energy landscape of a system can only ever be evaluated relative to the influence of the mechanical processes that occur within it. A system with a truly extreme energy gradient may, nevertheless, be entirely dominated by even more extreme mechanical interactions.

To illustrate this concept, consider that a billiard ball that is struck with great vigour will be less affected by variations in the surface landscape of the table than one that is struck very gently. This is because its momentum partially overwhelms the underlying surface variation. In the same way, the influence of both thermodynamics and mechanics on a system is dependent on one's effect relative to the other.

1.4.3 Time-Dependent (Mechanical) Processes Display Deterministic Chaos; This Causes Them to be Stochastic and Complex

As noted, a dramatic elaboration of Newton's laws of *mechanics* has been the recent development of a remarkable new field of research known as *chaos theory* [17]. Chaos theory derives from the property of *determinism*, which characterizes action–reaction sequences.

In a *deterministic* system [8], every action produces a reaction that becomes, in turn, the cause of subsequent reactions. The totality of these cascading events preordains the manner in which the system unfolds and allows one to predict accurately what the system will look like at any point in time. Thus, the outcome of starting the collapse of a line of dominoes seems to be predetermined by the fact that each domino has been set up to have a high probability of knocking over the next domino.

However, consider what happens when one has two or three *separate* sets of dominoes stacked in lines that intersect with each other multiple times. How will such dominoes fall?

Well, this will depend on which line of falling dominoes reaches each intersection point first (Fig. 1.5). The one that gets there first may block and halt the collapse sequence of the other line. Alternatively, it may prematurely set off the collapse of the other line at the intersection point, speeding its sequence. A third possibility is that both lines of falling dominoes may reach an intersection point at precisely the same instant, causing the two dominoes at the intersection point to fall up against each other and not knock over the next domino in their respective sequences. In such a case, the collapse sequence of both lines will be mutually halted. The presence of more than one action–reaction sequence within the same system may therefore lead to unexpected and unpredictable overall outcomes.

To illustrate just how unpredictable such a system could be, consider that all of the above possible events will be exquisitely sensitive to the *precise* arrangement of the dominoes in the respective lines. For example, if two dominoes in one of the lines were placed slightly further apart than the rest, this would cause them to take slightly longer to knock over their next domino. The collapse sequence of that line would thereby by slowed, albeit imperceptibly. Such a slowing could drastically alter the

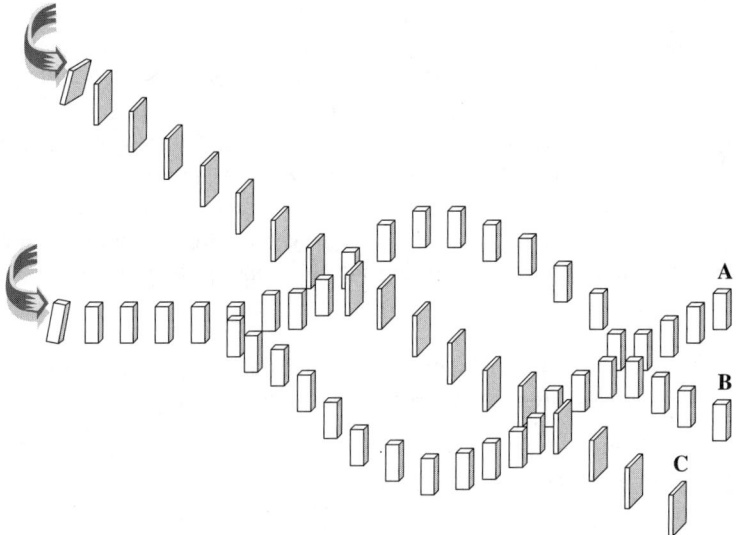

Figure 1.5. *Chaos theory: a chaotic time-dependent system*: Which of dominoes **A**, **B**, or **C**, will fall first? It depends on the precise spatial arrangement and spacing of the dominoes in each line. Even trivially small variation in these spatial arrangements may change the outcome of the process in a nonsimple way. Attempts to replicate this experiment in different venues may result in significant variations because of this extreme sensitivity to the initial conditions.

1.4 THE DIFFERENCE BETWEEN *ENERGY-DEPENDENT* AND *TIME-DEPENDENT*

outcomes at the intersection points. In other words, the way in which two, intersecting domino sequences interact with each other will be extraordinarily sensitive to the precise initial conditions, down to the *exact* spacing and arrangement of each domino.

Now take this example and imagine what will happen if there were 20, 100, or a 1000 intersecting lines of dominoes, instead of the 3 lines in Figure 1.5. The behavior of such systems would become increasingly impossible to predict. This unpredictability would occur despite the fact that the individual components of the system, the dominoes, would each act in a perfectly deterministic and apparently preordained way.

Moreover, if the system were run over and over again with even minutely small variations in the initial placement of the dominoes, wildly different, *chaotic* outcomes would result, because of the extreme sensitivity of the process to the initial conditions. Systems of this type are said to display *deterministic chaos* or simply *chaos* [17].

Chaotic systems are therefore characterized by *determinism* at the microscopic level; that is, a preordained predictability exists at the level of the individual components involved in any one set of interactions. However, the macroscopic properties of the system are *chaotic* or *complex*. Unpredictable macroscopic behavior of this type is termed *stochastic* [18]. Processes involving stochastic behavior are known as *stochastic processes* [18].

The property of an extreme sensitivity to the initial conditions is commonly known as the *butterfly effect* after a paper presented by Edward Lorenz in 1972, entitled *Predictability: Does the Flap of a Butterfly's Wings in Brazil set off a Tornado in Texas?* [19]. The point being made was that an extremely small change in the initial condition of a weather system, which involves a true time-dependent, chaotic system, may conceivably cause a chain of events leading to large-scale phenomena.

A good illustration of a *time-dependent* stochastic process in chemistry is the pressure of a gas [1]. Each molecule in a gas moves in a clearly defined path and undergoes collisions with readily predictable, deterministic outcomes. However, the overall motion of a large collection of such molecules is complex and unpredictable, because, although the individual gas molecules collide with each other in a simple manner, the sheer number of the resulting action–reaction chains and the intersections between such chains drastically complicates its overall, macroscopic behavior. A large enough set of gas molecules will, consequently, exhibit *stochastic* behavior, including filling the container, exerting equal pressure everywhere, and diffusing along a concentration gradient.

Energy-dependent processes are, by contrast, not stochastic in character even when they comprise extremely large collections of participating elements, because energy-dependent processes are not pathway-dependent. In other words, no critical pathways exist *per se*. Moreover, the intersections between the critical pathways that are needed to generate *complexity* are not present. Instead, the system gravitates to, and is distributed into, the lowest energy outcome regardless of the many and varied individual pathways that are followed and the manner in which they intersect with each other.

Time-dependent systems that comprise large collections of participants, such as chemical or gaseous systems, therefore typically involve many individual subsystems whose behavior is deterministic. However, the *overall* behavior of the system is *complex* and cannot be described in simple, classic terms. The study of such behavior forms a subdiscipline within the field of *complex systems science* [20], which has been created to study practical examples of chaotic systems.

Complex systems are generally considered to display the following distinctive features, among others [20]:

(1) *They are nonlinear*: A small perturbation may cause a disproportionately large effect (for example, the butterfly effect). In a linear system by contrast, the effect is *always* directly proportional to the cause.
(2) *They contain feedback loops*: The behavior of one element in the system may be fed back into the system in such a way as to amplify the effect drastically.
(3) *They are open and nonequilibrium*: No boundary of the type present between the system and the surroundings in thermodynamics exists. Thus, complex systems may exit in an energy gradient far from a thermodynamic equilibrium.
(4) *They may have a memory*: The history of a complex system may be important insofar as they change over time, and prior states may have an influence on current states.
(5) *They may have hierarchies of complexity*: The components of a complex system may be nested within subsystems that are themselves complex systems. For example, two intersecting lines of dominoes may form a separate subsystem of a large multiplicity of intersecting lines of dominoes. This subsystem may develop in a complex way and then interact chaotically with other, similar subsystems.

1.4.4 Time-Dependent (Mechanical) Processes Often Involve Synergies of Action

The term *synergy* refers to the phenomenon in which the whole of a system is greater than the simple sum of its parts [20]. For example, a collection of soccer players may, individually, have mediocre ball skills. Despite this failing they may, when combined with one another in a team, play very well and outdo another team whose players are, individually, more skilled. The team of mediocre players is then said to exhibit a good *synergy*. Therefore, the soccer players interact with each other in a way that cumulatively amplifies their skills. In other words, the whole of the team is better than the simple sum of the skills of its individual players.

Synergy is an extremely important effect in the time-dependent processes that characterize mechanics. Most mechanical devices, for example, involve one cog, which turns another cog, which turns another one, and so on, to thereby create an overall effect that is not available to any of those cogs individually (Fig. 1.6). Only in *correctly synchronized combination* do the components of the mechanical device bring about the desired outcome. They achieve this result by creating synergy between their components. This synergy is necessary because, at the end of the day, each component is typically just a bit of intricately machined metal. On

1.4 THE DIFFERENCE BETWEEN *ENERGY-DEPENDENT* AND *TIME-DEPENDENT*

Synchronized actions create synergy and utility   No synchronization: no synergy, no utility

Figure 1.6. *Mechanical systems* are characterized by *synchronization* of actions to yield an outcome that is more than the simple sum of the parts (*synergy*). With such synchronization, mechanical devices can perform astounding feats. Without it, they are just a collection of spare parts.

its own it is, frankly, useless. But together with the other parts it forms a machine with special properties.

Energy-dependent processes are, by contrast, far less reliant on synergy because they use a different approach, namely, to harness a large, intrinsic, energetic driving force to create the desired effect. This effect occurs spontaneously within the system as a response to its surroundings.

In essence, therefore, energy-dependent processes induce change by the application of overwhelming inputs of energy. Time-dependent processes, however, induce change by finessing impediments with clever synergies of action (Fig. 1.7).

To illustrate this distinction, consider that in a time-dependent system, each domino knocks over the next one because its spatial position is such that this is an inevitable outcome. If the spatial position of the first domino is changed somewhat, it may not knock over the next one. In the comparable energy-dependent system, however, a different tactic would be used. The system would, instead, rely on large size and bulk in the first domino to knock over the next one. Under these circumstances, spatial positioning would be much less important.

The history of engineering has largely involved using the synergies of mechanics to decrease the thermodynamic energies required in devices. Thus, the first steam engines were extraordinarily energy intensive and inefficient. But the subsequent introduction of clever mechanical systems led to a drastic reduction in their energy requirements. This process of more effectively harnessing synergy continues within the field of engineering to this very day.

1.4.5 Time-Dependent (Mechanical) Processes Characterize Numerous Aspects of Human Experience

The philosophical concept of *determinism*, which is exemplified by Newton's third law of motion, is observed in many facets of science [8]. In fact, *time-dependent* action–reaction sequences of the type observed in the domino examples are commonplace in human experience, being present in an astonishingly wide diversity of situations.

For example, in mathematics, such sequences are termed *cascade processes* [17]. All theoretical constructs in mathematics are arrived at by a series of steps that cascade in logical sequence from an initial assumption.

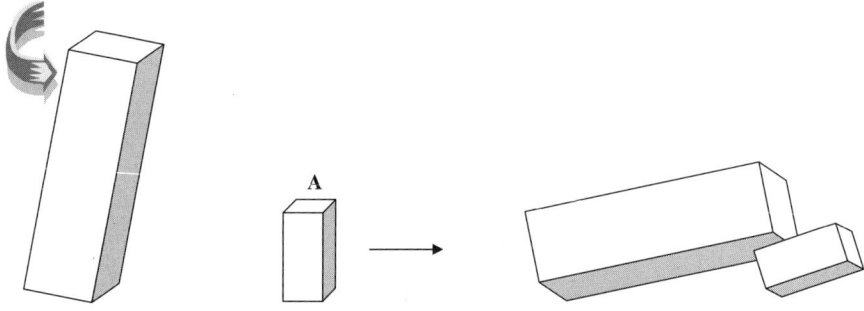

Thermodynamic systems induce change by applying an overwhelming input of energy

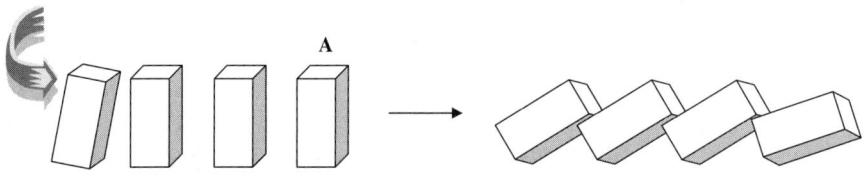

Mechanical systems finesse change by applying clever synergies of action

Figure 1.7. To knock down block **A**, a *mechanical system* will typically employ *synchronized* actions to create clever synergies to do the job. This process makes the system path-dependent and spatiotemporal in character. A *thermodynamic system* by contrast will typically employ a large input of energy, so that there is no critical pathway; the energy input is such that virtually any pathway will be sufficient.

In biology, numerous biochemical functions are believed to involve action–reaction sequences, including the feedback loops implicated in metabolic functioning [21]. The process of evolution is, arguably, itself an action–reaction sequence, as is the physical development of organisms during their lifecycle [20,21] Protein folding must, similarly, be dependent on time-dependent rather than on energy-dependent processes, because calculations reveal that even very simple proteins would literally take millions of years to find their energetically most stable folded conformation if they were to populate every possible combination of conformations [22].

Other examples of time-dependent processes include:

(1) The transmission of nerve impulses to and from the brain, and the fundamental operation of the brain itself (*neural networks*) [23]. Neural networks are based on parallel processing and on the recognition of patterns of "sensory" input from external sources. In other words, rather than sequential processing and execution, neural networks employ multiple, interacting pathways whose overall outcome is complex.
(2) Family trees, where each generation creates the next (*genealogy*) [20].
(3) Laissez-faire economic systems, which involve chains of mutually beneficial commercial transactions (*stock markets* and *capitalism*) [24]. The

eighteenth-century economist Adam Smith used the metaphor of a "hidden hand" to describe the overall outcome of free markets. This metaphor seeks to describe how the individual, mutually beneficial transactions synergistically create an overall outcome that is still more beneficial, but in a chaotic and complex way.

(4) Storm events such as cyclones and hurricanes, which involve mutually reinforcing confluences of subtle climatic effects (*weather systems*) [25].
(5) The operation of computers, which employ sequences of logic gates during data processing (*computer science*) [26].
(6) The movement of vehicular traffic on the intersecting freeways of, for example, Los Angeles, which involves numerous, critically intersecting pathways of movement (*traffic management*) [27].
(7) Fractionation plates in distillation columns, where each plate sets up the next one (*fractional distillation* and *refining*) [28].

Because of their ubiquity and importance, time-dependent processes are therefore of significant and topical interest in many different fields.

1.5 *TIME-* AND *ENERGY-DEPENDENCE* IN CHEMISTRY AND CATALYSIS

1.5.1 The Origin of Time- and Energy-Dependent Processes in Chemistry

In a chemical reaction, two or more *reactant* molecules or atoms initially collide with each other to form momentarily a combined species known as the *transition state* [Fig. 1.8(a)] [1]. The transition state is an unstable and high-energy moiety that exists for only an extremely brief period of time. Thereafter, its constituent atoms or molecules disengage either to yield the original reactants back again or to generate entirely new chemical species known as the *products* of the reaction. The products typically comprise new chemical arrangements of the atoms originally present in the reactants.

The energy of the collision determines its outcome. Products are formed if the collision is sufficiently energetic to overcome a certain threshold energy, known as the *activation energy* (or E_a) [1]. If this threshold is not overcome, the collision is unsuccessful and the reactants are reformed.

As the collisions involve a random variety of reactant orientations, E_a is formally defined as the *average* energy barrier that must be traversed during a random collision between the reactants. The lower the E_a threshold of a reaction, the greater will be the proportion of successful collisions. It will, in turn, result in a greater overall reaction rate.

The rate of a chemical reaction is given by Arrhenius's equation (1.2):

$$k = A \exp\left(\frac{-E_a}{RT}\right) \qquad (1.2)$$

(a) Uncatalyzed reaction

(i) *Mechanical action*

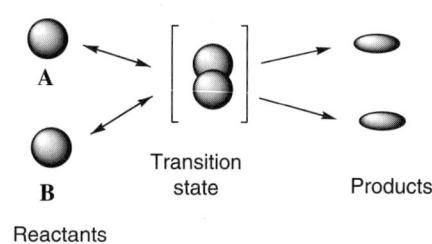

(ii) *Energy profile*

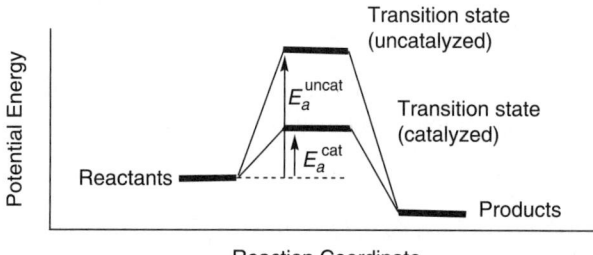

Figure 1.8a. Schematic representation of an uncatalyzed reaction. Figure 1.8(b) is several pages further on.

which relates the reaction rate (k) to the frequency with which the reactant molecules collide with each other (the preexponential term A known as the *collision frequency*) and the proportion of those collisions that are sufficiently energetic to result in product formation (the exponential term $-E_a/\mathrm{RT}$).

In reactions where the activation energy E_a is large, the exponential term will dominate the overall reaction rate k. In such a case, the process will be dependent on the *thermodynamics* of the collision and the reaction will be *energy-dependent*. That is, the progression of the reaction will depend on the activation energy E_a.

In reactions where the activation energy E_a is negligibly small; however, the preexponential term will dominate the reaction rate. That is, the rate will depend on the *collision frequency* of the reactants A. Since frequency is a measure of time, the reaction will then, in effect, be *time-dependent*. In other words, the reaction rate will depend only on the *mechanics* of the reactants colliding with each other.

Time-dependent reactions are today widely referred to as being *diffusion-controlled*. That is, they occur at a rate that is dependent only on how rapidly the reactants diffuse to each other in solution. In other words, the reaction rate is a function of the *collision frequency* of the reactants. We use the term "collision" here to refer to liquid- as well as gas-phase reactions.

The terminology of *diffusion control* is an excellent descriptor when the reactants are separate chemical entities that must find and collide with each other within a liquid or gaseous medium. However, it does not describe the situation when the reactants are, for example, part of the same, large chemical entity and therefore not separate molecules. In such a case, the reactants cannot *diffuse* to each other since they are mutually attached via an intermediate structure. Rather, they must be brought into collision by the conformational or other flexing of the intermediate chemical structure.

The term *diffusion controlled* does not, therefore, properly describe the concept of a reaction in which the activation energy is negligible. For this reason, we use the term *time-dependent* in this series.

In a following chapter we will discuss in detail the origin, implications, and outcomes of time- and energy-dependent reaction processes.

1.5.2 Examples of Time-Dependent Processes in Chemistry

Numerous examples of time-dependent processes exist in chemistry. Perhaps the best example involves the chain sequences that form the basis of nuclear reactions [29,30]. Although much energy is released during such reactions, they are not driven by this change in energy *per se*. Rather, they are impelled by a first fission or fusion event, which then sets off other such events much like the dominoes referred to earlier.

The challenge in creating a nuclear reaction is, consequently, similar to that involved in setting up a line of dominoes, namely, to create the conditions under which the chain process may take place [29,30]. In the case of nuclear reactions, this process requires extraordinary purification of the reactants followed by their engineering into carefully designed physical structures. These structures must then be further combined in very particular ways to initiate the reaction.

This need to identify and painstakingly set up the elements in a suitable spatial arrangement is highly characteristic of the design requirements for a time-dependent process. It originates because such processes rely on *synergy*. In the case of a nuclear reaction, the synergy is created by setting up the system so that every *one* fission or fusion event sets off *two or more* other events, thereby creating an accelerating cascade. If the system did not have this inherent synergy, a sustained nuclear reaction would not occur.

Other examples that may illustrate the distinction between energy-dependent and time-dependent processes in chemistry involve *thermodynamic* versus *kinetic* products [31].

Many reactions generate the *thermodynamic* or the energetically most favored product. Such reactions occur in the presence of a reversible thermodynamic equilibrium. The overall outcome is dictated by the stabilities of the various possible products. That is, the lowest energy option overall will be followed, yielding the most stable of the possible products in greatest proportions. This result occurs because its formation is accompanied by the greatest release of energy and is, consequently, subject to the largest thermodynamic driving force. Reactions that generate thermodynamic products, therefore, inevitably involve an *energy-dependent* reaction process.

(b) **Catalyzed reaction**

(i) *Mechanical action*

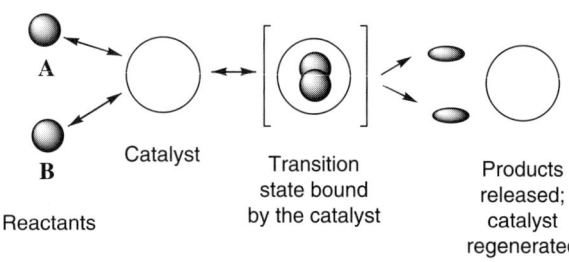

(ii) *Energy profile*

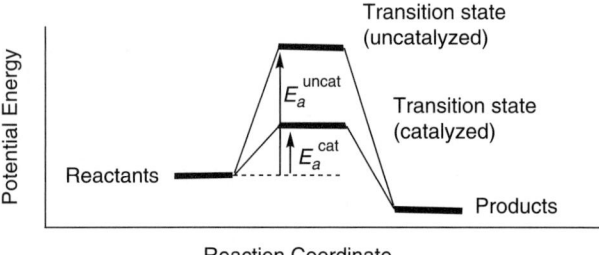

Figure 1.8b. Schematic representation of a catalyzed reaction, and its energy profile. Figure 1.8(a) is a few pages back.

Other reactions, however, deliver a so-called *kinetic* product. These reactions occur in the absence of a reversible thermodynamic equilibrium or when the equilibrium is slow. Kinetic products are obtained via the pathway with the *fastest reaction rate*. The outcome of such a reaction may be governed by energy considerations in cases where the kinetic product is the one with the lowest energy barrier to reaction. However, kinetic products may also form without formal regard to the underlying energy landscape of the collision. This process takes place when, effectively, no barrier to the reaction exists at all. An example in this respect is the reaction of H^+ and OH^-, whose rate is determined only by the rate at which these species diffuse to each other. In other words, as soon as H^+ and OH^- collide with each other, they react to form H_2O, which occurs regardless of their orientations during the collision or the nuances of their approach trajectories, their speed, or their energy of encounter. Effectively, no thermodynamic controlling factor exists to this reaction. It is governed only by the *mechanics* of the collision. To put it another way, the reaction is so thermodynamically favorable that its energy landscape about the collision is flat and its rate is determined only by how quickly the reactants collide with each other. Certain kinetic products can, consequently, also be characteristic of a *time-dependent* reaction process.

A feature of modern-day chemistry is its increasing use of cascade-type reactions that may involve time-dependent steps. For example, a developing theme in synthetic

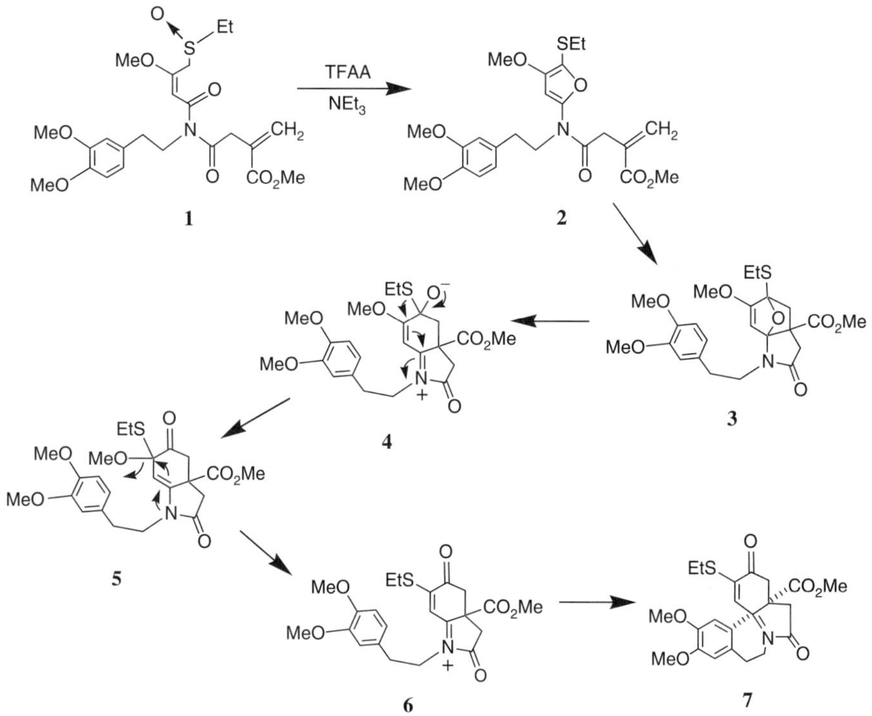

1.5 *TIME-* AND *ENERGY-DEPENDENCE* IN CHEMISTRY AND CATALYSIS

Scheme 1.1.

organic chemistry involves the creation of cascading, multistep, organic reactions. Such processes drastically simplify multistep organic synthesis. However, they must be setup with scrupulous and extreme care to ensure that each step initiates the next.

Scheme 1.1 illustrates an organic reaction recently described by Padwa that cascades through five intermediates before generating the final product **7** [32]. In normal organic synthesis, such a reaction would require multiple individual steps, with isolation and purification of the products of each step. However, in Scheme 1.1, it all happens in one, single process.

Consider what happens now if any one of the steps in Scheme 1.1 involves a time-dependent action, which is limited only by the mechanics of the interaction involved. Such a step cannot be said to be *diffusion-controlled*, because the reactants do not collide as a result of diffusion. Rather, they collide as a result of their mutual attachment to the chemical framework present and, possibly, its conformational flexing.

Thus, the term *diffusion-controlled* is inadequate to describe reactions that are *mediated by chemical frameworks* and that are limited only by the mechanics of the interaction involved. The term *time-dependent* is a more general and a better description of such reactions.

In conclusion, time- and energy-dependent reactions exist as two different types of chemical reactions. Time-dependent reactions have been known as *diffusion-controlled* reactions to date. However, this term is insufficiently broad to cover all aspects of the concept of a process governed by a mechanical interaction.

In the next section, we will briefly consider time- and energy-dependence in catalysis. In a following chapter, we will consider the theory of time- and energy-dependent reactions in detail.

1.5.3 Time- and Energy-Dependent Processes in Catalysis

Catalysts are species that create and mediate successful collisions between reactants. That is, they participate in and facilitate chemical reactions without themselves being transformed. They are generally believed to operate by intervening in and modifying the energy of the reaction (collision) process.

As schematically depicted in Figure 1.8(b), catalysts are thought to bind the reactants and to mediate their collision with each other, thereafter releasing the resulting species and regenerating themselves in an unchanged form. The catalyst is therefore not changed in the reaction process. But the process certainly is. In particular, the catalyst binds and stabilizes the transition state, thereby lowering the E_a barrier for the reaction and speeding up its overall rate.

Figure 1.8(b)(ii) displays a typical energy profile for an uncatalyzed and a catalyzed reaction [1]. As can be seen, the catalyst reduces the average energy barrier that must be overcome. In so doing, it speeds up the reaction rate.

At all levels of chemistry, everyone who ever studies catalysis learns that a catalyst speeds up a chemical reaction by decreasing the *energy barrier* involved in the reaction. It does so without itself being changed. The more the catalyst can reduce the energy barrier, the more rapidly the reaction will occur.

This scenario of an energy-related explanation for catalytic rate accelerations is currently the common wisdom and, indeed, the only wisdom in catalysis. But, does it describe *all* catalytic processes? Are there catalyzed reactions that do *not* fit this picture?

To put this another way, are there catalysts that operate in a time-dependent, not in an energy-dependent manner? In other words, do catalytic systems exist in which the reaction rate is limited only by the frequency with which the catalyst brings the bound reactants into collision with each other? In cases where the reactants are pulled apart, do catalytic processes exist that are determined only by the rate at which the catalyst pulls them apart? In other words, are there catalysts whose properties are determined by the mechanics of the interactions involved?

The answer is that there surely must be, given that we are aware of time-dependent, *diffusion-controlled* chemical reactions. If such reactions exist, then time-dependent catalysis must surely also.

Certainly many catalytic processes are *diffusion-controlled*. *Diffusion*, in this context, refers to the rate at which the reactants *bind with the catalyst*. That is, the rate in a *diffusion-controlled catalyst* is dependent on the rate of *catalyst–reactant binding*.

One may think that at least some diffusion-controlled catalysts may also be time-dependent. However, the situation is not so simple. Time-dependent chemical reactions are formally limited by the rate at which the reactants *collide with each other*. Thus, in catalysis, it is the rate at which *the catalyst brings the bound reactants into collision* with each other, in other words: the *catalyst-mediated collision frequency*. This concept is not the same thing as a *diffusion-controlled* catalyst, whose rate depends on the rate at which *the reactants collide with the catalyst*.

The *catalyst-mediated collision frequency* may, certainly, be affected by, or even be dependent on, the rate of reactant diffusion to the catalyst. But it may also not be.

For example, if the catalyst is inefficient at bringing the bound reactants into contact with each other, or if the reactants attach only very transiently on each occasion of binding, then *time-dependent catalysis* may occur even when there is an *excess of reactants* about the catalyst in solution. In such cases, the system will be dependent on the *catalyst-mediated collision frequency*. Such a catalyst will then be time-dependent regardless of the quantity or concentration of reactant molecules about the catalyst.

The term *diffusion-controlled* is therefore inadequate in describing time-dependent processes in catalysis. Diffusion-controlled catalysis is limited by the rate at which the reactants collide with the catalyst. Time-dependent catalysis is, however, limited by the rate at which the catalyst brings the bound reactants into physical contact with each other. This process is quite a different thing. It may be completely unaffected by the rate of reactant diffusion.

For this reason, again, we choose to use the term *time-dependent* rather than *diffusion-controlled* to describe catalyzed reactions whose rates are determined by the *catalyst-mediated collision frequency*.

In summary, then, does time-dependent catalysis exist? If so, where? Why has it not been properly recognized and characterized? More pertinently, what are its defining and observable macroscopic characteristics? Why do modern chemistry and catalysis science seem to consider only the possibility of energy-dependent catalysis?

Some answers to these questions may lie in the history of chemistry. The concept of catalysis controlled by the energy of its transition state has not always existed. In fact, it has a definite past, dating back mainly to the *transition state theory* of Henry Eyring in the 1930s [33]. Before that catalysis (and, indeed, chemical reactions in general) were not considered in purely energetic terms. The dominant hypothesis in the 1920s was the *Hinshelwood–RRK* theory, which is today still employed, in modified form, to describe high-temperature, gas-phase reactions [33].

Hinshelwood personally eschewed transition state theory, considering it narrow in its application [33]. The issue was more than merely personal. Transition state theory was, certainly, a very brilliant development of Hinshelwood–RRK theory [33]. But, as was pointed out, it actually describes only one special case of Hinshelwood–RRK theory, which has been called the *high-pressure limit* [33]. As such, Hinshelwood argued, it did not and could not possibly capture the full spectrum of chemical (or catalytic) action.

There must, therefore, be other types of catalytic action, involving time-dependent processes, that do not fall within the *energy-dependent* model described above.

But, what are they? In this volume, we will seek to illuminate and clarify this issue. We will try to identify and describe these other forms of catalysis.

1.5.4 Is There Such a Thing as a Time-Dependent Process in Catalysis?

Although chemistry is starting to come to terms with the properties and utility of time-dependent processes, the field of catalysis has not as yet formally considered the possibility that catalytic action may be controlled by a simple mechanical interaction. It has not explicitly asked, nor answered, the question: Does time-dependent catalysis exist? In the absence of a definitive answer, all we can do is look for clues in the scientific literature that address this question in an indirect way.

A more complicated, but perfectly equivalent, form of the question is as follows: Are there any types of catalysis that can be viewed as 1) *mechanical processes* that are 2) *path-dependent*, 3) *spatiotemporal*, and whose 4) *synergies* generate 5) *stochastic*, 6) *chaotic*, and 7) *complex* outcomes? Phrased in this way, it is clear that there are many distinctive keywords—all related to each other—with which to search for possible examples of time-dependent catalysis.

Do such catalytic systems exist? In fact, some catalytic systems have been characterized as displaying one or more of the above properties. But such reports have appeared only fairly recently. And they have been used only in respect of one class of catalyst: the natural catalysts of biology, known as enzymes.

In a 2006 study, Boehr and colleagues (with expansion by Dobson) has shown, very elegantly, that catalysis by the enzyme dihydrofolate reductase are *stochastic* in character [34]. In earlier work, Moss and others have described enzymes generally as *machine-like* "specialised combining centers" [35]. In a prescient and remarkable paper, Williams has characterized enzymatic action as being similar to a *mechanical device* [36]. Benkovic, Hammes-Shiffer, and others describe catalysis by enzymes as being critically dependent upon a series of *coupled protein motions* [37]. Menger has proposed a *spatiotemporal* hypothesis for enzymic action [38]. In a recent article, Willner and colleagues used a simple system involving two coupled enzymes (glucose dehydrogenase and horseradish peroxidase) to illustrate that enzymes may perform *logic gate* operations [40].

All of these descriptions depict time-dependent processes in one guise or another. Thus, mechanical devices and machines typically involve one cog moving another, which moves another, and so forth, in an engineered succession of actions stemming from a first action. The same can be said for coupled protein motions, which clearly involve a series of sequential actions in the enzyme that are unrelated to the formal activation energy of the reaction. In other words, if the catalytic rate of an enzyme is governed by coupled protein motions, how can it depend critically on the *activation energy* E_a of the collision? In similar vein, how can an energy-dependent catalyst rely on, and be affected by, its *spatiotemporal* properties or form a *logic gate*. Energy-dependent processes are, as we noted earlier, not influenced by the pathways employed in space and time.

All of the time-dependent processes alluded to above do not operate by seeking an energy minimum or by harnessing a drive to lower energy. Thus, many mechanical devices are used to *amplify* energy effects (as in, for example, hydraulic brakes). They are, consequently, independent of the underlying energy landscape of the process.

Although the action of some catalytic systems may therefore conceivably involve a time-dependent process, no theoretical or conceptual treatment has been proposed for catalysis of this type. Nor, indeed, has the nature of the catalytic action been described. The rational design and deliberate preparation of catalysts displaying time-dependent actions have also not been a subject of investigation.

The aim of this volume is to try to develop such a conceptual basis and to examine the incidence and operation of time-dependent processes in catalysis. In particular, we will aim to develop a methodology for designing species that undertake time-dependent catalytic action, that is, for making catalysts that act like mechanical devices.

1.6 THE AIMS, STRUCTURE, AND MAJOR FINDINGS OF THIS SERIES

In the remaining sections of this chapter, we will provide a brief summary of the contents and findings of this volume. This summary is intended to offer a drastically encapsulated version of the main topics for readers who are severely time-constrained. Other readers, who wish to consider the complete arguments, should skip over these next sections and go directly to Chapter 2.

1.6.1 Summary of the Key Finding: Many Enzymes Seem to be Time-Dependent Catalysts

We will show in this series that time-dependent catalysis does exist and that it has, astonishingly, been around us since time immemorial, being a form of catalysis employed by many enzymes. We will show that most theories of enzymatic catalysis and some of the overarching theories of heterogeneous catalysis actually describe time-dependent processes, albeit in an oblique and usually incoherent manner.

We will demonstrate that in *time-dependent* catalysis, the catalyst reduces the energy barrier of the collision between two or more reactants to such an extent that it, effectively, becomes irrelevant. It performs this process by carefully controlling the manner in which the reactants are brought into physical contact with each other. In effect, such catalysts, including many enzymes, act by driving their bound substrate functionalities into reactive contact (*collision*) with each other along near-optimum pathways during conformational flexing. The conformational limitations of the catalyst do not allow for deviations from this pathway; that is, the collision process is *path-specific*. The pathway, moreover, involves an exceedingly low energy barrier for reaction upon collision, meaning that the reaction process has a *flat energy landscape*.

In each cycle of conformational flexing, the possibility exists for a successful collision, which leads to product formation. But at the point of collision, all participating reactants must be correctly bound to their respective catalytic groups. That is, the process is not only *path-specific* but also *spatiotemporal*; everything must come together simultaneously, in the correct way, at the same instant in time.

This synergy is achieved in a highly repeatable and sustained way by ensuring that the individual catalyst–reactant binding interactions are dynamic, with the reactant/product functionalities being constantly bound and released. In the case of enzymes, this dynamism is a consequence of the weak interactions of biology that are employed (e.g., hydrogen-bonding, ion-pairing, and the like).

The overall rate of a time-dependent catalyst like many enzymes, therefore, depends on the extent to which its conformational flexing and its substrate binding are *synchronized* (Fig. 1.9). That is, the success of each collision action depends on the extent of overlap between the *time* during which the substrate is fully bound and the *time* during which the substrate functionalities are brought into contact by conformational flexing. The longer this overlapping *time*, on average, the larger will be the catalytic rate. The effect of this synchronization is that the catalytically active groups within a time-dependent catalyst like an enzyme act in a *synergistic* manner. That is, catalysis originates from coordinated, synchronized actions on the part of these groups.

Enzymatic rates are, of course, formally expressed in terms of the *inverse* of this overlapping time, namely as the average number of successful collisions per second. That is, they are expressed in terms of the *collision frequency* created by the catalyst [A in Equation (1.2)]. They are independent of the *activation energy* E_a.

This dependence on the *collision frequency* is the fundamental origin of the time-dependent character of many enzymes. It explains why such catalysis is *stochastic*, *chaotic*, and *complex*. Even exceedingly small variations in the physical makeup, conformational flexing, or binding properties of such an enzyme may drastically affect the required *synchronicity*, destroying or severely curtailing the catalytic effect.

Time-dependent catalysts like many enzymes are, consequently, like *machines* on a production line "stamping out" successful collisions between reactant functionalities. In each cycle, the same, optimum pathway is followed. On some occasions, the substrate functionalities are not bound to their respective catalytic groups at the instant of collision, which results in a failed collision attempt. Before the next collision cycle, however, that functionality will have left and another will have taken its place. This "stamping" and "ejection" process is perfectly analogous to a *mechanical device*. It is, arguably, also the true origin of Michaelis–Menten kinetics.

These conclusions explain numerous aspects of enzymatic catalysis. The large reaction rates displayed by many enzymes relative to similar but inactive proteins can, for example, be explained as a nonlinear structural effect. They derive from the fundamental path-dependent nature of the catalysis. Even very minor changes in enzyme structure that alter the approach pathways involved in the collision will, obviously, have a drastic influence on the catalysis. They may also change the synchronization of binding and flexing, with accompanying, disproportionately large changes in the overall rate. Much like the gears in a finely machined watch, any modification that affects the required synchronization will rapidly destroy the mechanical effect.

1.6 THE AIMS, STRUCTURE, AND MAJOR FINDINGS OF THIS SERIES 27

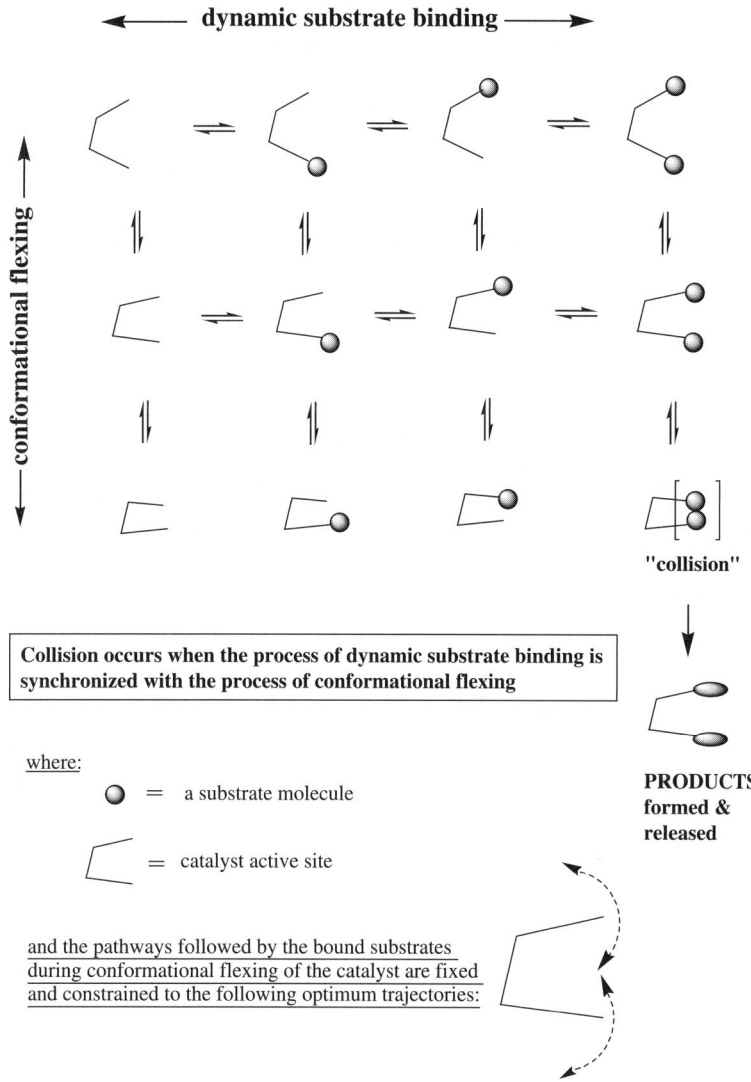

Figure 1.9. *Time-dependent catalysis*: Schematic representation of the two key processes present in many enzymes. Shown on the left, from top to bottom, is the process of *conformational flexing* of the enzyme active site. During conformational flexing, the groups in the enzyme that bind the substrate are repeatedly moved back and forth over the same optimum pathways in space, relative to each other. Shown at the top, from left to right, is the process of *dynamic substrate binding*. In this process, the substrate repeatedly binds and releases the relevant enzyme groups. A successful "collision" occurs when these processes of *conformational flexing* and *dynamic substrate binding* are synchronized, that is, when the substrates are fully bound at the instant that they are brought into physical contact with each other.

The propensity, first noted by Pauling [40], for enzymes to be structurally complementary to their transition states can also be explained. It is the only way to get any sort of synchronization between enzyme flexing and binding. In much the same way, the gears in a mechanical device must be structurally complementary to each other in order to act synchronously.

Finally, the ability of many enzymes to catalyze both the forward and the reverse directions of a reaction *in vitro*, even when the forward and reverse activation energies are very different, can be understood. The rates and catalytic action are not dependent on the energy barrier that must be traversed, and they are therefore unaffected by variations in this respect. Instead, the reaction rate is dependent on the enzyme-mediated collision frequency, and this may be the same in both the forward or the reverse direction. The entire concept of a pathway that may be followed in either the forward or the reverse direction is very typical and characteristic of mechanical systems. Machines can typically be run forward or backward with equal ease. Readers may recall an earlier statement to the effect that mechanical processes are like movies that may be played forward or backward. That is, they follow preordained courses that can be tracked in either the forward or the reverse direction.

1.6.2 The Aims and Structure of this Series. Summary: Other Major Findings of this Series

Chapter 2 of this series describes the different subdisciplines of catalysis; namely, *heterogeneous, homogeneous, and enzymatic catalysis*. The problem of conceptually uniting these fields into a single, unified theory of catalysis is raised. A key difficulty in this respect is that each subdiscipline employs a different set of concepts and terminologies to describe what are often the same fundamental properties. A key aim of Chapter 2 is to *translate* these different concepts and perspectives into simple, commonly understood terms that may form the basis of a shared understanding. The idea is to establish an agreed, underlying conceptual framework with which the reader may consider and fully appreciate the issues raised in the remaining chapters in this book. For reasons of simplicity, this framework has been kept to the minimum required.

This background discussion is followed in the final section of Chapter 2 by a practical example that seeks to cast the issues raised in this series into a sharper relief. The example discusses the problem of ensuring fidelity in the catalytic process. How this assurance can be achieved is a central theme of this book.

Chapter 3 examines and develops a *general theoretical basis for time-dependent reactions*, both catalyzed and uncatalyzed. We show that time-dependence and energy-dependence occur as a consequence of the requirements for reactants to:

(1) Collide with each other [A in Equation (1.2)]
(2) Collide with sufficient energy for reaction [E_a in Equation (1.2)]

Whichever of these requirements is the slower and least likely will determine the overall character and nature of the reaction.

1.6 THE AIMS, STRUCTURE, AND MAJOR FINDINGS OF THIS SERIES

Thus, in *time-dependent* reactions, process 1 is the slower and least likely overall; that is, the *kinetic dynamism of collision is the limiting characteristic*.

Energy-dependent reactions are, by contrast, controlled by process 2; that is, the *thermodynamic efficiency of the collisions is limiting*.

The pathway of the reaction is also determined by whichever of (1) or (2) above is the slowest and least likely.

Energy-dependent reactions are also shown to be described by Eyring's *transition state theory* [33]. However, time-dependent reactions fall outside of transition state theory because they are unaffected by, and do not depend on, the presence of a chemical equilibrium. Instead, they are best described by a special case in *Hinshelwood-RRK* theory, known as the *low-pressure limit* [33]. Energy-dependent reactions correspond to the *high-pressure limit* in *Hinshelwood-RRK* theory [33].

Chapter 4 discusses *time- and energy-dependent catalysis in heterogeneous catalysis*. Sabatier's Principle was first coined in heterogeneous catalysis about 100 years ago. It states that an ideal catalyst *binds its reactants strongly, but not too strongly*.

In this statement, Sabatier's Principle articulates the fact that catalysis involves an interplay between the *thermodynamic efficiency* of a catalyst and the *dynamism* of its catalyst–reactant interactions (especially with respect to reactant collision). The character of the catalysis is dominated by whichever of the above processes generates the slower overall reaction rate. This result occurs because the overall sequence of events in catalysis can only go as fast as the slowest participating step.

Thus, a system whose kinetic dynamism allows for a faster reaction than is possible according to its thermodynamic efficiency will be dominated by thermodynamics and will therefore be *energy-dependent*. The rate and character of such a reaction will depend on the exponential term $-E_a/RT$ in Equation (1.2) and on the activation energy (E_a) of the reaction.

However, a system whose thermodynamic efficiency allows for a faster reaction than is permitted by its kinetic dynamism will be dominated by this dynamism. That is, its rate (and other properties) will depend on the frequency with which the catalyst can bind reactants and then bring them into physical collision with each other. Such a catalyst will be controlled by the pre-exponential term in Equation (1.2) and on its *collision frequency A*. Since frequency involves a quantification of time, catalysis of this type will be *time-dependent*.

Sabatier's Principle therefore describes two independent catalytic realms: energy-dependent catalysis and time-dependent catalysis.

In Chapters 5 and 6, *the properties of homogeneous time-dependent catalysts* are discussed. Since several studies of enzymatic catalysis invoke concepts involving time-dependence [11–14], we examine enzymes and compare their general catalytic properties with those of comparable nonbiological catalysts.

We show that the general properties, which are most characteristic of enzymes are, in fact, consistent with a time-dependent catalytic action. They are, moreover, explicitly inconsistent with energy-dependence.

As such, enzymes seem to be generally, but certainly not exclusively, *time-dependent* catalysts. In other words, the most distinctive properties of enzymes as a class of catalyst seem to be also those of time-dependent homogeneous catalysts.

The catalytic activity of many enzymes is, consequently, implied to be limited by the dynamism of their enzyme–substrate interactions, not by the efficiency with which they stabilize their transition states. That is, it is determined by the rate at which they create successful collisions between bound substrate functionalities, namely, their *collision frequencies*.

By contrast, almost all known nonbiological homogeneous catalysts are *energy-dependent* in their mode of catalytic action. That is, their catalysis is limited by the energetic efficiency of the process.

Chapter 6 considers the *fundamental character of time-dependent homogeneous catalysis*. We show that the *activation energy* E_a does not provide a true measure of the threshold energy barrier that must be overcome in a time-dependent catalyst. This result occurs because E_a is formally the *average* energy barrier when the approach trajectories of the reactants (leading up to their collision) are randomly oriented. In enzymes, however, the bound reactants approach each other along highly constrained, optimum trajectories, so that the formal quantity E_a does not apply.

We also describe how many enzymes can harness the weak activation provided by amino acid groups to facilitate catalytic reactions. This process comes about because of the coordinated, synchronized nature of enzymatic catalysis. In effect, enzymes employ smart *synergies* of action to finesse the facilitation of a chemical reaction.

By contrast, most manmade homogeneous catalysts are energy-dependent. In such species, the catalyst also creates collisions between bound reactant functionalities during flexing. But their greater conformational freedom means that the approach pathways involved in the collisions are essentially random or close thereto. Each collision consequently involves a different energy barrier, with only those collisions having a low barrier being successful.

To maximize the likelihood of reaction during such collisions, energy-dependent catalysts must employ individual binding and activating interactions that are stronger and more sustained than those in enzymes (e.g., coordination bonds). The strength of these individual interactions keep the reactant functionalities fully bound to, and strongly activated by, the catalyst throughout the reaction process. They are, therefore, available for reaction during the many collisions required to form a product. In effect, therefore, most manmade homogeneous catalysts facilitate chemical reactions with an overwhelming input of energy.

Chapter 7 evaluates the many *theories proposed for enzymatic catalysis* and examines how they knit together to describe two distinct types of catalytic action: energy-dependent and time-dependent processes. The latter seem to describe many enzymes more correctly. The former, which is largely based on comparisons with manmade catalysts, seems to be flawed by the fact that most nonbiological catalysts are energy-dependent, not time-dependent.

Theories involving time-dependent catalysis such as *Coupled Protein Motions* [37], the *Proximity Effect* [41], and the *Spatiotemporal Hypothesis* [38], fully or partly describe enzymatic action. By contrast, theories invoking energy-dependence [40,42,43] are only partially appropriate or entirely inappropriate to enzymatic catalysis in general.

1.6 THE AIMS, STRUCTURE, AND MAJOR FINDINGS OF THIS SERIES

Chapter 8 discusses the issue of *synergy in catalysis*. Although both energy-dependent and time-dependent catalysts exhibit synergy, the character of this synergy is often quite different. Time-dependent, multifunctional catalysts display so-called *convergent synergies*, whereas their energy-dependent analogs display *complementary synergies*. Convergent synergies are dramatically more powerful than complementary synergies, because coordinated, synchronized action is an intrinsic and an essential part of a convergent system, whereas it can only ever be incidentally present in a complementary system.

Synergy often provides a useful and simple indicator of the type of action employed by a catalyst.

Synergy is important in another, more intellectual respect. It provides a conceptual connection with action–reaction chain processes outside of chemistry and catalysis. Many time-dependent processes, including *evolution* and *laissez-faire economics* among others, also often display convergent synergies [20]. The extraordinary catalytic power that is so characteristic of enzymes, therefore, has a profound and deeply fundamental origin that goes far beyond mere chemistry and catalysis.

Chapter 9 reconsiders the problem of *conceptually unifying heterogeneous, homogeneous, and enzymatic catalysis* in light of the previous chapters. We demonstrate that a common conceptual thread links these subdisciplines, namely, the existence of energy- and time-dependent catalytic actions. Nonbiological homogeneous catalysis is mainly energy-dependent. Biological catalysis is largely time-dependent. Heterogeneous catalysts seem to employ both time- and energy-dependent actions. Time-dependence in biology and certain heterogeneous catalysis seems to have developed because of the fundamentally combinatorial nature of their catalysis. A combinatorial approach has, by contrast, not been examined in nonbiological homogeneous catalysis.

The most fundamental catalytic processes that take place in heterogeneous, homogeneous, and enzymatic catalysis are therefore the same. They differ only insofar as they play out in different settings. In heterogeneous catalysts, they generally take place on two-dimensional (2-D) surfaces. In most homogeneous catalysis, they take place in three-dimensional (3-D) space. Because it is far more difficult to create synergies of action in 3-D space, most manmade homogeneous catalysts have, thus far, been energy-dependent. Enzymes, with their large and varied 3-D structures have, however, been able to create the necessary synergies and therefore bring about time-dependent catalysis.

In Chapter 10, the *rational design of time-dependent homogeneous catalysts* is addressed. We show that such catalysts rely on the simultaneous presence of two key attributes:

(1) Catalytic groups that are spatially disposed to flex about a structure that is complementary to the optimum transition state of the reaction, that is, an active site which is *structurally complementary* to the transition state (*structural complementarity*; this ensures the catalyst is *thermodynamically highly efficient*).

(2) Weak and highly dynamic individual catalyst–reactant interactions that are, nevertheless, sufficient to activate the bound reactants so that they will react when brought into contact with each other along optimum collision trajectories (*catalyst–reactant dynamism;* this ensures that *the collision frequency is limiting*).

Attribute 1 is needed to ensure a flat energy landscape during the reaction. This landscape is achieved by constraining to near optimum the approach pathways of the bound reactants immediately before the collision. Attribute 1 also ensures that the catalyst spends a significant proportion of its time during conformational flexing at or near the point of reactive collision.

Simultaneous with attribute 1, attribute 2 is needed to limit the duration of full reactant binding and thereby to limit the time available for a reactive collision between the reactant functionalities. This attribute is necessary to ensure fidelity in the reaction process and to eliminate the formation of unwanted products. It also ensures that catalyst–reactant dynamism is the limiting factor in the catalytic process.

The combined effect of these measures is to create a system whose thermodynamic efficiency allows for a faster reaction than its mechanical dynamism. The system is therefore time-dependent.

A key problem in the rational design of nonbiological, time-dependent catalysts is to replicate these attributes successfully. The trouble is that in a time-dependent species, the catalytic rate is dependent on the synchronization of catalyst flexing (attribute 1) and catalyst–reactant binding (attribute 2). If either of these attributes deviates even somewhat from the optimum, they rapidly become asynchronous and no catalytic effect is observed.

The rational design of a time-dependent homogeneous catalyst therefore involves *simultaneously optimizing* both attributes. This optimization process is problematic because it does not allow for systematic optimization of each variable separately while keeping the other arbitrarily constant. In effect, the two interconnected variables must be made *concurrently* ideal in order to observe even a minor effect. Only when both variables approach optimum will a measurable effect be obtained.

Achieving such an outcome is not a simple problem. Two major strategies are proposed. Chapter 10 also provides a literature survey of time-dependent, nonbiological homogeneous catalysts and their closely related energy-dependent analogs. The intention is to provide the reader with a framework and a toolbox of design elements with which to devise new time-dependent catalysts.

Chapters 11–13 employ the tools provided in Chapter 10 to discover and study several practical, efficient molecular catalysts. In Chapter 11, we demonstrate a *bio-inspired, solar water-splitting catalyst* and electrochemical cell that, arguably, employs a rationally designed, time-dependent catalyst. In Chapter 12, we describe a *combinatorial, hydrogen reduction electrocatalyst* that employs a time-dependent action within a bulk heterogeneous–homogeneous system. Chapter 13 discusses the development of a readily prepared, *combinatorial oxygen reduction catalyst* that may be time-dependent. These examples are intended to provide a practical

1.6 THE AIMS, STRUCTURE, AND MAJOR FINDINGS OF THIS SERIES

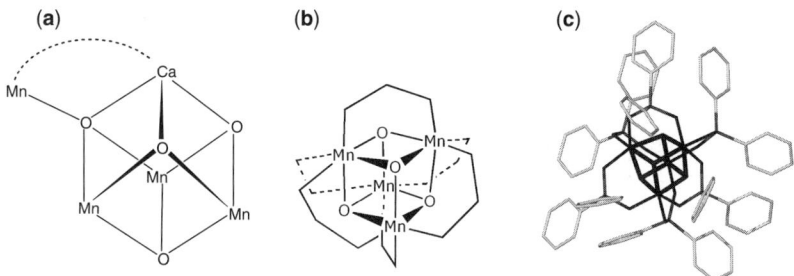

Figure 1.10. (a) Schematic depiction of the proposed cubical arrangement of the CaMn$_3$O$_4$Mn core in PSII-WOC. (b) Schematic depiction of its cubical Mn$_4$O$_4^{6+}$ core of Dismukes's cubane model complex Mn$_4$O$_4$(Ph$_2$PO$_2$)$_6$ and (c) XRD-derived structure of Mn$_4$O$_4$(Ph$_2$PO$_2$)$_6$.

illustration of the principles discussed in this series and the potential advantages associated with time-dependent catalysis.

Chapter 11 commences the above collection with the description of a fully functional, bio-inspired model of the water-oxidizing center in Photosystem II. This work was performed in the laboratories of Prof. G. Charles Dismukes at Princeton University, Princeton, NJ, and of Prof. Leone Spiccia at Monash University, Australia.

The Water-Oxidizing Complex (WOC) of Photosystem II (PSII) [Fig. 1.10(a)] is the only natural system that can efficiently photooxidize water using visible light according to

$$2H_2O \rightarrow O_2 + 4H^+ + 4e^- \qquad (1.3)$$

In this remarkable natural catalyst, an excited state of chlorophyll (P680*) is converted into a cation radical by charge separation (1.0 V vs. Ag/AgCl), which, in sequential photochemical steps, extracts four electrons from an oxo-bridged Ca$_1$Mn$_3$O$_x$ core, thereby releasing oxygen. The Ca$_1$Mn$_3$O$_x$ core has been proposed to involve a cubical structure.

Dismukes and coworkers at Princeton University have developed in recent years a range of model oxo-bridged, tetra-manganese complexes that approximately mimic in form and shape the proposed structure of the water-oxidizing center in Photosystem II. These *cubane* complexes have the formula Mn$_4$O$_4$L$_6$, where L$^-$ is a phosphinate ligand (p-R-C$_6$H$_4$)$_2$PO$_2^-$ (R=H, alkyl, methoxy). Each complex possesses a cubical Mn$_4$O$_4^{6+}$ core that assembles spontaneously from simple Mn^{2+} and MnO$_4^-$ salts in the presence of the phosphinate ligands, which form bridges between pairs of the Mn center [Fig. 1.10(b)]. The cubical core exists in a dynamic equilibrium with its constituents.

As noted, the active sites of many enzymes are structurally optimum for the formation of their respective transition states (attribute 1 above). If it can be assumed that the proposed cubical shape and structure of the PSII-WOC is also structurally optimum to its transition state for water oxidation, then Dismukes's *cubane* model complexes must, arguably, also fit this bill or at least come close to it. That is, a possibility exists that the *cubane* model complexes of Dismukes display attribute 1 in the section above on Chapter 10.

This result is significant because they undoubtedly also display attribute 2; as noted, the *cubanes* spontaneously self-assemble from Mn^{2+} in the presence of the phosphinate ligands. The *cubanes* clearly undergo dynamic and rapid catalyst–reactant binding.

According to the principles described in this series, the *cubane* model complexes are therefore likely to display both of the required attributes 1 and 2 and are, consequently, predicted to be candidate, time-dependent photooxidation catalysts of water.

When doped into a thin Nafion layer deposited on a suitable electrode and subjected to continuous illumination by white light, the *cubanes* were, indeed, found to facilitate sustained electrooxidation of water (at 1.0 V vs. Ag/AgCl), achieving a stable current and releasing O_2 continuously over a testing period of several days. This behavior occurred only when water was present in the electrolyte, indicating that the oxygen derives from the water, which is a reactant in the system. This result was confirmed by isotopic enrichment studies.

The catalytic assembly retained its activity when irradiated exclusively with visible light (>395 nm).

Chapter 12 then goes on to describe a family of readily prepared, *combinatorial, time-dependent, hydrogen-generating catalysts* with similarly remarkable properties. These species drastically accelerate the formation of dihydrogen from water (H^+) on the platinum surface of a normal hydrogen electrode (NHE). The catalyst generates hydrogen about seven times more rapidly than an equivalent Pt metal electrode. Pt metal is currently the best manmade hydrogen reduction catalyst. This work was largely performed by research fellows Drs. Jun Chen and Chee Too in the laboratories of Prof. Gordon G. Wallace at the University of Wollongong, Australia.

Chapter 13 expands on the concept of a combinatorial, time-dependent catalyst by describing the use of high concentrations of monomeric cobalt porphyrins within densely deposited polypyrrole to realize efficient *time-dependent oxygen reduction catalysis*. We show that the resulting catalysts are suitable for use in H_2/O_2 fuel cells. This work was also carried out at the University of Wollongong in Australia.

Although this series therefore initially delves deeply into the theory of catalysis, it becomes, in the end, a very practical account of how one could set about making highly active and specific new time-dependent catalysts capable of facilitating novel transformations.

REFERENCES

1. Atkins, P. W. *Physical Chemistry*, Oxford University Press, 1978, pp. 52–782.
2. Van Ness, H. C. *Understanding Thermodynamics*, Dover Publications Inc., 1969; ISBN 0-486-63277-6.
3. (a) Dunning-Davies, J. *Concise Thermodynamics: Principles and Applications*, Horwood Publishing, 1997; ISBN 1-8985-6315-2; (b) Waldram, J. R. *The Theory of Thermodynamics*, Cambridge University Press, 1985; ISBN 0-521-28796-0.

4. (a) Marion, J.; Thornton, S. *Classical Dynamics of Particles and Systems*, Harcourt College Publishers, 1995; ISBN 0-03-097302-3; (b) Tipler, P. *Physics for Scientists and Engineers: Mechanics, Oscillations and Waves, Thermodynamics (5th ed.)*, W. H. Freeman, 2005; ISBN 0-7167-0809-4.

5. Cardwell, D. S. L. *From Watt to Clausius: The Rise of Thermodynamics in the Early Industrial Age*, Heinemann, 1971; ISBN 0-435-54150-1.

6. (a) Kleppner, D.; Kolenkow, R. J. *An Introduction to Mechanics*, McGraw-Hill, 1973; ISBN 0-07-035048-5; (b) Goldstein, H.; Poole, C. P.; Safko, J. L. *Classical Mechanics (3rd ed.)*, Addison Wesley,2001; ISBN 0-201-65702-3.

7. Westfall, R. S. *Never at Rest: A Biography of Isaac Newton*, Cambridge University Press, 1983.

8. See, for example: (a) Pearl, J. *Causality*, Cambridge University Press, 2000; ISBN 0-521-77362-8; (b) Salmon, W. *Scientific Explanation and the Causal Structure of the World*, Princeton University Press, 1984.

9. Dugdale, J. S. *Entropy and its Physical Meaning*, Taylor and Francis, 1998; ISBN 0-7484-0569-0.

10. Lewis, G. N.; Randall, M. *Thermodynamics and the Free Energy of Chemical Substances*, McGraw-Hill Book Co. Inc., 1923.

11. Nash, L. K. *Elements of Statistical Thermodynamics (2nd ed.)*, Dover Publications Inc., 1974; ISBN 0-486-44978-5.

12. de Hemptinne, X. *Non-Equilibrium Statistical Thermodynamics applied to Fluid Dynamics and Laser Physics*, World Scientific Publishing, 1992; ISBN 981-02-0926-6.

13. Chandler, D. *Introduction to Modern Statistical Mechanics*, Oxford University Press, 1987; ISBN 0-19-504277-8.

14. Bate, R. R.; Mueller, D. D.; White, J. E. *Fundamentals of Astrodynamics*, Dover Publications, 1971; ISBN 0-486-60061-0.

15. Pierce, A. D. *Acoustics: An Introduction to its Physical Principles and Applications*, American Institute of Physics, 1989.

16. White, F. M. *Fluid Mechanics*, McGraw-Hill, 2003; ISBN 0072402172.

17. (a) Gleick, J. *Chaos: Making a New Science*, Penguin, 1988; ISBN 0-14-009250-1; (b) Alligood, K. T.; Sauer, T. D.; Yorke, J. A. *Chaos. An Introduction to Dynamical Systems*, Springer Verlag, 2000; ISBN 0-387-94677-2; (c) Strogatz, S. H. *Nonlinear Dynamics and Chaos: With Applications to Physics, Biology Chemistry and Engineering*, Addison Wesley, 1994; ISBN 0-201-54344-3; (d) Katok, A.; Hasselblatt, B. *Introduction to the Modern Theory of Dynamical Systems*, Cambridge, 1996; ISBN 0-521-57557-5; (e) Stewart, I. *Does God Play Dice? The New Mathematics of Chaos*, Penguin, 1997; ISBN 0140256024; (f) Peitgen, H-O.; Juergens, H.; Saupe, D. Eds., *Chaos and Fractals: New Frontiers of Science*, SpringerVerlag, 1992.

18. Karlin, S.; Taylor, H. E. *A First Course in Stochastic Processes, (2nd ed.)*, Elsevier Science, 1975.

19. The term *"Butterfly Effect"* was first employed by E. N. Lorenz in an address to the *Annual Meeting of the American Association for the Advancement of Science,* in Washington D.C. on December 29, 1979.

20. (a) Lewin, R. *Complexity: Life at the Edge of Chaos*, Macmillan, 1992; (b) Waldrop, M.M. *Complexity: The Emerging Science at the Edge of Order and Chaos*, Simon and Schuster,

1992; (c) Corning, P.A. *Systems Research* **1995**, *12*, 89 (published on the web at http://www.complexsystems.org/publications/pdf/synselforg.pdf).

21. (a) Stryer, L. *Biochemistry (3rd ed.)*, W. H. Freeman and Company, New York, 1988, p. 232; (b) Solé R. V. and Goodwin, B. C. *Signs of Life: How Complexity Pervades Biology*, Basic Books, 2001.
22. Stryer, L. *Biochemistry (3rd ed.)*, W. H. Freeman and Company, New York, 1988, p. 34.
23. For example: (a) King, C. C. *Prog. Neurobiol.* **1991**, *36*, p. 279; (b) Schwarz, W. *Acta Physol. (Amst)* **2003**, *113*, 231.
24. (a) Cunningham, L. A. *From Random Walks to Chaotic Crashes: The Linear Genealogy of the Efficient Capital Market Hypothesis*, George Washington Law Review, Vol. 62, 1994, 546; (b) e.g: Bacry, E.; Kozhemyak, A.; Muzy, J. F. J. *Phys. Rev. E* **2006**, *73(6)*, Art. No. 066114 Part 2.
25. For example: Sneyers, R. *Envirometrics* **1998**, *8*, 517.
26. Kocarev, L.; Vattay, G. (Eds.) *Complex Dynamics in Communication Networks* (Series: Understanding Complex Systems), Springer, 2005; ISBN: 978-3-540-24305-2.
27. Braha, D.; Minai, Al. A.; Bar-Yam, Y. (Eds.) *Complex Engineered Systems: Science Meets Technology (Understanding Complex Systems)*, Springer, ISBN 3-540-32831-19.
28. For example: Seader, J. D.; Henly, E. J. *Separation Process Principles*, John Wiley and Sons, 2006, p. 66.
29. For example: Bertulani, C. A.; Danielewicz, P. *Introduction to Nuclear Reactions*, Graduate Students Series in Physics, 2004.
30. For example: Shkolnikov, P. L.; Kaplan, A. E.; Pukhov, A.; Meyer-ter-Vehn, J.; *Appl. Phys. Lett.* **1997**, *71*, 3471.
31. March, J.; Smith, M. B. *March's Advanced Organic Chemistry (5th ed.)*, John Wiley and Sons, 2001, p. 284.
32. For example: Padwa, A. *Pure Appl. Chem.* **2003**, *75*, 47.
33. Pilling, M. J.; Seakins, P. W. *Reaction Kinetics*, Oxford University Press, 1996, p. 66.
34. (a) Boehr, D. D.; McElheny, D.; Dyson, H. J.; Wright, P. E. *Science* **2006**, 313, 1638; (b) Vendruscolo, M.; Dobson, C. *Science* **2006**, *313*, 1586.
35. For example: (a) Moss, D. W. *Enzymes*, Oliver and Boyd, 1968, p. 68; (b) Agarwal, P. K. *Microb. Cell Fact.* **2006**, *15*, 2 and references therein.
36. Williams, R. J. P. *Trends in Biochemical Science* **1993**, *18*, 115.
37. (a) Hammes-Schiffer, S.; Benkovic, S. J. *Annu. Rev. Biochem.* **2006**, *75*, 519; (b) Alper, K. O.; Singla, M.; Stone, J. L.; Bagdassarian, C. K. *Protein Sci.* **2001**, *10*, 1319; (c) Agarwal, P. K. *Microb. Cell Fact.* **2006**, *15*, 2, and references therein.
38. Menger, F. M. *Acc. Chem. Rev.* **1993**, *26*, 206, and references therein.
39. Baron, R.; Lioubashevski, O.; Katz, E.; Niazov, T.; Willner, I. *Org. Biomol. Chem.* **2006**, *4*, 989.
40. Pauling, L. *Nature* **1948**, *161*, 707.
41. Bruice, T. C. *Ann. Rev. Biochem.* **1976**, *45*, 331.
42. Fersht, A. *Enzyme Structure and Mechanism (2nd ed.)*, W. H. Freeman and Company, 1977, p. 331, and references therein.
43. Page, M. L.; Jencks, W. P. *Proc. Natl. Acad. Sci. USA* **1971**, *68*, 1678.

2

HETEROGENEOUS, HOMOGENEOUS, AND ENZYMATIC CATALYSIS. A SHARED TERMINOLOGY AND CONCEPTUAL PLATFORM. THE ALTERNATIVE OF TIME-DEPENDENCE IN CATALYSIS

GERHARD F. SWIEGERS

2.1 INTRODUCTION: THE PROBLEM OF CONCEPTUALLY UNIFYING HETEROGENEOUS, HOMOGENEOUS, AND ENZYMATIC CATALYSIS

The broad topic of catalysis is currently separated into three different subdisciplines based on the physical nature of the catalytic phenomenon:

(1) Nonbiological heterogeneous catalysis
(2) Nonbiological homogeneous catalysis
(3) Biological enzymatic catalysis

Each of these disciplines is generally considered to involve a philosophically distinct and separate set of fundamental catalytic principles that are largely the preserve of a particular group of specialist researchers. Heterogeneous catalysis is, for example, dominated by chemical engineers, whereas homogeneous catalysis

---

Mechanical Catalysis: Methods of Enzymatic, Homogeneous, and Heterogeneous Catalysis,
Edited by Gerhard F. Swiegers
Copyright © 2008 John Wiley & Sons, Inc.

lies mostly in the realm of chemistry. Enzymatic catalysis is largely studied by biochemists.

Catalysis science has long recognized the need to unify coherently the different catalytic phenomena into a single conceptual whole. Recent workshops and texts have specifically identified the recognition of general trends and principles that unify the different types of catalysis as an important opportunity (and challenge) [1,2].

A key difficulty in this respect has been to relate conceptually the catalytic action in heterogeneous, homogeneous, and enzymatic catalysis. What is the connection? A conceptual thread that coherently links these fields has not been discovered.

Unearthing this underlying commonality is made particularly problematic by the physical diversity of the catalytic actions in the different subdisciplines.

Another obstacle has been the assortment of terminologies involved. The various subdisciplines of catalysis employ different names, concepts, and perspectives to describe what must, in many cases, be the same fundamental phenomenon. These jargons dramatically complicate interdisciplinary dialogue between practitioners in the different subdisciplines.

Along with the need to *express* and *understand* the conceptual overlap between the different fields, therefore, a need to *translate* the different terminologies into a single, commonly understood language also exists. In other words, before considering catalysis as a single conceptual entity, one must first translate the vernaculars employed into a common vocabulary that may serve as a platform for cross-disciplinary dialogue. Only with such a common conceptual framework in hand can one get to grips with the issues raised in this series.

The aim of this chapter is therefore to develop a common basis for a mutual understanding of catalysis as it is manifested in homogeneous, heterogeneous, and enzymatic systems. Out of such an approach, we hope, in later chapters, to move toward a unified conceptualization of catalysis.

In this chapter, we will also introduce and describe several examples of catalytic systems that we will use, often repeatedly, in later chapters to illustrate issues and concepts. Our intention in employing the same set of practical examples over and over again is to minimize, as far as possible, the need for new learning by the reader.

In the final section of this chapter, we will reconsider and clarify the issue of time-dependent catalysis in light of the common concepts described herein.

2.2 BACKGROUND: WHAT IS HETEROGENEOUS, HOMOGENEOUS, AND ENZYMATIC CATALYSIS? TRENDS IN CATALYSIS SCIENCE

2.2.1 Homogeneous and Heterogeneous Catalysis

The history of catalysis has seen it develop into two major intellectual streams: *homogeneous* and *heterogeneous* catalysis.

Homogeneous catalysis involves processes that take place entirely in a single phase, such as a solution phase. Heterogeneous catalysis occurs at the interface between two phases, such as on the gaseous surface of a solid, metal catalyst. Although heterogeneous catalysis currently dominates the industrial applications of catalysis, homogeneous catalysis is increasingly employed, particularly in high-value-added processes.

The two fields differ fundamentally in that homogeneous catalysis is molecular in nature. Thus, homogeneous catalysts comprise individual molecules. Heterogeneous catalysts, by contrast, are typically large agglomerations of atoms, such as metals or metal oxides.

Because the catalytic process takes place on a solid surface, heterogeneous catalysis typically involves the simultaneous involvement of many different types of active site, including *face*, *edge*, *step*, and *defect* sites of the type depicted in Figure 2.1(a)–(d). Heterogeneous catalysis, therefore, has the property of being, fundamentally, *multisite* in character. That is, catalysis occurs concurrently at many different active sites.

Homogeneous catalysts, by contrast, are generally individual molecules solubilized in open solution. In such catalysts, only one type of active site is normally present, because the site is defined by the catalyst molecule, which is the same everywhere that catalysis takes place [Fig. 2.1(e)]. Homogeneous catalysis is therefore *single-site* and *molecular* in character.

This molecular and single-site nature endows homogeneous catalysis with important advantages over heterogeneous catalysis.

These advantages include [3]:

(1) Greater selectivity and reproducibility, with accompanying improvements in the properties of the products generated.
(2) Milder reaction conditions.
(3) Ready chemical modification.
(4) More easily studied reaction mechanisms.
(5) Greater inherent efficiency as all of the metal atoms participate in the catalysis. In a heterogeneous catalyst, most metal atoms lie within the bulk and not at the surface of the catalyst. They are therefore unable to participate in the catalysis.

However, homogeneous catalysts have several practical disadvantages. They are typically more expensive, less robust, and shorter lived than heterogeneous catalysts. Homogeneous catalysts are also generally more complicated to handle than heterogeneous catalysts. For example, because they are dissolved in solution, homogeneous catalysts may become incorporated into the products they generate. This process can be an expensive outcome if the catalysts are not cheap. By contrast, heterogeneous catalysts, in the form of gauzes or screens, can usually be physically withdrawn from the product stream without any difficulty.

For these reasons, heterogeneous catalysts remain more practical and widely used than homogeneous catalysts in most industrial applications. Much of the current development of homogeneous catalysis is focused on improving on heterogeneous catalysis or on developing catalysts for transformations that cannot be catalyzed in heterogeneous systems.

Heterogeneous catalysts: *multisite*

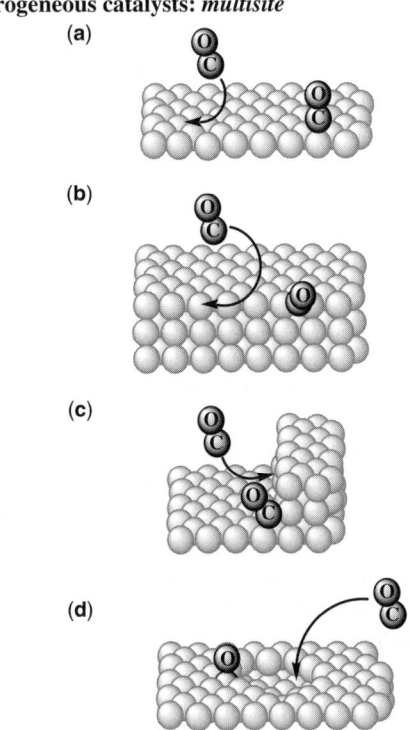

Homogeneous catalysts: *single site*

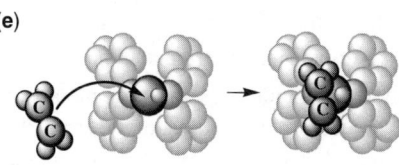

Figure 2.1. Schematic illustrating *heterogeneous and homogeneous single-site and multisite catalysts*. *Heterogeneous catalysis* typically involves the simultaneous presence of many different types of active site on which a reactant (CO) may undergo a catalytic transformation, like (a) face, (b) edge, (c) step, and (d) defect sites on the catalyst (Pt) surface (*multisite catalysis*). *Homogeneous catalysis* involves the presence of a single type of active site (e), created by molecules of the catalyst (*single site catalysis*). In solution, each catalyst intermediate HPd(PPh$_3$)$_2^+$ is, on average, identical, as far as the reactant ethylene is concerned. For clarity, one of the phenyl groups in each PPh$_3$ has not been shown in (e).

2.2.2 Hybrid Homogeneous–Heterogeneous Catalysts

As a consequence of the above factors, one of the major trends in catalysis science over the last few decades has involved developing systems that combine the advantages of homogeneous and heterogeneous catalysts while avoiding their disadvantages.

The principal way in which this trend has been brought about has been by attaching, or *supporting*, molecular, homogeneous catalysts on solid substrates, such as inorganic oxides, glasses, or polymers [4]. This approach has, in many cases, drastically improved the properties of the homogeneous catalyst because the support may act to [5]:

(i) Isolate and protect the metal atom/ion, often in close proximity to another such species, thereby preventing bimolecular and other deactivation processes
(ii) Promote coordinative unsaturation of the metal atom/ion, thereby speeding up the reaction
(iii) Improve the selectivity of the catalysis, with an accompanying improvement in the homogeneity of product mixture obtained
(iv) Retain the metal atom/ion strongly, allowing maximum product generation by the catalyst and as a consequence of the above factors
(v) Allow cooperative, multistep catalysis in which the product of one step is consumed in the next

As a result of these developments, homogeneous catalysts, in supported form, are now used in a wide variety of industrial processes [6]. For example, whereas Wilkinson's catalyst, $RhCl(PPh_3)_3$, loses its catalytic activity when exposed to air in open solution, it remains catalytically active if supported on a polystyrene resin [7].

2.2.3 Enzymatic Catalysis

Another, much more nascent trend in catalysis science has involved exploiting the molecular character of homogeneous catalysis to emulate enzymes [8].

Enzymes are large protein macromolecules that perform catalytic tasks in biology [9–13]. The active site of a typical enzyme contains amino acids, known as catalytic groups, which simultaneously participate in localizing, binding, and catalyzing transformations of reactant (substrate) molecules [14,15].

Enzymes have a molecular mode of action, albeit that they are rather large molecules. They operate in open solution. However, their catalytic capacities generally far exceed those of comparable manmade systems. Indeed, enzymes as a class are considered the most efficient type of catalysts, by a considerable margin.

To give just one example, the enzyme *carbonic anhydrase* is one of the most active catalysts known. One molecule of the enzyme can convert 600,000 carbon dioxide molecules *per second* (in our muscles) into aqueous carbonic acid (in our blood stream). It performs this conversion in the presence of a wide variety of other possible reagents at body temperature and at a partial pressure of less than 1 atmosphere [16]. Other enzymes operate under similar conditions [16].

By contrast, large-scale industrial processes employing heterogeneous catalysts typically operate at increased temperatures ($>250\,°C$) and extremely high pressures (e.g., 200 atmospheres), and they require highly purified reagent streams.

The dream of practitioners in catalysis is, consequently, to develop catalytic systems that have the properties of an enzyme (namely, high activity and selectivity) but also the advantages of heterogeneous catalysts (namely, low cost, large scale, continuous processes, ready separation of the catalyst from the products, direct electrical regeneration of redox catalysts).

Although many enzymes can and have been attached to substrates, their catalytic activity under such conditions is generally a fraction of that in open solution. The challenge is, consequently, to devise *nonbiological catalysts that mimic the catalytic properties of enzymes* [17]. Breslow has termed such species *Artificial Enzymes* [17].

Enzymatic catalysis is formally a type of homogeneous catalysis because it is molecular, single-site, and not based on physical properties unique to bulk aggregations of atoms as observed in heterogeneous catalysis [9–13]. In their extraordinary properties, enzymes can, indeed, be considered to embody the ultimate maximization of the advantages of homogeneous catalysis over heterogeneous catalysis.

But, how does one accurately replicate enzymatic catalysis?

2.2.4 Theories and Mimicry of Enzymatic Catalysis

The origin of enzymatic catalysis has been a subject of lively debate for many years. In Chapter 7, we will examine the many different theories of enzymatic catalysis that have been proposed. In summary, however, it is broadly agreed that the high catalytic activity of enzymes seems to derive from the proximity to each other and near-optimum spatial positioning of the catalytic groups within enzyme active sites. These groups consequently bind substrate molecules or atoms so that they are held very closely to each other, in arrangements that facilitate their combination in the transition state.

Pauling has, indeed, suggested that the structural arrangement of the catalytic groups in enzymes is complementary to the transition state of their reaction [18]. This complementarity is believed to stabilize the transition state greatly, thereby reducing the activation energy (E_a) of the reaction, and dramatically speeding up the overall conversion of the substrate to products [18–20].

Whereas enzymes seem to rely on the inherent, "in-built," structural arrangement in their catalytic groups to bring their substrates into reactive contact, heterogeneous catalysts typically depend on *adventitious proximity* in surface adsorbed reagents or intermediates. That is, heterogeneous catalysts present a large variety of possible active sites upon which the adsorbed reactants may migrate. During this migration process, the adsorbed reactants may, occasionally, happen to find themselves optimally disposed and proximate for reaction. Therefore, an element of statistical chance exists in the way that catalysis occurs in hetereogeneous catalysts, which is intrinsically combinatorial in character.

This reliance on statistical chance is arguably the fundamental cause of the comparative inefficiency of heterogeneous catalysts. It is also the reason that they require extreme conditions. High pressures and temperatures maximize the statistical likelihood that reactants will, at some point, be sufficiently closely and optimally arrayed to each other on the catalyst surface. Heterogenous catalysts are, of course,

2.2 WHAT IS HETEROGENEOUS, HOMOGENEOUS, AND ENZYMATIC CATALYSIS?

often also unselective in their binding, so that reagent impurities can easily poison the catalyst by binding to it irreversibly.

Various strategies have been employed to replicate the proximity effects of enzymes in nonbiological homogeneous catalysts. These strategies are considered in detail in Chapter 10. However, the most common approach has involved chemically attaching prospective catalytic groups close to each other, in binuclear or multinuclear homogeneous catalysts. Figure 2.2 depicts some successful examples of such catalysts, all of which involve two metallic catalytic groups (Co, Fe, Mn, Cu) held in close proximity to each other. The metals in these catalysts bind small molecules, which are then transformed in the space between them. The examples in Figure 2.2 involve:

(i) A co-facial metallodiporphyrin **A** that catalyzes the transformation of O_2 to H_2O [21]
(ii) A heterobimetallic capped porphyrin **B** that catalyzes the transformation of H_2O to O_2 [22]
(iii) An angled metallodiporphyrin **C** that catalyzes the transformation of O_2 to H_2O [23]

Other *biomimetic* catalysts of this type include, among others, bis(N_3-donor) dicopper complexes [24], covalently linked dinuclear Schiff base complexes [25], μ-carboxylate bridged diiron molecules [26], and various metal clusters [27].

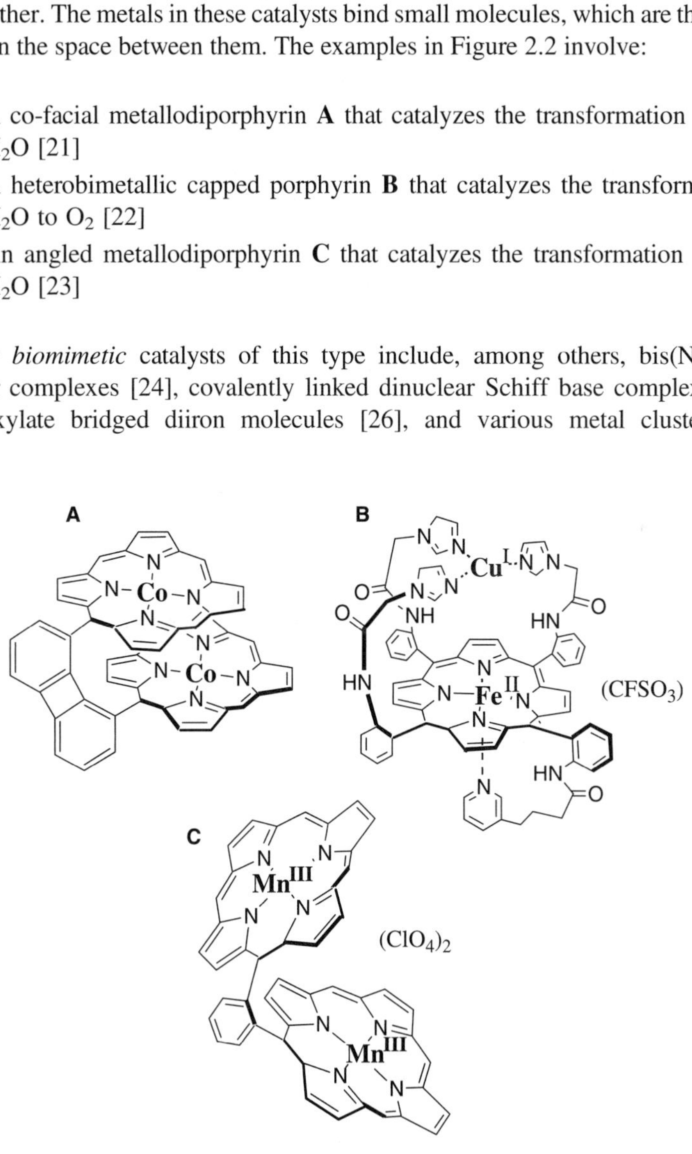

Figure 2.2. *Biomimetic catalysts.*

Several alternative approaches have also been used to create enzyme-like proximity effects in nonbiological catalysts [28,29].

2.3 DISTINCTIONS WITHIN HOMOGENEOUS CATALYSIS: *SINGLE-CENTERED* AND *MULTICENTERED* HOMOGENEOUS CATALYSIS

To avoid any confusion between practitioners of the different subdisciplines of catalysis, we need to now distinguish very carefully between catalyst *binding sites* and catalyst *active sites*. Both are *sites*, but what happens at these sites is quite different.

Active sites are collections of atoms about which catalysis takes place. *Binding sites* are atoms within the active site to which the reactants are bound during their catalytic transformation.

The following discussions are intended to illustrate this point.

2.3.1 Single-Centered Homogeneous Catalysts. Most Manmade Homogeneous Catalysts Are Single-Centered Catalysts

In most industrial homogeneous catalysts, a single atom binds the reactants and performs all of the direct catalytic functions. For example, consider the *metallocene* homogeneous catalysts **1** and **2** in Figure 2.3 [30–32]. In a solvent such as hexane or toluene, both catalysts spontaneously convert propylene into poly(propylene) in the presence of a co-catalyst, known as methyl aluminoxane or *MAO*.

The significance of the catalysis is that the polypropylene generated by **1** is exclusively *isotactic*, which means that each methyl group along the polymer chain has the opposite orientation to the previous one. In other words, the relative disposition of the pendent methyl groups as one moves along the polypropylene chain alternates between *up*, then *down*, then *up*, then *down*, and so on. This steric arrangement imparts the resulting *isotactic*-polypropylene with useful physical properties, including toughness and improved processing capacities.

By contrast, catalyst **2** converts propylene into so-called *syndiotactic* polypropylene, in which the orientation of the methyl groups along the polymer backbone does not alternate but rather stays the same: *up*, *up*, *up*, and so on. Syndiotactic polypropylene has remarkably different physical properties to *isotactic* polypropylene, being, for example, softer and more maleable. Both isotactic and syndiotactic polypropylene are, nevertheless, still polypropylene.

Thus, metallocene catalysts like **1** and **2**, offer an ability to *tailor* and *customize* the properties of the polymer. For a mature, commodity material like polypropylene, this approach has revolutionized the industry and has created an entire new niche manufacturing business in this respect.

How do these catalysts work? They are believed to operate by forming a positively charged Zr intermediate in solution. The Zr ion in this intermediate has a vacant coordination site to which the electron-rich propylene monomer readily binds in a side-on fashion. After binding, the propylene monomer undergoes an *insertion*

2.3 SINGLE-CENTERED AND MULTICENTERED HOMOGENEOUS CATALYSIS

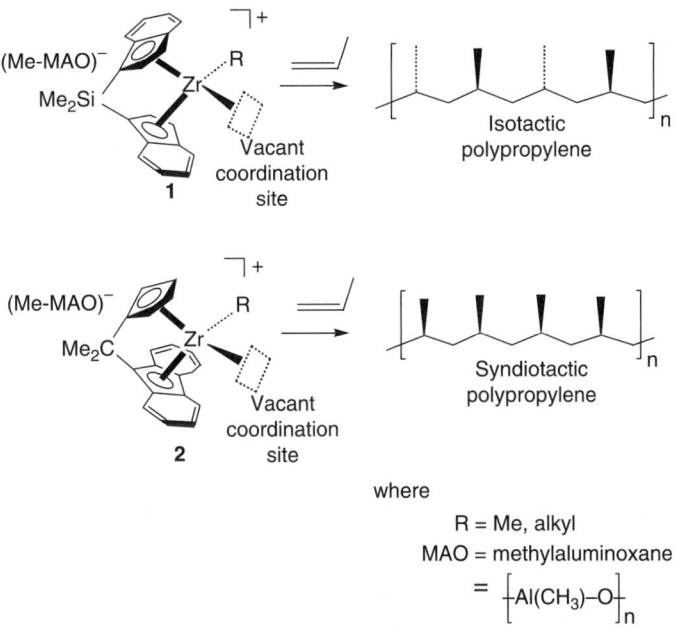

Figure 2.3. *Single-centered homogeneous catalysis.* Propylene polymerization by ansa-zirconocenes in toluene solution. Catalysis occurs at a single active site containing only one catalytically functional chemical group: the Zr ion [30–32].

reaction in which it is inserted into the growing polymer chain on the other side of the Zr ion. In so doing, the insertion reaction regenerates the vacant coordination site, starting the whole process again.

In the active intermediates depicted in **1** and **2**, the vacant Zr coordination sites are shown by the broken squares. It is to these sites that the propylene monomer binds in a side-on manner. In so doing, the alkyl group R bound to the opposite side of the Zr ion is disturbed, which causes it to insert the propylene into the Zr–alkyl bond [32,33]. A Zr coordination site now again becomes vacant, which allows for the above process to repeat. Many such repetitions on alternating sides of the catalyst may take place, thereby generating the polymer shown.

The important point to note here is that the reactants and products in this catalyst are only ever attached to the Zr ion. Although the other atoms present in **1** and **2** may be electronically or sterically essential to the catalysis, they are not directly involved in the process *per se*. Thus, although a *single site* is at work in this catalysis, in fact, only *one atom* (the Zr ion) exists within that site that is performing all the work.

Species in which the reactants bind and are transformed by a *single atom* in the catalyst are conveniently termed *single-centered* homogeneous catalysts [34]. It is important to distinguish single-centered species from other homogeneous catalysts in which more than one atom within the catalyst site facilitates the reaction.

Species of that type are termed *multicentered*. That is, they involve multiple atoms within the active site that bind and transform the reactants.

All homogeneous catalysts are therefore *single site* in character. That is, the catalysis occurs only at one, identically arranged collection of atoms. This occurs because, in solution, such catalysts offer only one arrangement of atoms in their active site. However, not all homogeneous catalysts are *single centered*. Only those species in which a *single atom* performs all of the catalytic work are single centered [34]. Homogeneous catalysts in which multiple atoms within the active site facilitate the catalysis are termed *multicentered* [34]. We will describe and illustrate *multicentered* homogeneous catalysts in the next section.

Beyond the zirconocene polymerization catalysts shown in Figure 2.3, other prominent examples of single-site, single-centered catalysts include various chiral Pd arylation catalysts [35,36].

2.3.2 Multicentered Homogeneous Catalysts: Most Enzymes Are Multicentered Homogeneous Catalysts

Some homogeneous catalysts involve two or more atoms binding and transforming reactants during catalysis [37]. Species **A**, **B**, and **C** in Figure 2.2, for example, contain two catalytic groups (the Co, Mn, Fe, and Cu atoms). Both of these groups are actively involved in binding and transforming the reactants during catalysis.

Another example is depicted in Figure 2.4, which shows how the two Cr ions in **3a–c** direct the catalytic ring-opening of an epoxide (the O< species at the bottom left of the figure) [38]. As can be seen in the square brackets at the middle and bottom of Figure 2.4, one Cr ion in **3a–c** binds the epoxide (O<), whereas the other binds an azide (N_3). The catalyzed reaction involves the reaction of the bound epoxide with the bound azide in a "head-to-tail" manner as shown at the bottom of the figure. The resulting product has a specific stereochemistry, as shown at the bottom right of the figure, with the azide substituent (N_3) oriented downward relative to the five-membered ring and the hydroxide (OH) oriented upward.

Two monomers **4** may also catalyze the reaction in a similar way, but this does not generate the same stereospecificity in the product that is obtained with the tethered Cr ions in **3a–c**.

The active site in these catalysts, as well as in **A**, **B**, and **C** in Figure 2.2, is the space *between* the two metal centers.

Species in which two or more groups within the active site bind and transform reactants are conveniently termed *multicentered* homogeneous catalysts [34]. It is important to realize that such species are nevertheless still *singlesite* catalysts because all of the active sites in solution are identical.

Figure 2.5 depicts in a schematic form a *single-centered* and a *multicentered* catalyst. As noted, enzymes are homogeneous catalysts and therefore *single-site* in character. However, enzymes typically bind and transform their substrates using many individual binding interactions, which makes them *multicentered* catalysts.

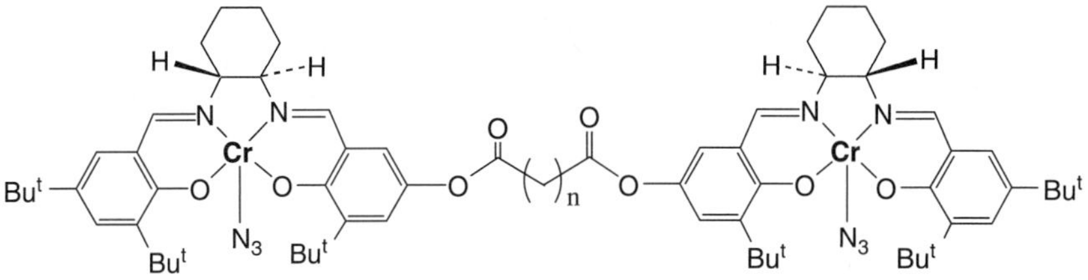

2.3 SINGLE-CENTERED AND MULTICENTERED HOMOGENEOUS CATALYSIS

a: n = 5 **b:** n = 2 **c:** n = 10

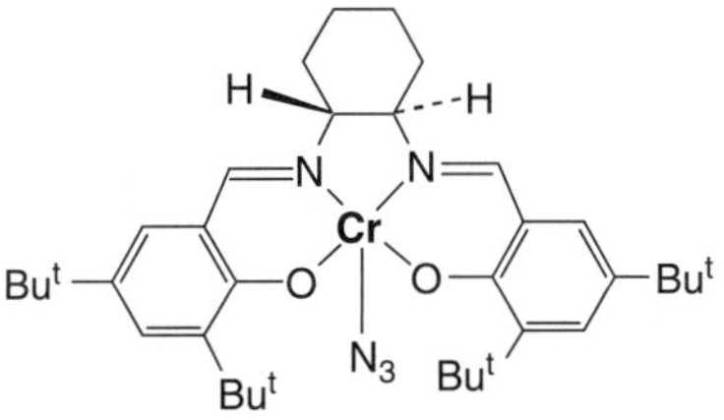

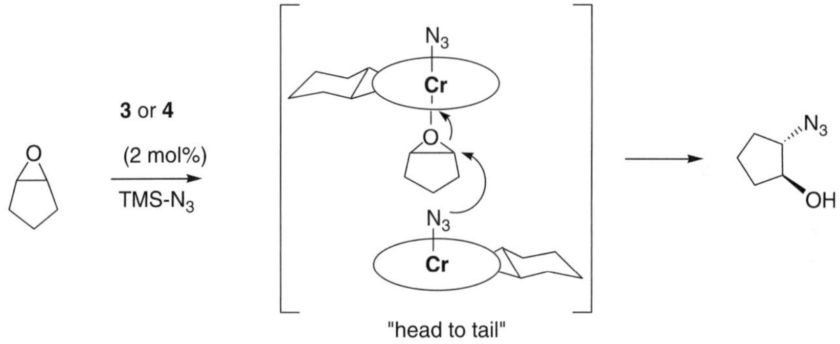

Figure 2.4. *Multicentered homogeneous catalysis.* Catalytic ring-opening reaction of an epoxide. The reaction involves two catalytically functioning Cr groups [37].

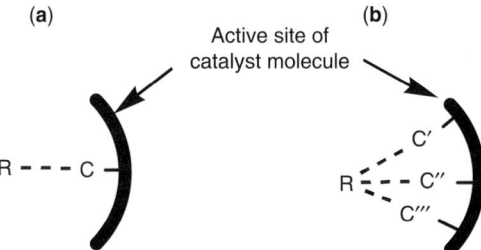

Figure 2.5. Schematic illustrating *single-centered and multicentered homogeneous catalysis*. The transformation of a reactant R within the active site of the catalyst may involve the direct actions of (a) one catalytic group C (*single-centered catalysis*) or (b) more than one catalytically functioning chemical group C′, C″, and C‴ (*multi-centered catalysis*).

2.4 THE DISTINCTION BETWEEN *SINGLE-SITE/MULTISITE* CATALYSTS AND *SINGLE-CENTERED/MULTICENTERED* CATALYSTS IN *HETEROGENEOUS CATALYSIS*: AN IMPORTANT CONVENTION USED IN THIS SERIES

The reason that we needed to distinguish clearly between single-site/multi*site* and single-centered/multi*centered* catalysts in the preceding discussion is that the discipline of heterogeneous catalysis makes no formal distinction between the *active site* (that is, the location where catalysis take place) and *binding sites* (that is, the atoms within the active site to which the reactants bind and are transformed). Within heterogeneous catalysis, all such centers and sites are simply and generically termed *catalytic sites*.

This terminology is likely a consequence of the relative scarcity of structural and mechanistic information that is available in heterogeneous catalysis. In the absence of such information, it is usually impossible to say which specific atoms on the surface of a heterogeneous catalyst are binding sites and which collections of atoms are active sites.

Although this issue has not been a major impediment to heterogeneous catalysis thus far, it does mean that the discipline cannot currently formally define and conceptually distinguish between single-centered and multicentered catalysis. As more structural and mechanistic information becomes available, such a distinction may, in future, be possible.

2.4.1 A Key Convention Used in This Series: A Catalytic Site Is a Collection of Atoms about Which a Reaction Is Catalyzed. A Catalytic Center Is an Atom Within that Site Which Binds and Facilitates the Transformation of a Reactant

To avoid any confusion in this respect, and for the benefit of practitioners of heterogeneous catalysis, we will, in this series, clearly and explicitly distinguish between the active sites of a catalyst, which will be termed *catalytic sites*, and the individual binding sites within the active site, which will be termed *catalytic centers*. This distinction is crucial to the arguments made in subsequent works within this series.

2.5 THE ALTERNATIVE OF *TIME-DEPENDENCE* IN CATALYSIS

As noted, the key feature of enzymatic catalysis is currently considered to be that enzymes are structurally complementary to their transition states. It is believed to reduce the activation energy E_a and to impart them with a powerful *thermodynamic* inducement for reaction, thereby explaining their remarkably high catalytic activities. Although this conceptualization is fine in theory, it does not address the practical question of what actually happens during an enzyme-catalyzed reaction. In particular, it does not answer the question: How do enzymes induce the many catalytic groups in their active sites to act in a *concerted, coordinated* fashion? Such a *synchronized*

2.5 THE ALTERNATIVE OF *TIME-DEPENDENCE* IN CATALYSIS

interplay between the different catalytic groups is essential to achieving the great specificities for which enzymes are famous.

Perhaps the best illustration of this issue can be found in the field of *biomimetic chemistry*, which seeks to emulate the principles on which enzymes operate. Numerous nonbiological *biomimetic* catalysts have been prepared that are, to varying extents, complementary to their transition states (including, arguably, **A–C** in Figure 2.6). Yet, these catalysts typically do not display anything like the sort of activities and specificities that are common to enzymes.

For example, consider Breslow's famous *biomimetic* Mn porphyrin catalyst **6**, which is laterally (*trans*-) appended with two β-cyclodextrins, as shown in Figure 2.6 [39].

When treated with a linear olefin substrate **7** of appropriate length, the long-chain tails of the olefin are strongly favored to be included in the two opposite cyclodextrin cavities, with the C–C double bond then held atop the Mn ion (as depicted in the schematic in square brackets in Figure 2.6). Oxidative epoxidation consequently occurs mainly at the central C=C bond. However, it does not occur *exclusively* at this bond. It also occurs elsewhere.

In other words, **6** has been designed to be approximately complementary to its transition state in the same way that enzymes are said to be complementary to their transition states. But this has not led to activities or selectivities that remotely approach those observed with enzymes. It has certainly improved the selectivity of the catalysis; the comparable *cis*-appended **8** catalyzes the same reaction with no selectivity whatsoever.

Key Point. *The key issue here is that one can create a catalyst that is complementary to its transition state, but how does one ensure that the substrate is fully bound to it at the instant of reaction?* In other words, how does one create an *absolute requirement* for the substrate in Figure 2.6 to bind the cyclodextrins, *in order for* the Mn ion to catalyze the reaction? Such a requirement does not exist in the example in Figure 2.6. The Mn ion in **6** can and does act on its own on some occasions in a single-centered fashion. In effect, **6** has its catalytic groups arranged fairly ideally, but it does not have a way of *ensuring* that the substrate is correctly and *simultaneously* attached to all of the catalytic groups at the precise moment of reaction. Thus, the substrate is sometimes attached to only one, or even none, of the cyclodextrins at the instant of reaction.

Key Point. *The point being made here is that complementarity to the transition state is insufficient to create a notable catalytic effect on its own. Something more is needed to ensure concurrent actions.*

Enzymes, which also employ multiple catalytic groups, clearly do not have this problem. They somehow have a means of ensuring that at the precise instant of reaction, all of the substrate functionalities are correctly and simultaneously attached to their respective catalytic groups. This approach is the only way in which they can achieve their high selectivities. They meet these goals despite the fact that their substrates are bound by multiple, weak and dynamic binding interactions, such as hydrogen bonds, hydrophobic–hydrophilic interactions, and the like. How do they ensure

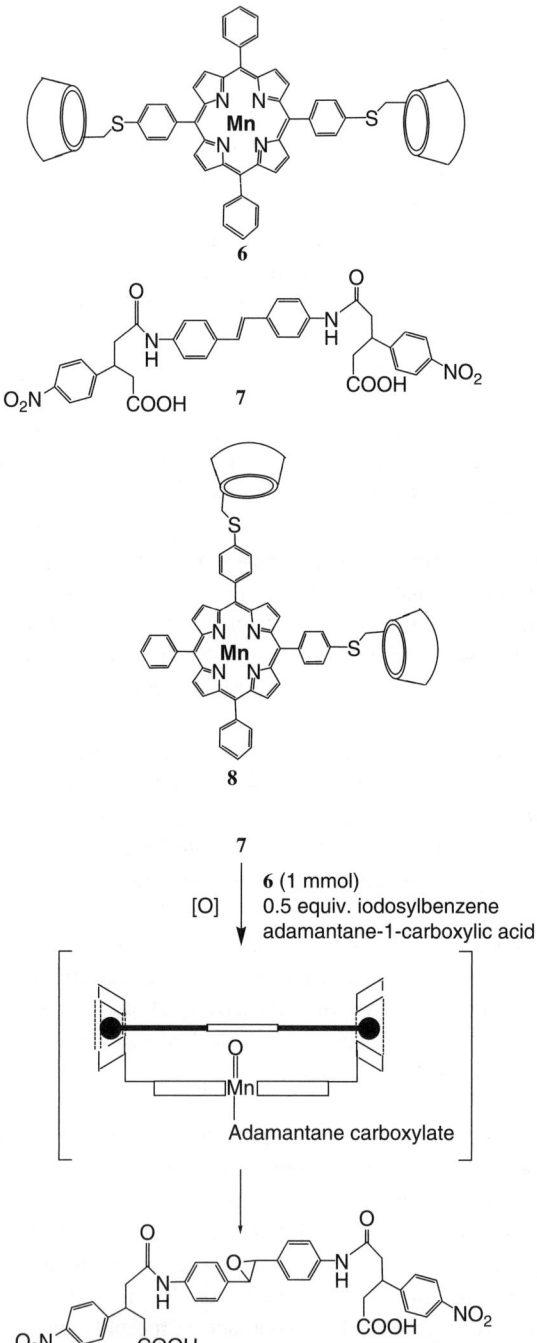

Figure 2.6. Biomimetic epoxidation catalyst. Mn porphyrin **6** oxidizes substrate **7** with high, but not perfect, selectivity. Porphyrin **8** shows little selectivity under comparable conditions, because it has no method to ensure that the substrate is simultaneously attached to all of the catalytic groups at the precise moment of reaction.

that every one of these weak and dynamic individual binding interactions are in place at the exact moment of reaction?

This problem has been recognized in biochemistry. The current thinking is that enzymes are not only complementary to their transition states but also to the substrates, albeit less so. In other words, enzymes bind their substrates strongly by a process of molecular recognition, but then they bind their transition states more strongly still [19,20]. This process is proposed to create a powerful thermodynamic driving force for reaction.

Thus, biochemists would argue that the above result occurs because **6** is *not complementary enough* to the substrate **7**. In other words, **7** can bind to **6** in several different ways, and this is reflected in the product distribution. If **6** were more complementary to **7**, and more strongly induced to bind **7** correctly, those options would be reduced.

Undoubtedly, there is merit to this argument given the rudimentary nature of catalysis by **6** relative to enzymes. The goodness-of-fit of **6** and **7** is certainly not nearly as precise as is the case with enzymes and their substrates. However the fact that other products are obtained using **8** means that these other products will *always* also be present, albeit in increasingly tiny proportions, as **6** is made more complentary to its transition state and its substrate. In other words, *some* reactant will always exist that does not bind properly at the instant of reaction. This must surely also be the case with enzymes. Indeed, the problem should be greater with enzymes given the weak individual interactions they employ with their substrates.

To put this another way, as noted in Chapter 1, a thermodynamic system will always distribute itself over the energetically most-stable possible outcomes. Thus, in a chemical reaction, the thermodynamically most stable product will be formed in greatest proportion. However, *other products* will also be generated in proportion to their stabilities *relative* to the most-favored product. In other words, what counts is not the *absolute* stability of the product or the energy barrier of the pathway but their stability *relative* other competing products and pathways.

In effect then, if enzymes employ a powerful thermodynamic driving force, then their extraordinary specificities can only develop if an enormous *differential* exists with the thermodynamic driving force for other, competing products. The current thinking in biochemistry is that such a differential comes about because of strong enzyme–substrate binding, which is then amplified by still stronger enzyme–transition state binding.

A purely thermodynamic driving force of this type is, of course, a valid conceptual possibility that we will examine in some detail in this series.

Key Point. ***However, another alternative is available. And that is that enzymes employ a time-dependent process of some sort in which reaction is only possible if all of the elements are simultaneously present and in place at the precise instant of reaction. In other words, the reaction can only take place if all of the substrates are fully and "correctly" bound, activated, and in optimum, physical, reactive contact at the precise moment of reaction. If any one, or even any one part of those elements is not present, then no reaction can occur.***

In effect, the reaction occurs because of an *advantageous confluence of circumstances*. That is, it occurs because, at a precise instant in time, everything is right for it to occur. Put another way, a limited slice of time is available during which all of the necessary elements of binding, activation, and physical contact may be in place for a reaction to take place. If they are synchronized at this precise instant in time, the reaction proceeds. If not, then no reaction occurs.

This is the essence of *time-dependent* processes. They do not take place because of a thermodynamic or other driving force, but because the physical confluence of circumstances is such that there is *no alternative but for them to take place*. Moreover, if the confluence of circumstances is not perfect—if any of the required parameters are not optimum—then they do not occur.

In precisely the same way, a line of dominoes of the type described in Chapter 1 does not fall because of gravity, although gravity is certainly essential to the process. Rather, they fall because there is no alternative *but* for them to fall once the first domino is knocked over. Moreover, if any one domino in that line is incorrectly placed, the whole process halts or does not occur in the first place.

The same is true of a finely tuned mechanical device or a machine. The machine will only work if all of its key parts are operating correctly and in unison at the same time. If any one part is nonoptimal, at any one instant in time, everything will grind to a halt.

If such a time-dependent mechanism is operative in certain enzymes, then it would potentially explain their remarkable specificities. Moreover, it could also explain their high catalytic activities if it occurred regularly and often. However, these features would then develop not out of a thermodynamic driving force but out of an advantageous confluence of circumstances in which all of the necessary elements happened to be simultaneously in place at one precise instant in time. At that exact moment, a thermodynamic influence may exist that completes the reaction in the same way that gravity takes over when a domino in a line is knocked over. However the fundamental impetus for the reaction would not be the thermodynamic gradient but the confluence of circumstances that led to it.

How, though, would such an action come about? How would it operate? And, perhaps more pertinently, how could it be replicated in a manmade system?

In the remainder of this series, we will examine the possibility of a time-dependent catalytic process in the various forms of catalysis, both homogeneous and heterogeneous. We will particularly examine enzymes in this respect and reflect on the current thermodynamically based explanation for their efficiency.

REFERENCES

1. (a) Author not stated. *Catal. Lett.* **2001**, *76*, 111, and references therein; (b) Bercaw, J. *Opportunities for Catalysis in the 21st Century, A Report from the Basic Energy Sciences Advisory Committee* (Workshop Chair: White, J. M.), May 14–16, 2002, U.S. Department of Energy.
2. Thomas, J. M.; Williams, R. J. *Phil. Trans. R. Soc.* **2005**, *363*, 765.

REFERENCES

3. Iwasawa, Y. Chap 1 in *Tailored Metal Catalysts*, Ed. Iwasaw, Y., D. Reidel Publishing Company, 1986, p. 87 and references therein.
4. Basset, J. M.; Smith, A. K. In *Fundamental Research in Homogeneous Catalysis*, Eds. Tsutsui, M.; Ugo, R.; Plenum Press, 1977, p. 69.
5. Hirai, H.; Toshima, N. Chap 2 in *Tailored Metal Catalysts*, Ed. Iwasawa, Y., D. Reidel Publishing Company, 1986, p. 87 and references therein.
6. Keim, W. In *Fundamental Research in Homogeneous Catalysis, Vol 4.*, Eds. Tsutsui, M.; Ugo, R.; Plenum Press, 1984, p. 131.
7. Capka, M.; Svoboda, P.; Cerny, M.; Hetflejs, J. *Tetrahedron Lett.* **1971**, 4787.
8. Parshall, G. W.; Ittel, S. D. In *Homogeneous Catalysis (2nd ed.)*, John Wiley & Sons, 1992, p. 1.
9. Kluger, R. In *Enzyme Chemistry: Impact and Applications (3rd ed.)*, Eds. Suckling, C. J.; Gibson, C. L.; Pitt, A. R.; Blackie Academic & Professional, 1998, p. 9–50.
10. Leadlay, P. F. *An Introduction to Enzyme Chemistry*, The Chemical Society, 1978, p. 24–36.
11. Lipscomb, W. N. In *Structural and Functional Aspects of Enzyme Catalysis*, Eds. Eggerer, H.; Huber, R.; Springer-Verlag, 1981, p. 17–23.
12. Yonaha, K.; Soda, K. In *Molecular Aspects of Enzyme Catalysis*, Eds. Fukui, T.; Soda, K.; Kodansha, VCH, 1994, p. 1–13.
13. Moss, D. W. *Enzymes*, Oliver & Boyd Ltd. 1968, p. 65–79.
14. Fersht, A. In *Enzyme Structure and Mechanism (2nd ed.)*, W. H. Freeman, 1984, p. 311–346.
15. Palmer, T. In *Understanding Enzymes (4th ed.)*, Prentice Hall/Ellis Horwood, 1995, p. 67–75.
16. Stryer, L. *Biochemistry (3rd ed.)*, W. H. Freeman and Company, 1988 p. 191.
17. Breslow, R. *Acc. Chem. Res.* **1995**, *28*, 146.
18. Pauling, L. *Nature* **1948**, *161*, 707.
19. Stryer, L. *Biochemistry (3rd ed.)*, W. H. Freeman and Company, 1988, p. 182–183, 191.
20. Williams, D. H.; Stephens, E.; Zhou, M. *Chem. Commun.* **2003**, 1973, and references therein.
21. For example, Collman, J. P.; Wagenknecht, P. S.; Hutchison, J. E. *Angew. Chem. Int. Ed. Engl.* **1994**, *33*, 1537, and references therein.
22. For example, Collman, J. P.; Rapta, M.; Bröring, M.; Raptova, L.; Schwenninger, R.; Boitrel, B.; Fu, L.; L'Her, M. *J. Am. Chem. Soc.* **1999**, *121*, 1387; Ricard, D.; Andrioletti, B.; L'Her, M.; Boitrel, B. *Chem. Commun.* **1999**, 1523.
23. For example, Naruta, Y.; Sasayama, M.; Sasaki, T. *Angew. Chem. Int. Ed. Engl.* **1994**, *33*, 1839.
24. For example, (a) Tolman, W. B. *Acc. Chem. Res.* **1997**, *30*, 227 and references therein; (b) Liang, H.-C.; Dahan, M.; Karlin, K. D. *Curr. Opin. Chem. Biol.* **1999**, *3*, 168, and references therein.
25. For example, Watkinson, M.; Whiting, A.; McAuliffe, C. A. *J. Chem. Soc. Chem. Commun.* **1994**, 2141.
26. For example Lee, D.; Lippard, S. J. *J. Am. Chem. Soc.* **2001**, *123*, 4611, and references therein.

27. For example, in Ruettinger, W.; Dismukes, G. C. *Chemical Review* **1997**, *97*, 1.
28. For example, (a) Oyaizu, K.; Haryono, A.; Yonemaru, H.; Tsuchida, E. *J. Chem. Soc. Faraday Trans.* **1998**, *94*, 3393; (b) Yagi, M.; Kaneko, M. *Chem. Rev.* **2001**, *101*, 21.
29. For example, Okada, T.; Katou, K.; Hirose, T.; Yuasa, M.; Sekine, I. *Chem. Lett.* **1998**, 841.
30. Brintzinger, H. H.; Fischer, D.; Mulhaupt, R.; Rieger, B.; Waymouth, R. M. *Angew. Chem. Int. Ed. Engl.* **1995**, *34*, 1143.
31. Ewen, J. A.; Elder, M. J.; Jones, R. L.; Curtis, S.; Cheng, H. N. *Int. Symp. of Catalytic Olefin Polymerisation*, Elsevier, 1989.
32. Bochmann, M. *J. Chem. Soc., Dalton Trans.* **1996**, 255.
33. Grubbs, R. H.; Coates, G. W. *Acc. Chem. Res.* **1996**, *29*, 85.
34. Steinhangen, H.; Helmchen, G. *Angew. Chem. Int. Ed. Engl.* **1996**, *35*, 2339.
35. Blaser, H. *Chem. Rev.* **1992**, *92*, 935.
36. Tietze, L. F.; Ila, H.; Bell, H. P. *Chem. Rev.* **2004**, *104*, 3453.
37. Molenveld, P.; Engbersen, J. F. J.; Reinhoudt, D. N. *Chem. Soc. Rev.* **2000**, *29*, 75.
38. Konsler, R. G.; Karl, J.; Jacobsen, E. N. *J. Am. Chem. Soc.* **1998**, *120*, 10780.
39. Breslow, R.; Zhang, X.; Xu, R.; Maletic, M.; Merger, R. *J. Am. Chem. Soc.* **1996**, *118*, 11678.

3

A CONCEPTUAL DESCRIPTION OF ENERGY-DEPENDENT ("THERMODYNAMIC") AND TIME-DEPENDENT ("MECHANICAL") PROCESSES IN CHEMISTRY AND CATALYSIS

GERHARD F. SWIEGERS

3.1 INTRODUCTION

In the opening chapter of this book, we distinguished between *energy-dependent* and *time-dependent* processes [1,2]. *Energy-dependent* processes were said to be fundamentally thermodynamic in character, being driven by underlying energy gradients (e.g., objects reacting to a gravitational field or to an exchange of heat with the surroundings). *Time-dependent* processes are, by contrast, fundamentally mechanical in character, because they concern physical interactions between objects. These interactions include chain sequences [3] involving particular spatial arrangements of the participating elements in time (e.g., billiard balls colliding).

Time-dependent processes are distinguished from their energy-dependent analogs in being *path-dependent*, and they have to do with events in time and space (*spatiotemporal*). Another distinction is that time-dependent systems composing of extremely large collections of participants (e.g., chemical/catalytic systems) may display *chaotic* and *complex* macroscopic properties, sometimes termed *stochastic* behavior.

Mechanical Catalysis: Methods of Enzymatic, Homogeneous, and Heterogeneous Catalysis,
Edited by Gerhard F. Swiegers
Copyright © 2008 John Wiley & Sons, Inc.

In the context of chemistry, time-dependent chemical reactions have, to date, typically been observed in the interaction of exceedingly reactive species, whose reaction is dependent only on the rate at which they physically collide with each other. Such reactions have been termed *diffusion-controlled* or *diffusion-limited*. However, this terminology does not cover instances in which the reactants do not diffuse to each other. For example, it is not appropriate in cases where the reactants are formed *in situ* while they are mutually attached to a single, large chemical framework. For this reason, we employ the more general term "time-dependent" to describe such reactions.

Within the field of catalysis, several researchers have invoked notions related to time-dependence to describe catalytic action. Most have involved the action of certain of the biological catalysts known as enzymes. For example, enzymes have been, individually and/or generally, portrayed as machine-like "specialized combining centers" [4], analogous to mechanical devices [5], dependent on "coupled" protein motions [6], and *spatiotemporal* in their action [7]. Catalysis by the enzyme dihydrofolate reductase has been shown to be *stochastic* [8]. All of these representations refer, fundamentally, to time-dependent, and not to energy-dependent, processes.

Many catalytic systems are, moreover, known to be *diffusion-controlled* [4(c)]. Some such cases may well involve time-dependent reactions. *Diffusion control* in catalysis is, however, formally limited by the rate at which the reactants bind to the catalyst. This process is not the same as *time-dependence*, which is limited by the rate at which the catalyst mediates the collisions of bound reactants.

What, then, is the origin of time-dependence in catalysis, and how can it be properly and fully described? To answer this question, one must find a way to relate time-dependence to the existing conceptual framework of energy-dependence in catalysis. In this chapter, we will seek to do that. We will try to develop a theoretical conceptualization for time-dependent catalysis. We will show that time- and energy-dependence are fundamental properties of chemical reactions. When they are translated into catalysis, it generates time- and energy-dependent catalysis. In the field of catalysis, these distinctions therefore mirror, in a conceptually precise way, the choices and pathways available in uncatalyzed chemical reactions.

Finally, it should be noted that in liquid solution, the gas-phase term "collision" is formally called an encounter [1,2]. For the purposes of clarity and continuity in argument, however, we will in the remainder of this series continue to use the terms "collision" and "collision frequency" to describe reactive contact between species in the liquid phase.

3.2 THEORETICAL CONSIDERATIONS: COMMON PROCESSES IN UNCATALYZED REACTIONS

3.2.1 Reactions as Collisions Between Molecules

According to collision theory, two molecules can only react with each other when they physically collide with sufficient energy to form a so-called "transition state"

3.2 COMMON PROCESSES IN UNCATALYZED REACTIONS

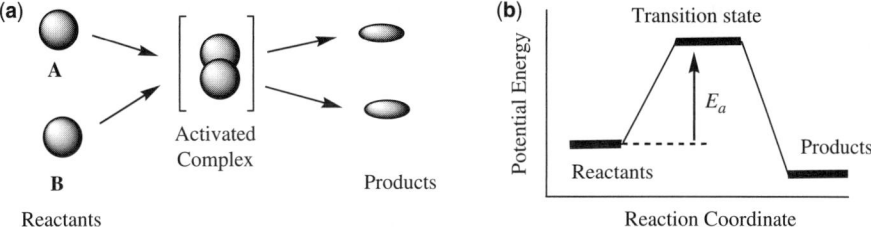

Figure 3.1. Schematic depiction of (a) MECHANICS: a collision between two molecules, A and B, leading to a chemical reaction in which products are formed, and (b) THERMODYNAMICS: the energetic profile followed during the collision, showing the minimum threshold energy needed to bring about product formation (termed: the activation energy E_a).

(formally known as the activated complex) [1,2]. Figure 3.1(a) illustrates this process. The reaction proceeds to completion if the transition state successfully separates into the products. The activation energy E_a represents the average threshold energy that must be overcome in the activated complex for the reaction to take place [Figure 3.1(b)]. The overall rate k of such a reaction is given by the Arrhenius equation [1,2].

$$k = A \exp\left(\frac{-E_a}{RT}\right) \qquad (3.1)$$

The pre-exponential term A is the average frequency of collisions between the reactant molecules (the *collision frequency* or *frequency factor*). The exponential term $(-E_a/RT)$ describes the proportion of those collisions that are sufficiently energetic to result in product formation [1,2].

An uncatalyzed reaction therefore involves two processes that occur in sequence:

(1) A collision between the reactants
(2) The formation of products when the transition state of the collision has sufficient energy

3.2.2 The Fundamental Origin of Energy-Dependent and Time-Dependent Reactions

It is important to note that any process involving a sequence of steps can only go as fast as the slowest step. Thus, the overall rate of an uncatalyzed reaction depends on which of the above steps, (1) or (2), is the slowest or the least likely. Two possible extremes exist in this respect.

On one extreme lies the situation where only a small proportion of collisions are sufficiently energetic to result in product formation. That is, the activation energy E_a

in Figure 3.1(b) is very large. The overall reaction rate is then dominated by the exponential term in Equation (3.1). That is, the reaction rate will, effectively, be controlled by the activation energy E_a and the thermodynamic profile depicted in Figure 3.1(b). The key determinant of the reaction is therefore an energy-based quantity, so that the reaction rate is *energy-dependent*.

On the other extreme lies the situation where virtually *every* collision leads to a reaction. That is, E_a is so small that it, effectively, constitutes no real barrier at all. In that case, the overall reaction rate will be dominated by the pre-exponential term A in Equation (3.1). That is, the reaction rate will, effectively, be controlled by the *frequency of collisions* and the tangible, physical interaction of the reactants as shown in Figure 3.1(a). Since the key determinant is then a frequency-based quantity and frequency involves a quantification of time, the reaction is considered *time-dependent*, not energy-dependent. That is, it will be governed by how the reactants physically (mechanically) interact with each other and is unaffected by the energy barrier involved in that interaction.

The above extremes will be influenced in different ways by changes in the physical conditions. For example, energy-dependent processes will be very strongly affected by changes that increase the likelihood of overcoming the E_a threshold. This will typically include small increases in temperature, for example. Time-dependent processes will, however, be most influenced by changes that alter the collision frequency, such as small increases in reactant concentration. An increase in the temperature of a time-dependent reaction will, of course, also lead to an increase in the reaction rate, but this will be because of an increase in the collision frequency, not because of an increase in the rate at which the E_a threshold is overcome (because all collisions already overcome E_a).

3.2.3 Time-Dependent and Energy-Dependent Domains Were First Observed in Unimolecular Gas-Phase Reactions

In the early part of the twentieth century scientists struggled to explain why certain gases decomposed with first-order kinetics at high pressure but with second-order kinetics at low pressure [2(e)]. An example in this respect is the decomposition of azomethane. At low pressure, this reaction is second order. At high pressure, however, the reaction proceeds via a first-order process as follows [2(e)]:

$$CH_3N_2CH_3 \rightarrow C_2H_6 + N_2 \qquad (3.2)$$

The heart of the conundrum lay in the fact that all reactions were, at that time, thought to involve collisions between molecules. How, then, could the collision of two or more molecules generate simple first-order kinetics at high pressure; first order kinetics must necessarily involve only a single reacting species? Moreover, why does this revert to second-order kinetics at low pressure?

3.2 COMMON PROCESSES IN UNCATALYZED REACTIONS

An explanation was provided by Lindemann and later experimentally verified by Ramsperger in the 1920s [2(e)]. Lindemann proposed that there were two steps in the process:

$$A + M \xrightleftharpoons{K} A^* + M \quad (3.3)$$

$$A^* \xrightarrow{k_2} P \quad (3.4)$$

where A is a reactant molecule that is excited to form an energised molecule A^* upon collision with M, which is either another A molecule or an added diluent gas molecule. The activated molecule A^* undergoes transformation into a product molecule P according to Equation (3.4).

At high pressures, the frequency of collisions [depicted in Equation (3.3)] is high, so that the rate of transformation into products [shown in Equation (3.4)] is rate-determining. As the rate-determining step contains only one reactant, the observed kinetics is first order [2(e)]. At low pressure however, the collision frequency is low, so that Equation (3.3) becomes rate-determining [2(e)]. In that case, the slowest step involves bimolecular collision with accompanying second-order kinetics. Thus, the observation was explained.

Although Lindemann's proposal qualitatively explained the observations, subsequent modifications by Hinshelwood along with Rice, Ramsperger, and Kassel (RRK) brought greater quantitative accuracy [2(e)]. Like Lindemann, the resulting *Hinshelwood–RRK* theory explicitly recognized two distinct realms in gas phase reactions [2(e)]:

(1) The so-called **high-pressure limit** in which the reaction rate is effectively determined by the rate at which the threshold activation energy E_a is overcome [Equation (3.4)]. As the reaction is dependent on E_a in the high-pressure limit, it is therefore *energy-dependent* in character.
(2) The **low-pressure limit** in which the reaction rate is determined by the *collision frequency* of the reactants with each other [Equation (3.3)]. As the reaction is then dependent on the frequency of collisions and is independent of E_a, it is then *time-dependent* in character.

3.2.4 The Pathway of the Reaction Is also Controlled by the Least-Likely Step in the Sequence

It is not only the reaction rate that is affected by the slowest step in the reaction sequence. The major pathway of the reaction is also dependent on the slowest process.

Consider the hypothetical situation where there are several different possible products of the reaction, depending on the relative orientations of the reactants during the collision. Figure 3.2(a) schematically depicts four hypothetical transition states, TS_1–TS_4, which are formed during collisions between reactants C and D. Each transition state generates a different product, P_1 to P_4, respectively. Each transition state also has its own activation energy, E_a^1 to E_a^4, respectively [Figure 3.2(b)]. To evaluate the

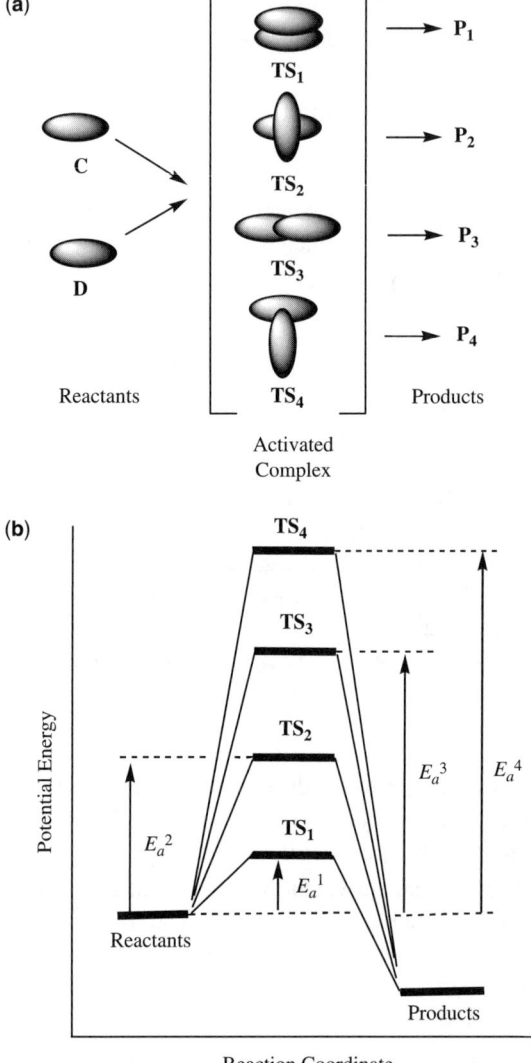

Figure 3.2. Schematic depiction of (a) four possible reactive collisions between two molecules, C and D, leading to chemical reactions generating products P_1 to P_4, from transition states TS_1 to TS_4, respectively, and (b) the corresponding energetic profiles followed during the collisions, showing the minimum threshold energies needed to bring about reaction in each case (the activation energies E_a^1 to E_a^4).

effect of the activation energy on the reaction pathway, we will consider the forward reaction only; that is, this is a kinetically slowly reversible reaction. Under these circumstances, it does not matter what the relative thermodynamic stabilities of the products P_1 to P_4 are; these are, consequently, set to have the same overall potential energy.

3.2 COMMON PROCESSES IN UNCATALYZED REACTIONS

In an energy-dependent reaction, the overall pathway followed and the nature of the products obtained will be determined by the transition state with the smallest E_a. This is because most collisions do not result in product formation. But, of those collisions that do, the transition state with the lowest E_a will generate the most products. In effect, the pathway involving the transition state of lowest energy will be most populated. The least populated pathway will involve the transition state of highest energy. The final mixture of products obtained will reflect this distribution.

To illustrate, if the example shown in Figure 3.2 involves an *energy-dependent* reaction, it will generate product P_1 in greatest numbers because its transition state TS_1, has the smallest activation energy E_a^1. Product P_4 will be produced in smallest numbers because its transition state TS_4 has the largest activation energy E_a^4.

If, however, the reaction in Figure 3.2 is *time-dependent* and dominated by collision frequency, the most favored transition state will be the one that is formed *first* and *most often* in collisions between the reactants. This is because under the physical conditions employed, the activation energies of the possible transition states are so small in relative terms that they no longer influence the reaction rate. Instead the rate-limiting step involves merely achieving a suitable collision.

In other words, the reaction rate will then be dependent on which *type of collision* occurs most often. This type of collision will also occur most often between reactants that are sufficiently energetic to overcome E_a during the collision. Thus, the populations of the different pathways depicted in Figure 3.2 will be controlled by the *frequency* with which these collisions are generated.

In a time-dependent reaction, the most favored pathway consequently involves the transition state associated with the *most common type of collision*. This transition state need not have the lowest E_a since all of the E_a's are so low that they do not affect the reaction rate. Indeed, the pathway of the reaction will be entirely unconnected with the relative E_a's of the various transitions states. Instead, the mixture of products will reflect the *statistical likelihood* of forming the various possible transition states during reactant collisions.

3.2.5 Transition State Theory (TST) Describes the Pathway and Rate of Energy-Dependent Reactions. Transition State Theory Corresponds to the High-Pressure Limit of Hinshelwood–RRK Theory

In the 1930s, Henry Eyring (with contributions from Polanyi, Evans, and others) developed an elegant simplification of the Hinshelwood–RRK theory [2(c)]. By assuming that the transition state, the reactants, and the products were all in equilibrium with each other and by the application of quantum theory, Eyring developed equations that *quantitatively* predicted reaction rates [2(c)]. Moreover, potential energy surfaces of the type depicted in Figure 3.3 could be calculated for the process of transition state and product formation. These surfaces indicated the lowest energy pathway for the reaction and consequently predicted its progression [2(c)]. The LEPS (London–Eyring–Polanyi–Sato) energy surface is, for example,

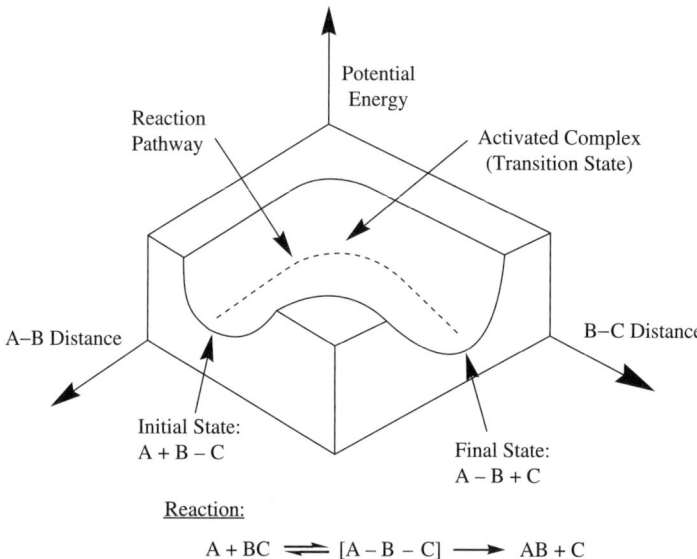

Figure 3.3. Illustrative potential energy surface for the hyothetical reaction of A and BC to produce AB and C. The reaction pathway is the dotted line on the surface.

today widely used in reaction dynamics [2(c)]. Eyring's theory subsequently came to be known as *transition state theory* (or *TST*) [2(c)].

As noted above, transition state theory assumes and depends on the maintenance of an *equilibrium* distribution of reactants and products. Such an equilibrium exists at high pressures in unimolecular gas reactions of the type described by Hinshelwood–RRK, [2(e)] because at high pressures, the rate of collisions [Equation (3.3)] far outweighs the rate of product formation [Equation (3.4)]. Although the reactant concentration falls during the reaction, a Boltzmann distribution of reactant energies is nevertheless maintained as depicted in Figure 3.4(a) [2(f)].

However, in the other special case of the *low-pressure limit* in Hinshelwood–RRK theory, fewer collisions occur. Moreover, only collisions between reactants that are sufficiently energetic result in a reaction. The population of sufficiently energetic reactants therefore becomes selectively depleted relative to the remaining reactants [2(e)]. A Boltzmann distribution of reactant energies cannot be maintained [Fig. 3.4(b)] [2(f)]. Kinetics in this realm is consequently *nonequilibrium* in character; that is, a thermodynamic equilibrium is not maintained. Under these circumstances, the key assumption of transition state theory is not valid. *Time-dependent* reactions are, therefore, not subject to, or described by, transition state theory [2(c)].

As such, transition state theory describes only the special case of the *high-pressure limit* in Hinshelwood–RRK theory. This case also describes *energy-dependence* in reactions [2(e)]. Thus, transition state theory stands as a definitive description of *energy-dependent* reactions [2(c)]. Such reactions involve *equilibrium* kinetics.

The theories of Hinshelwood and others today form a vital part of especially high-temperature gas kinetics that cannot be described accurately by transition state theory [2].

3.2 COMMON PROCESSES IN UNCATALYZED REACTIONS 63

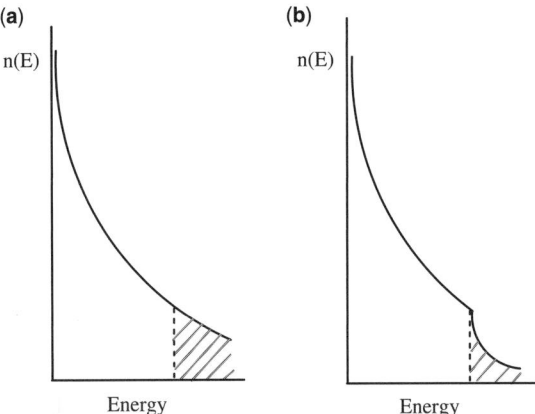

Figure 3.4. Boltzmann distribution of reactants under (a) high-pressure conditions and (b) low-pressure conditions. In the latter case, the higher (reactive) energy levels of the reactant distribution become significantly depleted [2(f)].

3.2.6 Time-Dependent Reactions in the Liquid Phase: Some Examples

Although *time-dependent* reactions have been documented in the gas phase, most traditional organic, inorganic, and physical chemistry is carried out at, or near, ambient temperature in the liquid phase. Given that the kinetic energies of the reactant molecules are substantially lower under these conditions, it is not unexpected that overcoming the E_a threshold will be the prime impediment to reaction. That is, most uncatalyzed, liquid-phase reactions will be *energy-dependent*.

However, not all liquid-phase reactions are energy-dependent. As noted, some liquid-phase reactions are *diffusion-controlled*, which means that the only impediment to reaction is actually getting the reactant molecules to collide with each other. How they collide with each other is, essentially, not relevant; that is essentially no energy barrier exists to their reaction once they collide. Such reactions are time-dependent, not energy-dependent. In effect, the reaction occurs when the reactant molecules physically encounter each other, regardless of their relative orientations, trajectories of approach, or other features of the collision. In other words, the underlying energy landscape about the collision is flat.

An example in this respect is the reaction of H^+ with OH^-, which is extremely rapid. It occurs at a rate (ca. 10^{-10} s) that is limited only by the rate at which the reactants diffuse to each other in solution. The overall rate is therefore dependent only on the *collision frequency* and is independent of E_a.

An example in which time-dependence determines both the rate and the reaction pathway can be observed in the Wurtz cyclization depicted in Figure 3.5 [9].

The Na-carbanion end of intermediate **1** is exceedingly reactive toward alkyl halides such as the $-CH_2-Br$ group at the other terminus of the molecule. This reactivity is such that the Na-carbanion will, effectively, react with the *first* alkyl bromide that it encounters in an appropriate collision.

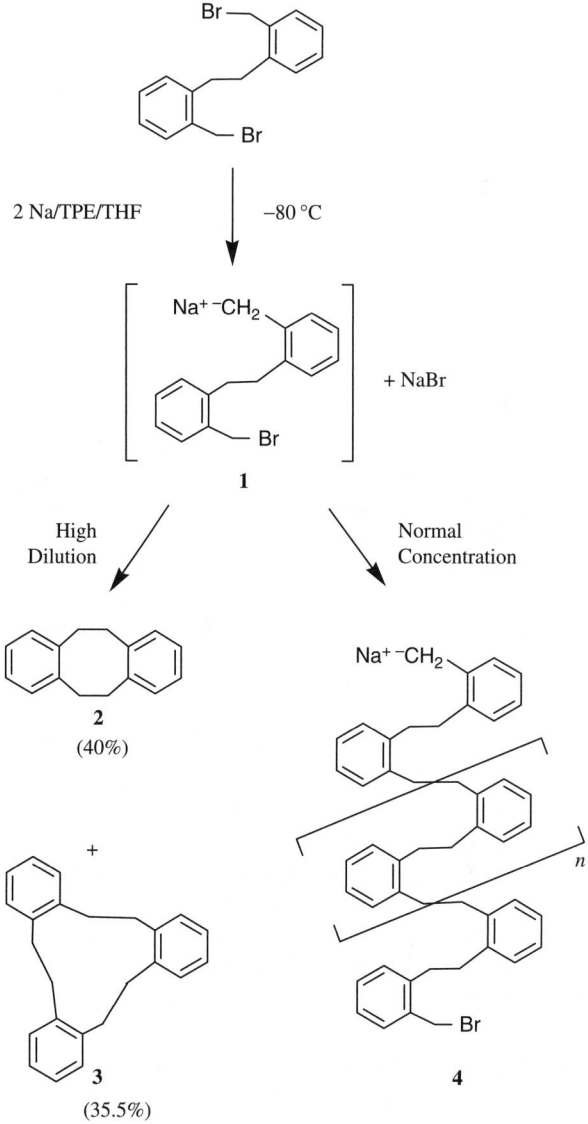

Figure 3.5. Formation of cyclophanes by Wurtz C–C coupling [9]. The reactivity of intermediate **1** necessitates the use of high-dilution conditions to form **2** and **3** selectively. Polymers of type **4** are obtained at normal concentrations.

Under normal conditions of concentration, the first bromide encountered will reside on a separate molecule, because polymers of type **4** are the major products of the reaction (Fig. 3.5). However, under high dilution conditions, the cyclized product **2** is obtained in high yield (Fig. 3.5). Even under such conditions, other cycles, such as **3**, are still formed as major by-products.

3.2 COMMON PROCESSES IN UNCATALYZED REACTIONS

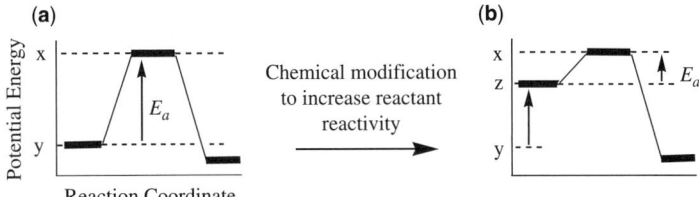

Figure 3.6. One means of making a reaction time-dependent is to modify chemically the reactants to be highly reactive (as in **1** in Figure 3.5). This increases the potential energy of the reactant [from y in (a) to z in (b)], with an accompanying decline in E_a.

The need for high-dilution conditions in the preparation of **2** can therefore, undoubtedly, only be for one reason: to *decrease the overall collision frequency* and thereby increase the likelihood that the *first* alkyl bromide encountered by the Na-carbanion will be the $-CH_2-Br$ group at the other terminus of the molecule.

The reaction by intermediate **1** must involve a very low activation energy. The reason E_a is low is because the potential energy of **1** is high; that is, the intermediate is highly reactive, even at the $-80\,°C$ temperature employed [9]. This property of **1** is depicted in Figure 3.6; the high potential energy of **1** results in a low E_a. The E_a must be small because this is the only way that such a Wurtz coupling could lead to a *statistical mixture* of products.

In this example, we see the typical characteristics of time-dependent processes, including a reliance on the statistical likelihood of spatial interactions in time. To control this statistical likelihood, it is necessary to design and set up the reaction conditions carefully (high-dilution conditions). Such processes are also unaffected by, and independent of, the underlying energy landscape of the reaction, which is, to all intents and purposes, too flat to have an influence on the overall progress of the reaction.

3.2.7 The Transition between Energy-Dependence and Time-Dependence as a Function of Temperature. Curvature in Arrhenius Plots

If both time- and energy-dependent reactions exist in the liquid phase, it must be possible to manipulate an energy-dependent reaction into becoming a time-dependent one. Moreover, it must be possible to observe the point at which that transition occurs. In other words, some change must be made in the physical conditions to induce an energy-dependent reaction to become time-dependent. As we have noted, in the gas phase, this change involves high and low pressure. At high pressure, many gas reactions are energy-dependent. At low pressure, they are typically time-dependent. But what change will induce a similar transformation in the liquid phase? And how could we observe this transition?

The most obvious candidate in this respect is temperature. As the temperature is increased, more collisions in a liquid-phase reaction will become successful. Presumably, at some stage—if one could continue to a suitably elevated temperature without evaporating the liquid—one must reach a point where the activation energy becomes essentially irrelevant to the rate of the reaction. At that stage, the reaction will have gone from being fundamentally energy-dependent to being time-dependent. If this is the case, then some physical indication of such a transition must surely exist? Where can one observe such an indication?

Arrhenius plots are a standard technique for measuring the activation energy of a reaction. They involve plotting the natural logarithm of the reaction rate ($\ln k$) against the reciprocal of the temperature ($1/T$). According to Equation (3.1), such a plot should yield a linear graph, with a slope of $-E_a/R$ and an intercept that corresponds to the natural logarithm of the collision frequency $\ln A$. By measuring the slope of the graph, one can calculate the activation energy of the reaction. Implicit in this method is an assumption that the collision frequency A is unaffected by, and independent of, temperature.

A curious feature of Arrhenius plots is the fact that they are often linear at or near ambient temperatures, but at high temperatures, they become curved [2(a)], as depicted in Figure 3.7 [2(a)].

Such plots indicate that the rate of most reactions at or near ambient temperature depends on the rate at which E_a is overcome; that is, most reactions are *energy-dependent* at or near ambient temperature. However, a new and different relationship is established at higher temperatures, which causes the curvature. The new relationship conceivably commences once overcoming the E_a threshold is no longer the prime impediment to reaction (that is, when the collision frequency and not E_a becomes the main determinant of the reaction rate). This new relationship may comprise the temperature-dependence of the collision frequency, not of E_a.

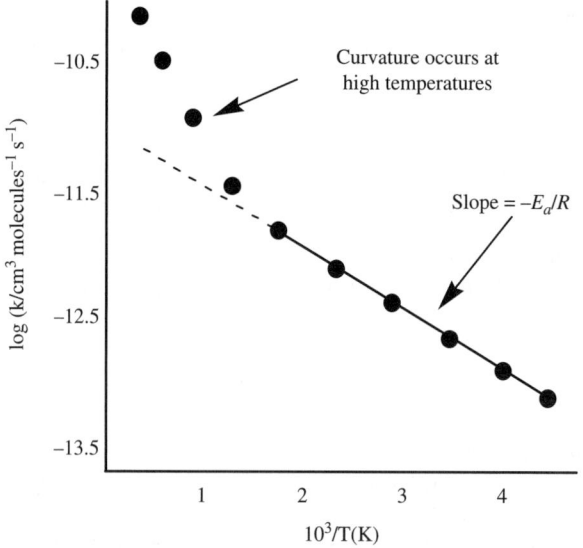

Figure 3.7. Illustrative Arrhenius plot for the reaction of OH radicals with ethane over the temperature range 222–500 K, showing substantial curvature at high temperatures [2(a)].

3.2 COMMON PROCESSES IN UNCATALYZED REACTIONS

To elaborate on this argument, consider the following: As the temperature is raised, reactant molecules move faster. Therefore, more collisions occur per unit time. However, a greater proportion of the reactant molecules also become sufficiently energetic to overcome the activation energy during collisions. That is, more collisions are unaffected by the energy landscape. Thus, not only more collisions occur, but a greater proportion of them are successful.

The collision frequency A is, consequently, not constant and independent of temperature under these circumstances. Moreover, it will be increasingly affected as the temperature is raised. It stands to reason then that, at some elevated temperature, overcoming the threshold E_a will no longer constitute the slowest step. Instead the frequency of collisions will ultimately become the rate-determining process. The slope of an Arrhenius plot at that stage will then reflect the temperature-dependence of the collision frequency rather than of the activation energy. In effect the reaction rate will become independent of E_a and dependent on the collision frequency. This process must necessarily show up as curvature in the Arrhenius plot.

Of course, not all curvature in Arrhenius plots is caused by a change in the fundamental nature of the reaction. Such curvature is, however, an expected outcome of the temperature-induced transition from an energy- to a time-dependent realm.

To illustrate these points, consider again the diffusion-controlled reaction of H^+ with OH^-. As noted, it occurs at a rate of ca. 10^{-10} s. That is, as soon as the reactant ions collide, they react. The overall rate is therefore dependent only on the *collision frequency* and is independent of E_a. Consider what a plot of $\ln k$ versus $1/T$ will give for this reaction. The slope of such a plot will certainly not be $-E_a/R$. Rather, it will give the temperature dependence of the collision frequency.

Behavior of this type in Arrhenius plots has not been widely recognized in the chemical literature. Thus, innumerable activation energies have been reported that are extremely low, even to the extent of sometimes lying within the thermal background. Activation energies of, for example, 5–7 kJ/mol, which have been determined by Arrhenius plots, can clearly not be meaningful given that the reaction must then be diffusion controlled and therefore not subject to an Arrhenius plot. The slope of such a plot must necessarily represent something else, including, possibly, the temperature dependence of the collision frequency.

We should note here that Arrhenius plots can be misleading in several other ways too. The presence of rapidly equilibrating intermediates in a reaction may, for example, also drastically distort Arrhenius plots. In extreme cases, Arrhenius plots may even suggest "negative" apparent activation energies [10].

In conclusion, one way of changing an energy-dependent, liquid-phase reaction into a time-dependent one would be to increase the temperature of the reaction progressively. This approach assumes, of course, that such a transition can be achieved before the liquid phase evaporates.

3.2.8 Methods of Creating Time-Dependent Reactions

In liquid-phase reactions at or near ambient temperature, there are therefore two ways in which a reaction may be made time-dependent.

As depicted in Figure 3.6, the potential energy of the reactants may be increased relative to that of the transition state by chemical modification to make the reactant more reactive. This process necessarily increases the potential energy of the reactant and decreases its activation energy. It is the reactivity of **1** in Figure 3.5 that makes its reaction time- and not energy-dependent.

The second method is to elevate the temperature. It has the effect of increasing the proportion of reactant molecules that have high kinetic energies, with the result that there are both more collisions and a greater proportion of them will be successful (that is, which overcome E_a). Eventually, overcoming E_a will no longer constitute the slow step in the reaction sequence. Instead the collision frequency will become rate determining, thereby rendering the reaction time-dependent.

3.2.9 Summary: The Key Properties of Time-Dependent and Energy-Dependent Reactions

In the previous sections, we have shown that energy-dependent reactions are as follows:

1. Reliant on the activation energy E_a and the underlying energy landscape for their overall rate and pathway.
2. Theoretically described by transition state theory and the high-pressure limit in Hinshelwood–RRK theory.
3. Typically equilibrium processes; that is, they occur in the presence of a thermodynamic equilibrium.

By contrast, time-dependent reactions are as follows:

4. Reliant on the collision frequency of the reactants for their rate and pathway. They are independent of the activation energy E_a.
5. Theoretically best described by the low-pressure limit in Hinshelwood–RRK theory. They are not subject to transition state theory.
6. Nonequilibrium processes; that is, they may occur in the absence of a thermodynamic equilibrium.

3.3 THEORETICAL CONSIDERATIONS: COMMON PROCESSES IN CATALYZED REACTIONS

3.3.1 Catalyzed Reactions Are More Likely to be Time-Dependent than Are Uncatalyzed Reactions

Although uncatalyzed solution-phase organic chemistry is dominated by energy-dependent reactions, time-dependence is likely to be more common among their catalyzed counterparts. There are two main reasons.

First, catalysts by their nature decrease the activation energy of reactions. They do so by binding and stabilizing the transition state of the reaction. Very good catalysts may conceivably decrease E_a to insignificant levels, thereby rendering the reaction time-dependent. But even catalysts that do not produce such dramatic declines in E_a can nevertheless turn a reaction that is energy-dependent, when it is uncatalyzed, into one that is time-dependent when catalyzed, because all that the catalyst has to do is reduce the activation energy *sufficiently* that reactant collision and not product formation is the slowest process in the reaction sequence.

Second, many heterogeneous catalysts, including a wide variety of metals and metal oxides, are not limited to operating at or near room temperature (as is the case in solution phase organic chemistry). The elevated temperatures available to such catalysts reduce the size of the activation barrier relative to the background thermal energy and therefore facilitate its crossing. This process necessarily speeds up the energetic step in the reaction sequence, increasing the likelihood that it will be the faster process.

For these reasons, it is not unreasonable to expect to encounter more time-dependent reactions in catalysis.

3.3.2 Catalysis Changes the Reaction Processes

How, then, does the involvement of a catalyst change the sequence of a reaction? All catalysts, regardless of their type, must necessarily undertake some common steps. Typical steps in this regard are depicted in Figure 3.8, namely [11]:[1]

(1) The catalyst must bind and "activate" the reactants. *Activation* involves the catalyst polarizing or "stretching" bonds within the bound reactant to, thereby, make them more conducive to reaction. This result is typically achieved by electron withdrawal from reactant bonding orbitals or electron donation into reactant antibonding orbitals [11(c)]. Figure 3.9 schematically illustrates the activation of a representative reactant, dioxygen. The practical effect of activation is to *increase the potential energy* of the bound reactants along the lines of that illustrated in Figure 3.6. That is, in binding the catalyst, the reactant becomes more reactive because the catalyst modifies its electronic structure,

(2) While bound to the catalyst, the activated reactants must be brought into physical contact with each other (collide). This process may involve, in the case of a heterogeneous catalyst, migration of the reactant from one catalyst atom to the next over the surface of the catalyst, until it contacts another bound reactant. Alternatively reactants may reorientate themselves on, or immediately after, binding, thereby coming into contact with an adjacent reactant. In the

---

[1]Langmuir–Hinshelwood and so-called "precursor" mechanisms involve transition states formed by reactants that are attached to the catalyst. Eley–Rideal mechanisms involve collisions where some, but not all, of the reactants are attached to the catalyst. Such mechanisms are generally seen only at high pressures in gas-phase reactions on solid catalysts.

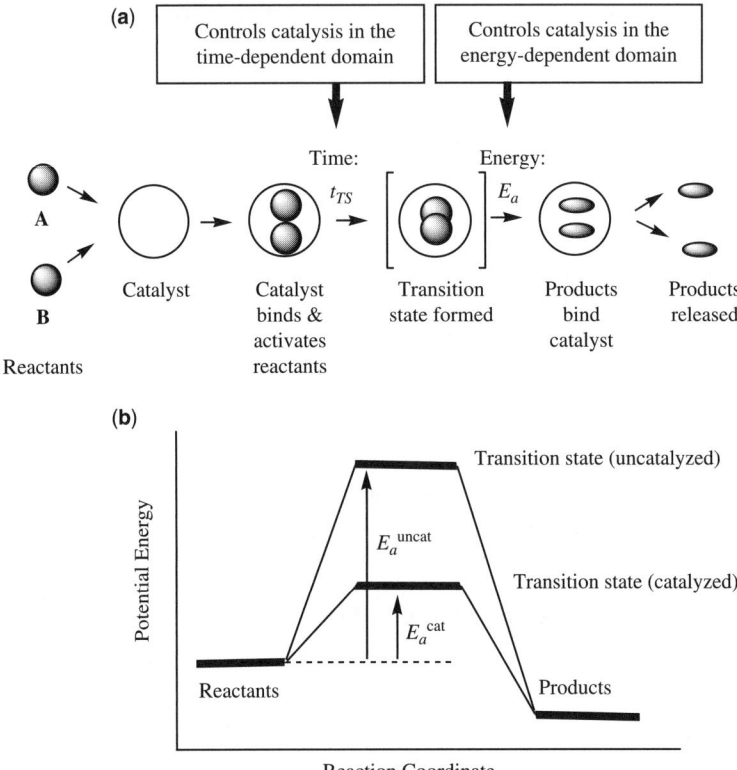

Figure 3.8. Schematic depiction of (a) a catalyzed, reactive encounter between two molecules, A and B, leading to the formation of products, and (b) the energetic profile followed during the catalyzed reaction, showing the minimum threshold energy (the activation energy E_a^{cat}) needed for reaction, relative to the same threshold for the uncatalyzed reaction (E_a^{uncat}). In an *energy-dependent* catalyzed reaction, the step that comprises the greatest impediment to reaction involves overcoming the activation energy E_a^{cat}. In a *time-dependent* catalyzed reaction, the greatest impediment involves collision of the bound reactants (with the average time required to bring about collision being t_{TS}).

case of a homogeneous catalyst, a collision may be brought about by a conformational change within the catalyst or by a rearrangement within the catalyst upon or immediately after reactant binding.

(3) In these collisions, transition states are formed. These transition states are stabilized, relative to the uncatalyzed reaction, by the binding of the catalyst.

(4) If a transition state overcomes E_a, it separates to form products that are, initially, also bound to the catalyst.

(5) The catalyst must release the products, thereby regenerating itself to undertake another cycle.

3.3 COMMON PROCESSES IN CATALYZED REACTIONS

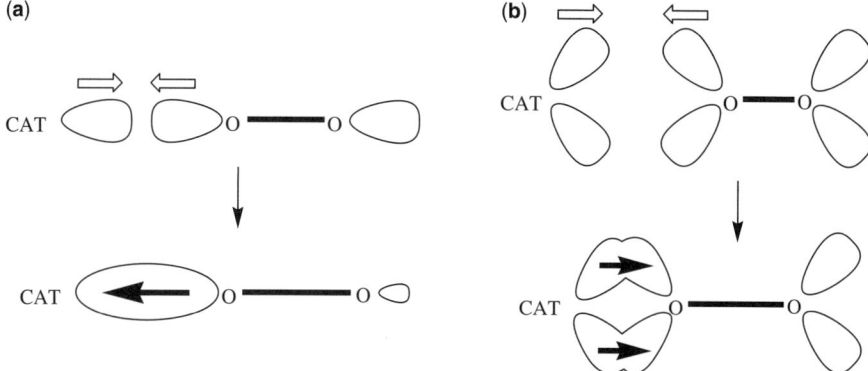

Figure 3.9. *Reactant activation by a catalyst.* Schematic illustrating a catalyst (CAT) activating a representative reactant (O_2) by increasing the O—O bond length in order to facilitate its cleavage. The activation is achieved by either or both of (a) withdrawing electron density from its σ-bonding molecular orbital and (b) increasing electron density in its π^*-antibonding orbital.

It should be noted that the catalyst schematically depicted in Figure 3.8 can be *any* species capable of catalysis, which includes a molecular species dissolved in open solution (a homogeneous catalyst), a biological species (an enzyme), as well as the surface of a bulk metal or metal oxide (a heterogeneous catalyst). All these species are catalysts and must necessarily undertake similar basic steps to facilitate reactions.

For heterogeneous catalysts, we should note that step (2) above describes a *Langmuir–Hinshelwood* or *Precursor* mechanism. Qualitatively similar arguments can be made for *Eley–Rideal* mechanisms [these mechanisms are described briefly in Footnote 1].

A key difference between a catalyzed and an uncatalyzed reaction is therefore that the reactants are, first, localized near to each other by binding with the catalyst. During a very brief, subsequent time period (labeled t_{TS} in Fig. 3.8), they come into reactive contact with each other. In one such collision, they form the transition state. For convenience, we will term t_{TS} the "*time required for transition state formation.*" It should also be noted that the transition state in a catalyzed reaction is very different to that in the uncatalyzed analog, being stabilized by binding to the catalyst.

The time t_{TS} involves the period during which the catalyst-bound reactants are ready and available for transition state formation but have not yet contacted each other and formed the transition state. Before the transition state can form, the activated reactants must, *inter alia*:

(i) Physically contact with each other at van der Waals distance or less
(ii) Form an intermediate, new, high-energy electronic structure

3.3.3 Physical Manifestation of Time- and Energy-Dependence in Catalysts

The function of a catalyst is to bind reactants and to mediate successful collisions between them, thereby accelerating their reaction with each other. Good catalysts will create many successful collisions and thereby drastically speed up reactions. Catalysis, at the most fundamental level, *is therefore all about creating successful collisions*. The ways in which a catalyst may do that is, in fact, the central subject of this volume.

In an energy-dependent catalyst, the collision frequency is high relative to the energetic efficiency. That is, the activation energy E_a comprises the greatest impediment to reaction and therefore determines the overall reaction rate. In other words, the reactants bind the catalyst and collide numerous times with each other, until they are finally transformed into products. The properties of such a catalyst depend on the rate at which the many collisions produce transition states that overcome E_a.

Key Point. In a time-dependent, catalyst, however, the collision frequency is low relative to the energetic efficiency. That is, getting the reactants into physical collision with each other is the key impediment to reaction; the activation energy E_a is sufficiently low to not influence the overall rate. The reactants therefore typically bind the catalyst and form products in the first collision that are properly mediated by the catalyst. Their reaction rate is then dependent on the average time t_{TS} taken to set up the transition state. That is, it is typically dependent on the time it takes the catalyst to bring the bound and activated reactants into contact with each other.

3.3.4 The Distinction Between Time-Dependent Catalysis and Diffusion-Controlled Catalysis

It is important to note that the intermediacy of a catalyst changes the nature of a time-dependent reaction relative to its uncatalyzed counterpart.

In an uncatalyzed reaction with independent reactant molecules, time-dependence is, effectively, equivalent to diffusion control. That is, the collision frequency is determined by the speed at which the reactants diffuse to each other in solution. However, when a catalyst mediates the reaction process, the collision frequency is determined by the rate at which *the catalyst binds the reactants and then brings the bound reactants into contact with each other*. This may be, but need not be dependent on the rate of reactant diffusion to the catalyst.

Thus, for example, if the catalyst is inefficient at bringing the reactants into contact, or if the reactants bind only very transiently on each occasion of binding, then *time-dependent catalysis* may occur even when there is an *excess of reactants* about the catalyst in solution. Any physical condition that inhibits the ability of the catalyst to bring suitably bound reactants into physical contact with each other will favor time-dependence.

Moreover, an energy-dependent catalyst may be diffusion controlled. This will occur when the catalyst creates successful collisions between bound reactants at a very rapid rate relative to the rate of reactant diffusion to the catalyst. Even though there are many collisions after binding, the overall catalytic rate could then still be dependent on the rate at which the reactants diffuse to the catalyst.

3.3 COMMON PROCESSES IN CATALYZED REACTIONS

Diffusion control in a catalytic system consequently often has to do with the presence of dilute reactant solutions relative to the efficiency of the catalyst. In such cases, the rate becomes dependent on the rate of diffusion of the reactant to the catalyst. This dependence may happen regardless of the type of action employed by the catalyst, which can be either energy- or time-dependent.

In other words, essentially any catalyst can be turned into a diffusion-controlled system if the concentration of reactant in the solution or gas phase about the catalyst is made sufficiently low. In some cases, where the catalyst turns over bound reactants into products extremely quickly, catalysis will be diffusion-controlled at any reactant concentration in liquid solution. That is, the speed of the catalyst is then so great that the overall rate, at any reactant concentration, will depend on how fast the reactants can get to the catalyst. Several enzymes, including triosephosphate isomerase and carbonic anhydrase, are diffusion-controlled at all available substrate concentrations. Catalysis of this type is generally termed *kinetically perfect*; it typically involves k_{cat}/K_m factors in the order of 10^8 to 10^9 M^{-1} s^{-1} [4(c)].

Diffusion control in a catalytic system, therefore, provides no information *per se* about the nature of the catalytic action. For this reason, it is not of interest and will not be considered in this work. Instead, we will consider only *reaction-controlled* catalysis; that is, systems whose rates are determined by the rate of the catalytic reaction, not by the rate of reactant diffusion to the catalyst. This focus allows us to distinguish between energy-dependent and time-dependent processes.

To summarize, although diffusion control is usually equivalent to time-dependence in uncatalyzed reactions, it is inequivalent in catalyzed reactions and says nothing about the catalytic action. We will, in this volume, consider only *reaction-controlled* systems, not *diffusion-controlled* ones.

3.3.5 Energy-Dependent and Time-Dependent Control of Catalysis

To illustrate better the respective roles of the time required for transition state formation t_{TS} and the activation energy E_a, consider the hypothetical, general case in which several different products may be obtained from the catalyzed reaction as depicted in Figure 3.10.

Each product originates from a different relative orientation of the reactants in the transition state. As in the uncatalyzed case, each of the possible transition states, TS_1 to TS_4, will have a different activation energy, E_a^1–E_a^4, respectively. Unlike the uncatalyzed case however, each transition state will be formed over a different average time period after catalyst–reactant binding, t_{TS}^1–t_{TS}^4, respectively.

If the overall reaction in Figure 3.10 is energy-dependent, the most favored transition state will be the one with the lowest E_a. This transition state will determine both the rate and the pathway of the reaction, in complete analogy with the uncatalyzed case. Thus, the reaction rate will be a function of the E_a of the most stable transition state. The products initially obtained will reflect the relative E_a's of the available transition states, with the lowest energy transition state being most highly populated and its product formed in greatest numbers.

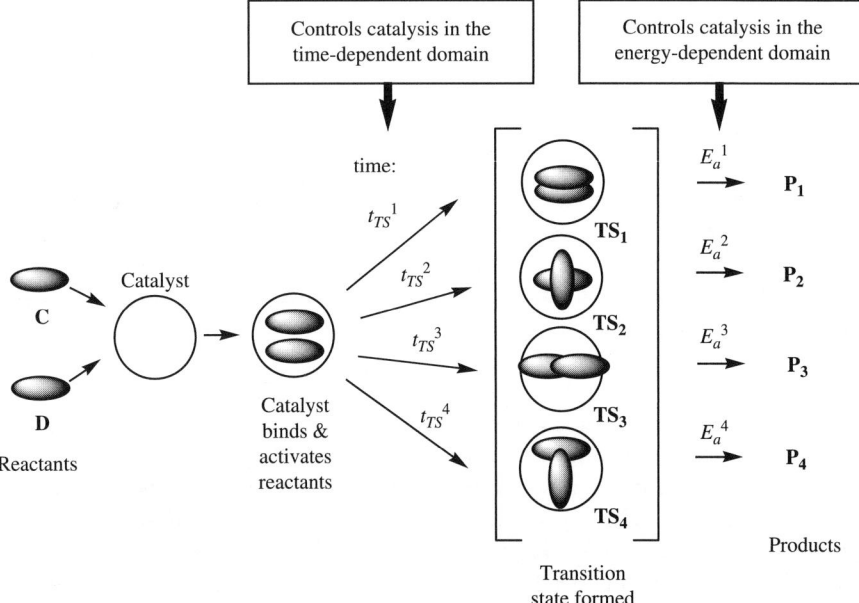

Figure 3.10. Schematic depiction of four hypothetical catalyzed reactions between two reactant molecules, C and D, generating the products P_1 to P_4, from transition states TS_1 to TS_4, respectively. Each transition state takes a certain average time, t_{TS}^1 to t_{TS}^4 to create and must achieve a minimum threshold energy (the activation energies E_a^1 to E_a^4) in order to proceed to products.

Key Point. In a time-dependent catalytic reaction, however, the rate is dependent on the catalyst-mediated collision frequency. The key variable in this regard is the time t_{TS} between reactant binding/activation and transition state formation, namely, the time required for transition state formation. The shorter this time, the greater will be the collision frequency (all other variables being maintained constant). The most favored transition state will therefore be the one involving the shortest average time t_{TS}. This transition state will be statistically favored because, under otherwise invariant conditions, it will generate more products per unit time than the other transition states, which are formed more slowly. In other words, the *first-encountered* transition state will be favored because this species has the highest overall, catalyst-mediated collision frequency. The shorter t_{TS} is for a particular transition state relative to the other transition states, the more favored this pathway will be.

3.3.6 The Influence of the Product Release Step

As a final note in this chapter, we should mention that the intermediacy of a catalyst introduces an additional, new step to the reaction process: the step of product-release from the catalyst (Fig. 3.8). Although we will not consider the influence of product release in this work, it is important to recognize that this step may dramatically affect the rate and pathway of a reaction. For instance, a catalyst that will release only one particular product will be governed, entirely, by the product release step.

3.4 CONCLUSIONS: ENERGY- AND TIME-DEPENDENT CATALYSIS

In Chapter 3, we have reviewed theories of uncatalyzed and catalyzed reaction kinetics. We have demonstrated that uncatalyzed reactions may be either energy-dependent or time-dependent and that this affects both their rate and pathway. In the former case, the reaction proceeds according to the intangible, underlying energy landscape that exists about the point of collision. In the latter case, the reaction follows the tangible, statistical occurence of spatial interaction in time.

Put more formally, energy-dependent reactions are controlled by the entity of the reaction transition state and its activation energy. Energy-dependent reactions typically involve a thermodynamic equilibrium among the reactants, the transition state, and the products. As such, they are described by *transition state theory*.

Time-dependent reactions are, however, controlled by the collision frequency of the reactants and are therefore independent of the nature and structure of the activated complex. Such processes are fundamentally nonequilibrium in character and are therefore not subject to *transition state theory*.

The identical situation exists in catalyzed reactions, except that the reaction is mediated by the intermediacy of a catalyst. Thus, two distinct catalytic domains are theorized: time-dependent and energy-dependent catalysis.

The energy-dependent domain is characterized by a reliance on E_a and the structure and energy of the catalyst-mediated transition state. As with uncatalyzed reactions, *transition state theory* provides a theoretical basis.

The time-dependent domain involves a dependence on the catalyst-mediated collision frequency. Time-dependent catalysis is not described by *transition state theory*.

The two catalytic domains differ in the manner with which the most favored transition state is selected. In the energy-dependent domain, the most favored transition state is the *lowest energy transition state*. This transition state has the smallest activation energy E_a. In the time-dependent domain, however, it is the first transition state that is formed, that is, the *first-encountered transition state*.

Energy-dependent catalysis provides the dominant paradigm of the catalytic process in chemistry and is therefore well known. Numerous chemistry textbooks teach this model of catalysis.

However, time-dependent catalysis is also theoretically possible. Time-dependence of this type has, to date, not been properly recognized or systematically studied in catalysis. Indeed, the field of catalysis has simply assumed that all catalysts must be energy-dependent. As such, time-dependent catalysis remains an unexplored area of research.

ACKNOWLEDGMENTS

The author gratefully acknowledges the involvement of Prof. R. J. P. (Bob) Williams (Oxford University) with whom he had a wide-ranging exchange regarding the contents of this chapter. In the process of these discussions, several of the foundation ideas in this series were clarified (Sections 3.2.3 and 3.2.5). The author also thanks

Prof. G. Charles Dismukes (Princeton University) and Prof. Fred Menger (Emory University) for useful and insightful comments.

REFERENCES

1. Atkins, P. W. *Physical Chemistry*, Oxford University Press, 1978, p. 52–782. Collision Theory: p. 897–928.
2. Pilling, M. J.; Seakins, P. W. *Reaction Kinetics*, Oxford University Press, 1996, p. 3–142: (a) Curvature in Arrhenius Plots: p. 54, p. 85, (b) Collision Theory: p. 61–66, (c) Eyring's Transition State Theory: p. 66–85, (d) Reaction Dynamics and Potential Energy Surfaces: p. 106–120, (e) Lindemann and Hinshelwood-RRK Theory: p. 121–138, (f) Boltzmann Distributions in the High- and Low-Pressure Limit of Hinshelwood-RRK Theory: p. 128–130.
3. See for example: (a) Peitgen, H-O.; Juergens, H.; Saupe, D., Eds. *Chaos and Fractals: New Frontiers of Science*, Springer Verlag, 1992; (b) A very simple primer on cascade processes is provided by Devaney, R. L. *The Cascade Model of Morphological Flows*, which is published at http://www.mbscientific.com/1_The_Cascade_Process.html; (c) Karlin, S.; Taylor, H. E. *A First Course in Stochastic Processes (2nd ed.)*, Elsevier Science, 1975; (d) Jouault, B.; Lipa, P.; Greiner, M. *Phys. Rev. E.* **1999**, *59*, 2451.
4. For example: (a) Moss, D. W. *Enzymes*, liver and Boyd, 1968, 68, and references therein; (b) Agarwal, P. K. *Microb. Cell Fact.* **2006**, *15*, 2, and references therein; (c) Kinetic "perfection" in an enzymatic system is described in: Knowles, J. R.; Albery, W. J. *Acc. Chem. Res.* **1977**, *10*, 105.
5. Williams, R. J. P. *Trends Biochem. Sci.* **1993**, *18*, 115.
6. Hammes-Schiffer, S.; Benkovic, S. J. *Annu. Rev. Biochem.* **2006**, *75*, 519; (b) Benkovic, S. J.; Hammes-Schiffer, S. *Science* **2003**, *301*, 1196; (c) Alper, K. O.; Singla, M.; Stone, J. L.; Bagdassarian, C. K. *Protein Sci.* **2001**, *10*, 1319; (d) Tousignant, A.; Pelletier, J. N. *Chem. Biol.* **2004**, *11*, 1037; (e) Agarwal, P. K. *Microb. Cell Fact.* **2006**, *15*, 2, and references therein.
7. (a) Menger, F. M. *Acc. Chem. Rev.* **1993**, *26*, 206, and references therein; (b) Menger, F. M. *Acc. Chem. Rev.* **1985**, *18*, 128, and references therein.
8. (a) Boehr, D. D.; McElheny, D.; Dyson, H. J.; Wright, P. E. *Science* **2006**, *313*, 1638; (b) Vendruscolo, M.; Dobson, C. *Science* **2006**, *313*, 1586.
9. Mueller, E.; Roscheisen, G. *Chem. Ber.* **1957**, *90*, 543. See also: Rossa, L.; Vogtle, F. *Top. Curr. Chem.* **1983**, *113*, 1.
10. See, for example: Tanaka, A. *Analyt. Sci.*, **2001**, *17* (Supplement) i797; Frank, R.; Greiner, G.; Rau, H. *Phys. Chem. Chem. Phys.* **1999**, *1*, 3481; Becerra, R.; Boganov, S. E.; Egorov, M. P.; Nefedov, O. M.; Walsh, R. *Mendeleev Commun.* **1997**, *7*, 87.
11. (a) Masel, R. I. *Chemical Kinetics and Catalysis*, Wiley-Interscience, 2001, and references therein. (b) See p. 287–290. (c) See p. 707–742.

4

TIME-DEPENDENCE IN HETEROGENEOUS CATALYSIS. SABATIER'S PRINCIPLE DESCRIBES TWO INDEPENDENT CATALYTIC REALMS: TIME-DEPENDENT ("MECHANICAL") CATALYSIS AND ENERGY-DEPENDENT ("THERMODYNAMIC") CATALYSIS

GERHARD F. SWIEGERS

4.1 INTRODUCTION

In previous chapters, we have discussed and distinguished between *energy-dependent* and *time-dependent* chemical processes [1,2].

In energy-dependent reactions, the critical impediment that must be overcome is the energy barrier involved in the reaction, namely, the *activation energy* E_a. In time-dependent reactions, however, the activation energy is sufficiently low that the key impediment to reaction involves merely bringing the reactants into physical contact with each other ("collision"); that is, it involves the *collision frequency* of the reactant molecules. Time-dependent processes are observed in so-called "diffusion-controlled" reactions. However, we will use the term "time-dependent" in this series, because "diffusion control" does not describe all possible chemical reactions of this type.

Mechanical Catalysis: Methods of Enzymatic, Homogeneous, and Heterogeneous Catalysis,
Edited by Gerhard F. Swiegers
Copyright © 2008 John Wiley & Sons, Inc.

In the field of catalysis, two distinct realms have been theorized to exist along these lines: 1) *energy-dependent catalysis* and 2) *time-dependent catalysis*.

In realm 1) the catalytic rate and pathway was proposed to be a function of the thermodynamic efficiency of the catalyst. That is, it depends on how well the catalyst stabilizes its transition state and its resulting *activation energy* E_a. In realm 2), however, catalysis depends on the ability of the catalyst to bring the reactants into physical or mechanical contact with each other. That is, it is a function of the *catalyst-mediated collision frequency*. In this chapter, we will examine the existence of such realms and their observable, physical manifestations. We will restrict this study to the field of *heterogeneous catalysis*.

Sabatier's Principle is a well-known and long-established tenet of the general theory of catalysis. As proposed by the 1912 Nobel Prize winner, the principle states that an ideal catalyst must *bind its reactants strongly but not too strongly* [1]. This assertion alludes to the presence, in most catalysts, of processes that are contrarily influenced by variations in the strength of catalyst–reactant binding. Maximum efficiency is achieved when these processes are perfectly balanced, that is, at a particular, intermediate catalyst–reactant binding strength.

What are the processes? In its simplicity, the principle makes no pronouncement on the underlying mechanisms involved.

What happens when the processes are not perfectly balanced? Here again, the principle is silent.

Finally, what is the principle really saying about catalysis at the most fundamental level? Broad, but accurate generalizations like Sabatier's Principle inevitably originate from specific and clearly definable physical causes. What is the physical phenomenon at the origin of Sabatier's Principle?

Despite its long history, the fundamental character of Sabatier's Principle has never been fully explored. Indeed, it could be said that the principle has not even been considered in a truly comprehensive manner. The insights that it may potentially provide have, therefore, not been clearly articulated.

In this work we will review Sabatier's Principle and examine in detail the proposition laid out in the cumulative studies in the scientific literature. We will show that excellent catalysis must combine energy efficiency with highly dynamic catalyst–reactant interactions. These two fundamental processes are, however, contrarily influenced by the strength of catalyst–reactant binding. That is, increasing energetic efficiency is necessarily accompanied by decreasing dynamism in catalyst–reactant binding, and vice versa. Catalysis, therefore, involves a balance of these two imperatives. This juxtaposition comprises the underlying physical origin of Sabatier's Principle.

We will also demonstrate that the key properties of most heterogeneous catalysts, including rate and pathway, depend on which of the above processes dominate the catalytic process. That is, which of energy efficiency or kinetic dynamism comprise the key impediment to catalytic reaction?

Catalysis whose catalyst–reactant dynamism allows for a faster reaction than is permitted by its thermodynamic efficiency is governed by the transition state and its activation energy (E_a). Such catalysis is *energy-dependent*.

Catalysis limited by the dynamism of the catalyst–reactant binding is, however, directed by the time available for catalysis on each occasion of binding and is not influenced by E_a per se. Such catalysis is *time-dependent*.

Sabatier's Principle, consequently, conceptually relates energy-dependent and time-dependent processes in catalysis.

At the most fundamental level, this chapter will show that energy-dependent catalysts select the transition state with the lowest activation energy, that is, the *lowest energy transition state*. Time-dependent catalysts, however, select the first transition state that is formed, that is, the *first-encountered transition state*. Although the former involves the pathway with the lowest available thermodynamic energy, the latter is intrinsic to a mechanical-type process.

4.2 SABATIER'S PRINCIPLE IN HETEROGENEOUS CATALYSIS

4.2.1 Volcano Plots

In the 1950s, Balandin and others noticed a distinctive relationship between the relative rates at which formic acid is decomposed in the gas phase on various metal catalysts and the strength of catalyst-formate binding [2–4]. This relationship is

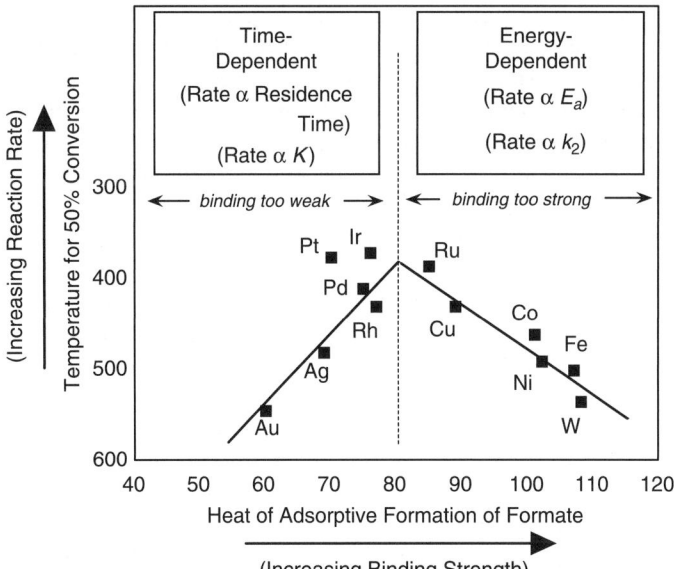

Figure 4.1. Volcano plot for the decomposition of formic acid by a range of heterogeneous metal catalysts. The constants K and k_2 are those of Equations (4.1) and (4.2a–b). The "Residence Time" is the average time that the reactants are present on the catalyst on each occasion of binding (see Section 4.2.3.1). The labels "binding too weak" and "binding too strong" are taken from Reference 5.

depicted in Figure 4.1. As can be seen, the reaction rate initially rises with increasing binding strength (quantified as the heat of adsorption of formate), after which it falls, giving a characteristic volcano-like shape. Such graphs are thefore termed "volcano plots."

Based on various physical studies [2–4], it was concluded that the formate ion is an intermediate in this reaction, which has the mechanism ($K = k_1/k_{-1}$):

$$HCOOH \xrightleftharpoons{K} HCOO^-_{\text{adsorbed}} + H^{(+)}_{\text{adsorbed}} \qquad (4.1)$$

$$H^{(+)}_{\text{adsorbed}} + HCOO^-_{\text{adsorbed}} \begin{array}{c} \xrightarrow{k_2} CO_2 + H_2 \quad (4.2a) \\ \\ \xrightarrow{k_2} CO_2 + H_2O \quad (4.2b) \end{array}$$

The importance of these equations is that they, effectively, depict the two general steps involved in a catalyzed reaction. Thus, Equation (4.1) involves the rate at which the reactants are brought together on the catalyst. That is, it illustrates the process in which the reactants bind to the catalyst and are thereby brought together in a catalyst-mediated collision. Equations (4.2a–b) describe the rate at which the transition state, once formed, is converted into products. That is, they depict the process of overcoming their respective activation energies (E_a's). As such, these equations are, effectively, the catalyzed analogs of the uncatalyzed steps shown in Equations (4.3 and 4.4) in Chapter 3.

The trends depicted in the volcano curve must represent these underlying processes. As summarized by Thomson and Webb [6]: On one side of the plot, Equation (4.1) (reactant localization and collision) is rate-limiting, whereas on the other side, Equation (4.2a) or (4.2b) (overcoming E_a in the transition state) is rate-limiting. At the peak of the volcano plot, the two processes are balanced and their rates are identical.

Fahrenfort, van Reyen, and Sachtler have put it thus [4]: "[the reason for the shape of the volcano plot is that] on one side of the maximum, the *activation energy* becomes unfavorably high; on the other side, the *frequency factor* [collision frequency] drops to low values."

Thus, the volcano plot graphically depicts the relationship between time- and energy-dependent catalysis in a range of heterogeneous catalysts.

On the left-hand side of Figure 4.1, catalyst–reactant binding interactions are weak, so that the slowest and least-likely process involves reactant localization and collision. Thus, the reaction is limited by the rate at which $HCOO^-_{\text{adsorbed}}$ is formed, that is, by Equation (4.1) and the value of K.

4.2 SABATIER'S PRINCIPLE IN HETEROGENEOUS CATALYSIS

On the right of Figure 4.1, catalyst–reactant binding interactions are strong, so that the slowest and least-likely step involves the successful separation of the transition state into the products, that is, by Equation (4.2a) or (4.2b), and k_2 [1].

On the left-hand side of Figure 4.1, the collision step [Equation (4.1)] is therefore slower than the energetic step [Equation (4.2a–b)]. Thus, the system involves *time-dependent* catalysis.

On the right-hand side of Figure 4.1, this process is reversed: The energetic step [Equation (4.2a–b)] is slower than the collision step [Equation (4.1)]. The system consequently describes *energy-dependent* catalysis in this domain.

As depicted in Figure 4.2, the volcano plot displays, in a physically observable form, the slower of the underlying processes involved in the reaction. As noted in Chapter 3, the slowest process in a reaction sequence determines the properties of the reaction and hence establishes its character.

The optimum catalyst lies at the peak of the volcano plot where the best balance is achieved between the competing requirements of highly dynamic kinetics (the weakest reasonable binding) and highly favorable thermodynamics (the greatest reasonable transition state stabilization). For the decomposition of formic acid, the optimum catalyst is Pt or Ir.

Incidentally, the ratio of products in the decomposition of formic acid was estimated to be ca. 10 H_2:1 H_2O [Equation (4.2a):(4.2b)] for all catalysts in Figure 4.1 [4]. This estimation was complicated, however, by the water gas shift reaction, which may alter the H_2:H_2O ratio *in situ*, according to Equation (4.3). Complications of this type are typical in heterogeneous catalysis.

$$CO_2 + H_2 \leftrightarrow H_2O + CO \qquad (4.3)$$

In the following sections, we will examine how volcano plots get their distinctive shape. Before we do that, however, we need to make a few important points about them.

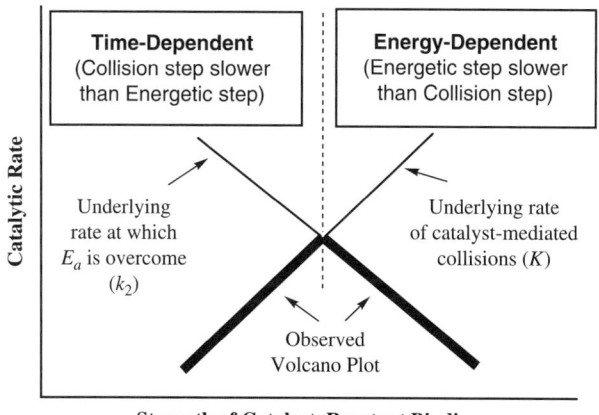

Figure 4.2. Origin of the distinctive shape of volcano plots [5].

4.2.2 Some Important Points about Volcano Plots

As can be seen, the y-axis in Figure 4.1 employs the temperature for 50% conversion as a measure of the relative reaction rate for each of the catalysts [1]. John Meuring Thomas and W. John Thomas have made the point that one should relate activity to activation energies and pre-exponential factors [7]. For this reason, it is preferable to plot the reaction rate of the catalysts at a fixed temperature rather than to plot the reaction temperature at a fixed conversion, as is done in Figure 4.1. This process is important because a change in the temperature may prematurely change the rate-limiting step. In practice, however, it is easier to measure accurately the temperature at fixed conversion than it is to measure the conversion at fixed temperature. Moreover, the elevated temperatures typically used in gas-phase catalysis are such that distortions in the plot are likely to be relatively unimportant, at least in respect of the practical need to select the most active catalyst for an industrial reaction.

A second important point is one made by Masel [1], who noted that almost all heterogeneous catalytic systems display volcano plots, including gas-phase, liquid-phase, and electrochemical reactions [8]. For this reason, volcano plots are routinely used to select the optimum catalyst for heterogeneous, industrial reactions [1]. This fact confirms that volcano plots reflect a set of underlying reaction dynamics that is common to all catalyzed reactions. That is, the sheer prevalence of volcano plots means that they cannot generally originate from some other effect, such as an adventitious change in the catalytic sites involved. This contention is supported by the fact that other, related plots, such as so-called "Sachtler-Fahrenfort [9]" and "Tanaka-Tamaru [10]" plots, are also generally volcano-shaped (as will be later discussed in Section 4.2.6). Moreover, experimentally observed volcano plots may be accurately modeled as a function of bond strength using theoretical equations for the collision frequency and the activation energy (as later discussed in Section 4.4.7).

Finally, it is important to note that the experimental data in volcano plots compare the rate of reaction *at the catalyst only*. They do not consider the rate at which reactants diffuse to the catalyst, which is maintained invariant in the experiments. That is, volcano plots do not involve diffusion-controlled systems. Rather, they directly compare time- and energy-dependent catalysis in the absence of diffusion constraints.

4.2.3 Time-Dependent Catalysis in Volcano Plots

4.2.3.1 How Is Time-Dependence Created on the Left-Hand Side of the Volcano Plot? As noted, catalysts on the left-hand side of Figure 4.1 are time-dependent. The manner in which this time-dependence is created is different, however, to that previously described for uncatalyzed reactions as discussed in Section 3.2.8 of Chapter 3. As shown in Figure 4.2, it originates because the catalyst-mediated collision frequency falls below the rate at which E_a is overcome. How does this reduction occur?

The key feature of catalysis on the left-hand side of Figure 4.1 is that the formate intermediate is bound and activated by the catalyst only very briefly and transiently on each occasion of binding [2–7]. The average time during which this intermediate is bound and available for transition state formation is termed the *residence time* τ [7,11,12].

4.2 SABATIER'S PRINCIPLE IN HETEROGENEOUS CATALYSIS

As the strength of catalyst–reactant binding declines, this residence time also declines. This occurs because, the weaker the catalyst–reactant binding is, the shorter will be the time that the reactant remains bound to the catalyst and available to be brought into collision on each occasion of binding.

As the residence time becomes shorter, there is therefore less and less of an opportunity for the bound reactant to be involved in reactive collisions while attached to the catalyst. The catalyst-mediated collision frequency therefore also declines.

At some relatively low residence time, the collision frequency becomes rate determining and the catalyst is then time-dependent. This point is marked in Figure 4.2 by the dotted line running through the peak of the volcano plot. To the left of this line, catalyst–reactant dynamism allows for only a slower overall reaction than is possible according to the energy efficiency of the catalysis. Thus, time-dependence is established.

But it is not only the collision frequency that varies according to changes in the residence time. The pathway of the reaction is also affected.

In Sections 3.3.2–3.3.5 of Chapter 3, we showed that the properties of a time-dependent catalyst are determined by the time needed for the bound reactants to form the transition state t_{TS} in Figure 4.3 and t_{TS}^1 to t_{TS}^4 in Figure 4.4. The statistically most favored transition state is the one with the shortest value of t_{TS}. But the extent to which it is favored depends on the time available for the bound reactants to react. Thus, as the residence time τ is made more and more brief, the transition state with the shortest t_{TS} becomes increasingly favored.

To illustrate this effect, we will use the hypothetical example in Figure 4.4. We will set the times for transition state formation (t_{TS}) to be $t_{TS}^4 > t_{TS}^3 > t_{TS}^2 > t_{TS}^1$. Figure 4.5 graphs the relative transit times to depict the influence of variations in the residence time τ relative to t_{TS}^{1-4}.

If the average residence time of the reactants on the catalyst is longer than t_{TS}^4 (that is, $\tau > t_{TS}^4$), all of the possible transition states are available to be formed [Fig. 4.5(a)].

However, if the residence time becomes shorter than t_{TS}^4 (that is, $t_{TS}^1 < t_{TS}^2 < t_{TS}^3 < \tau < t_{TS}^4$), the corresponding transition state TS_4 cannot be formed because the average time taken to set it up is greater than the average time that the reactants are bound and activated by the catalyst [Fig. 4.5(b)]. Thus, at this residence time, no products deriving from TS_4 will be generated.

If the residence time is made progressively more fleeting, it will eventually also become shorter than t_{TS}^3, thereby eliminating the availability of the corresponding transition state TS_3 [Fig. 4.5(c)] ($t_{TS}^1 < t_{TS}^2 < \tau < t_{TS}^3 < t_{TS}^4$). No products deriving from TS_4 or TS_3 will then be generated.

Ultimately, only one transition state will be available for catalysis—the transition state with the shortest t_{TS}; namely, TS_1 [Fig. 4.5(d)] ($t_{TS}^1 < \tau < t_{TS}^2 < t_{TS}^3 < t_{TS}^4$). The first-formed transition state therefore becomes increasingly favored as the residence time declines.

In the extreme situation shown in Figure 4.5(e), the residence time will be shorter than all of t_{TS}^1 to t_{TS}^4, meaning that no transition state can be formed during the short-lived occasions that the reactants are bound by the catalyst. In that case, the reactants will simply dissociate from the catalyst without reaction. In other words, the catalytic effect will be lost.

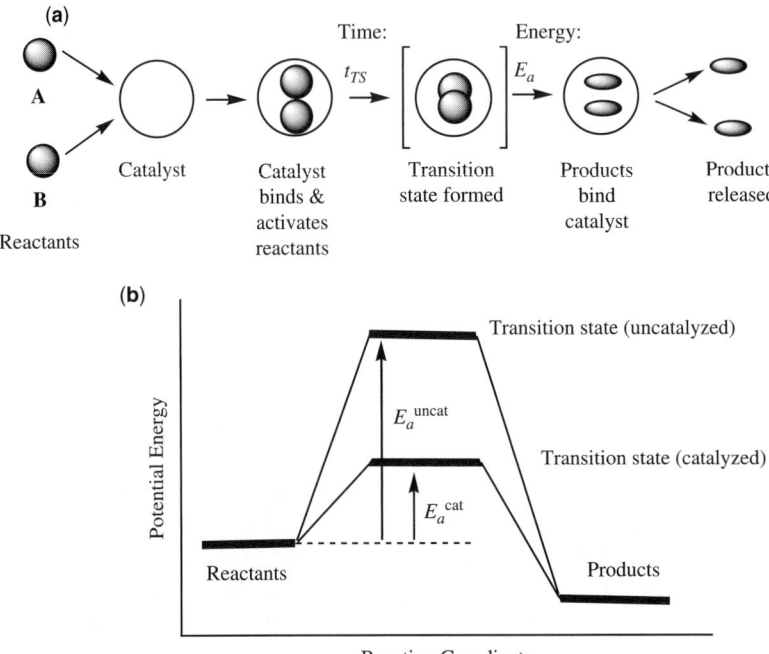

Figure 4.3. Schematic depiction of (a) a catalyzed, reactive encounter between two molecules, A and B, leading to the formation of products, and (b) the energetic profile followed during the catalyzed reaction, showing the minimum threshold energy (the activation energy, E_a^{cat}) needed for reaction, relative to the same threshold for the uncatalyzed reaction (E_a^{uncat}). In an *energy-dependent* catalyzed reaction, the step that comprises the greatest impediment to reaction involves overcoming the activation energy, E_a^{cat}. In a *time-dependent* catalyzed reaction, the greatest impediment involves bringing the bound reactants into collision with each other (average time required is t_{TS}).

Of course, all of these times, including the residence time τ and the times t_{TS}, are *average* values only. The observed rate may also be affected by surface turbulence and other factors [12]. So this analysis demonstrates only a general trend. But the outcome is clear: The brevity of reactant binding and activation creates time-dependent catalysis in a volcano plot and controls the product mixture generated.

4.2.3.2 Why Do Volcano Plots Slope Upward on the Left. The graph slopes upward on the left side of Figure 4.1 because increasing strength in reactant binding leads to an increase in the residence time [1]. Thus, the catalyst-mediated collision frequency is increased, with an accompanying escalation in the overall reaction rate.

Chemical engineers put it as follows: The physical manifestation of brevity in the residence time is that time-dependent heterogeneous catalysts display an incomplete surface coverage at any one time [1,11]. That is, many catalytic centers are not

4.2 SABATIER'S PRINCIPLE IN HETEROGENEOUS CATALYSIS

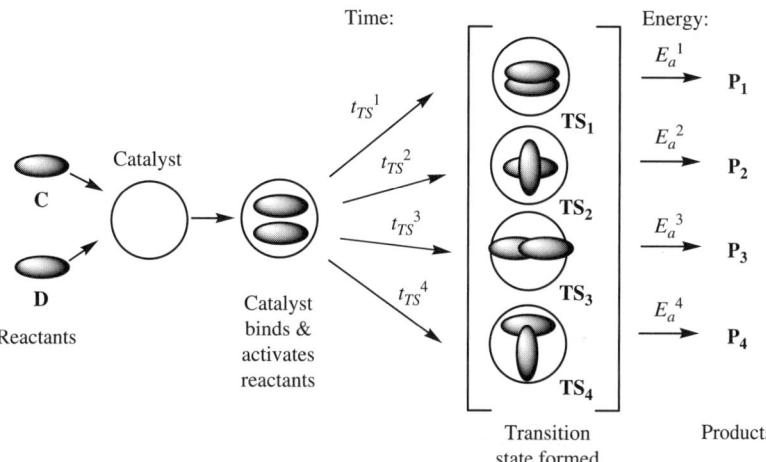

Figure 4.4. Schematic depiction of four hypothetical catalyzed reactions between two reactant molecules, C and D, generating the products P_1 to P_4, from transition states TS_1 to TS_4, respectively. Each transition state takes a certain average time t_{TS}^1 to t_{TS}^4 to create and must achieve a minimum threshold energy (the activation energies, E_a^1 to E_a^4) in order to proceed to products.

occupied because the reactants are not attached for very long on each occasion of binding. The reaction rate is then determined by the fraction of the total catalytic centers that are occupied at any one instant θ [1,11]. This fraction increases with increasing catalyst–reactant binding strength, thereby increasing the reaction rate.

In summary, the shorter the residence time, the lower will be the surface coverage and the smaller the fraction of catalytic centers that are occupied, thereby diminishing the overall rate. For very weak binding, the reactants are attached to the catalyst and activated too briefly on each occasion to allow for much transition state formation. But the stronger the binding becomes, the greater the average residence time and the larger the rate.

In mathematical terms, the overall rate exhibited by a time-dependent catalyst is proportional to K [Equation (4.1)], which is a function of the free energy of reactant adsorption ΔG_{ads} [equation (4.4a)] [1]. An increase in ΔG_{ads} (or in its constituent, ΔH_{ads}) leads to an increase in K and therefore an increase in the reaction rate.

$$K = \exp\left(\frac{-\Delta G_{ads}}{kT}\right) \quad (4.4a)$$

According to Ertl [7,12], simple molecules adsorbed on metal surfaces have residence times:

$$\tau = \tau_o \exp\left(\frac{-E_{des}}{RT}\right) \quad (4.4b)$$

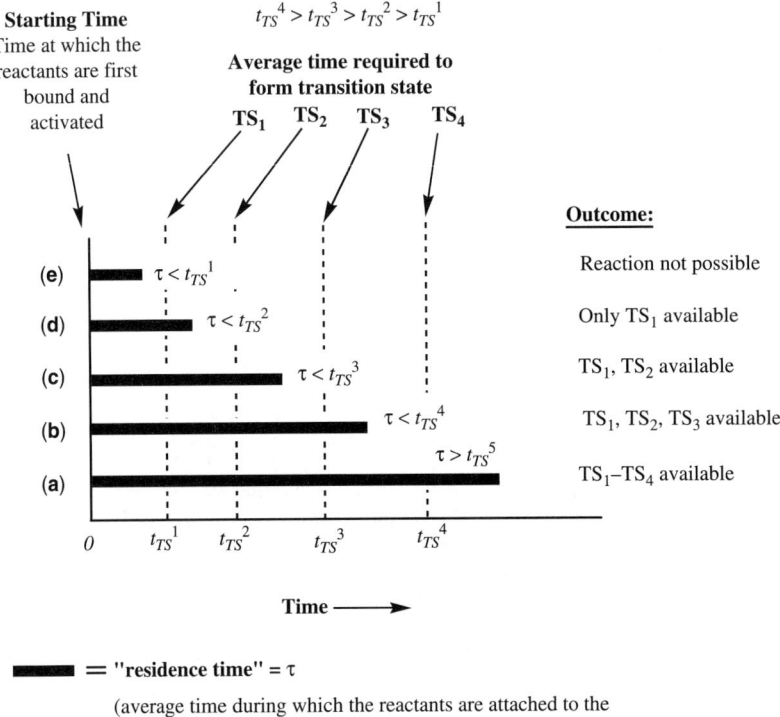

Figure 4.5. Illustrative effect of residence time (τ) on the transition states for catalysis in Figure 4.4. As the residence time (thick lines) decreases in the order shown in (a) to (e) above, it becomes progressively shorter than the time needed to form certain of the transition states ($t_{TS}^1 - t_{TS}^4$). This decreases the number of transition states available for catalysis until, ultimately, it is too short for any transition state to form [example (e)]. In this case, reaction is not possible on the catalyst because the catalyst-mediated, collision frequency falls to zero.

where E_{des} = activation energy of desorption. The residence times of the formate ion on the catalysts in Figure 4.1 have not, to the best of our knowledge, been determined. However, by way of example, residence times in the order 0.1–10 ms are measured for CO on Pd(111) at 580–700 K [7].

4.2.3.3 The Rate-Determining Step in a Time-Dependent Catalyst.
The overall reaction rate of a time-dependent catalyst is therefore determined by the catalyst-mediated collision frequency, which involves two components:

(1) The average duration of the binding events (the residence time)
(2) The frequency of catalyst–reactant binding events (which may be a function of the rate of diffusion of the reactants to the catalyst)

As both of these processes influence and determine the collision frequency, both formally comprise the rate-determining steps, with (1) above being the more critical because it exerts ultimate control of the rate when (2) is maximized [13].[1] Constant K in Equation (4.1) incorporates both of these processes.

Key Point. The key distinguishing feature of a time-dependent catalyst is, consequently, that its rate-determining step involves a rapidly forming and dissociating intermediate like the HCOO$^-_{adsorbed}$ species depicted in Equation (4.1). The lifetime of this intermediate, and the frequency with which it is formed, determines the overall rate.

4.2.3.4 The Physical Manifestation of Time-Dependent Catalysis. "Saturation" of a Time-Dependent Catalyst.

In an archetypal time-dependent catalyst, the reactants are constantly and rapidly binding and dissociating from the catalyst. During some of these binding events, the residence time is sufficiently long that the reactants are able to collide and form products.

What happens however, when the reactant concentration is systematically increased until the catalyst is *saturated*? How does saturation occur?

As the reactant concentration is increased relative to the catalyst concentration, the frequency of the catalyst–reactant binding events increases. It is accompanied by an increase in the proportion of catalytic centers that are occupied at any one time. Eventually, however, the frequency of catalyst–reactant binding will be maximized; that is, it simply cannot increase any more. At this point, the catalyst is *saturated*. A proportion of the catalytic centers will then still be unoccupied because, at any one instant, most catalytic centers are involved in receiving or releasing reactants, not in physically holding them.

Key Point. At saturation, the overall rate is determined only by the average residence time of the reactants on the catalyst because the frequency of catalyst–reactant binding is then at its maximum and unable to increase. In other words, the lifetime of the rapidly forming and dissociating intermediate becomes the sole determinant of the catalytic rate at saturation.

Rapidly forming and dissociating intermediates in systems at saturation are distinctive of several catalytic systems, including, most prominently, enzyme catalysts that display so-called Michaelis–Menten kinetics. In the following chapter, we will examine whether enzymes are generally time-dependent or energy-dependent catalysts. Several heterogeneous catalysts have been shown to also display the Michaelis–Menten kinetics that is so characteristic of enzymes [14].

---

[1] The rate-determining step (also known as the rate-limiting or rate-controlling step) is defined by the IUPAC as the "elementary process" whose rate constant exerts the strongest effect of all rate constants on the overall reaction rate.

4.2.4 Energy-Dependent Catalysis in Volcano Plots

4.2.4.1 How Is Energy-Dependence Created on the Right-Hand Side of the Volcano Plot? A very different catalytic regime exists on the right-hand side of Figure 4.1. Because the catalyst–reactant binding interactions are strong in this region, the overall reaction rate depends on the rate at which the bound and activated reactants successfully traverse the transition state to form products; that is, on the rate of Equation (4.2a) or (4.2b). The more stable the transition state, relative to the reactants, the lower will be the activation energy (E_a) and the faster will be the rate. The most favored transition state in this regime is therefore the one that is *energetically most stable*. This need not be the first transition state encountered, although, in practice, many catalysts involve only one possible transition state, so that these are often the same.

If multiple transition states are available, they will be populated in proportions that reflect their relative energetic stabilization, with the transition state having the lowest E_a being most highly populated. All of the available transition states can be populated only because the residence time during which the reactant is bound and activated by the catalyst is sufficiently long.

4.2.4.2 Why Do Volcano Plots Slope Downward on the Right? The graph on the right-hand side of Figure 4.1 slopes downward, because the potential energy of the bound reactants declines as catalyst–reactant binding becomes stronger, but the potential energy of the transition state remains unchanged. Thus, the activation energy, which is the difference between these two potential energies, becomes larger with increasing catalyst–reactant binding strength. This result is illustrated in Figure 4.6. The larger E_a becomes, the slower will be the rate of catalysis. In the extreme case, very strong binding effectively halts transition state formation, causing the catalyst to be "poisoned."

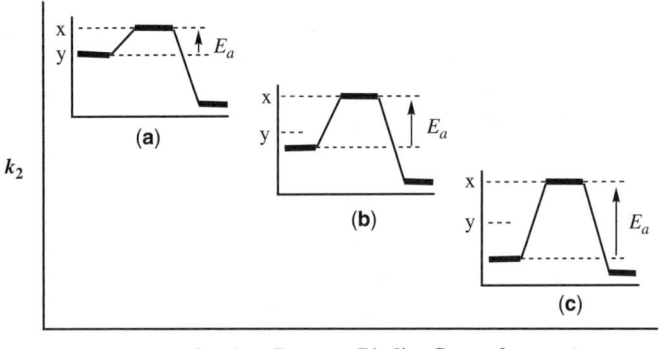

Figure 4.6. Activation energy (E_a) as a function of catalyst–reactant binding strength. In increasing the catalyst–reactant binding strength [going from (a) to (b) to (c)], note that the potential energy of the transition state (depicted as x) does not change. However, the potential energy of the bound reactants (depicted as y) declines.

4.2.4.3 The Rate-Determining Step in an Energy-Dependent Catalyst.
The overall rate of an energy-dependent catalyst is consequently governed by k_2 [Equation (4.2a–b)], which is dependent on the activation energy E_a. E_a gets larger with increasing binding strength according to

$$k_2 = k_0 \exp\left(\frac{-E_a}{kT}\right) \tag{4.5}$$

The rate-determining step in an energy-dependent catalyzed reaction is therefore the rate at which E_a is overcome.

4.2.4.4 The Physical Manifestation of Energy-Dependence. Saturation in an Energy-Dependent Catalyst.
In an archetypal energy-dependent catalyst, the reactants bind and stay on the catalyst until they are transformed into products. During the period that they are attached to the catalyst, they constantly collide with each other, but only a small proportion of these collisions results in product formation. Most catalytic centers will be occupied at any one instant, because the residence time of the reactants on the catalyst is long.

The properties of such a catalyst consequently depends on the rapidity with which the reactants are transformed into products once they are attached to the catalyst.

The catalytic rate is, however, also affected by the average time between a reactant or product leaving the catalyst and the next one attaching to it. An increase in the concentration of the reactants will cause this time to decline. The catalytic rate therefore increases with increasing reactant concentration up to the point where the time between product release and reactant binding approaches zero. At that stage the catalyst will be *saturated*. All catalytic centers will then be occupied. This result will occur only at a very high, sometimes inaccessible, *absolute concentration* of the reactant.

Key Point. **The key distinguishing feature of an energy-dependent catalyst is therefore that its rate-determining step involves the process of overcoming the activation energy as depicted in Equation (4.2a–b). It does not involve a kinetically detectable, rapidly forming, and dissociating intermediate.** Such a catalyst is, theoretically, only capable of becoming saturated when all catalytic centers are occupied. It will only occur at a high absolute concentration of reactants.

4.2.5 The Physical Origin of Sabatier's Principle

The overall reaction rate achieved by a catalyst therefore comprises the *slower* of the two processes involving the stablization of transition state (k_2) and the kinetics of catalyst–reactant binding, activation, and collision (K). These processes are contrarily influenced by the strength of catalyst–reactant binding. Whereas the latter increases with binding strength, the former declines with binding strength. In the ideal catalyst, the two processes will be perfectly balanced, with identical rates.

This, then, is the physical origin of Sabatier's Principle. The underlying processes described above are necessarily common to *all catalysts* requiring both kinetic dynamism and energetic efficiency. It is this commonality that makes Sabatier's Principle so widely true and accurate.

The two realms of time-dependence and energy-dependence described by Sabatier's Principle are conceptually identical to those that exist in uncatalyzed reactions as previously described in Sections 3.2.1–3.2.9 of Chapter 3.

Thus, energy-dependent catalysis is characterized by *equilibrium kinetics* displaying a *dependence on E_a*. It is described by *transition state theory*.

By contrast, the time-dependent domain involves nonequilibrium kinetics displaying a dependence on the collision frequency. Time-dependent catalysis is not described by transition state theory.

4.2.6 Other Plots Illustrating Sabatier's Principle

A problem with drawing up volcano plots is that the heat of adsorption of a particular reactive intermediate is often not available. However, it is generally true that the enthalpy of formation per mole of oxygen of the most stable oxide is *directly proportional* to the heat of adsorption of an intermediate [1]. Thus, one can construct volcano plots that mimic the desired curve relatively accurately by plotting the heat of formation of the oxides *per mol oxygen* on the *x*-axis. Such graphs are known as *Sachtler–Fahrenfort* curves [9].

An alternative approach is to plot the enthalpy of formation *per metal atom* of the highest oxide. Such graphs are known as *Tanaka–Tamaru* plots [15]. Figure 4.7 depicts illustrative plots of these types for the hydrogenation of ethylene. Although certainly more scatter exists in the data, such graphs are generally also volcano-shaped, supporting the universality of Sabatier's Principle.

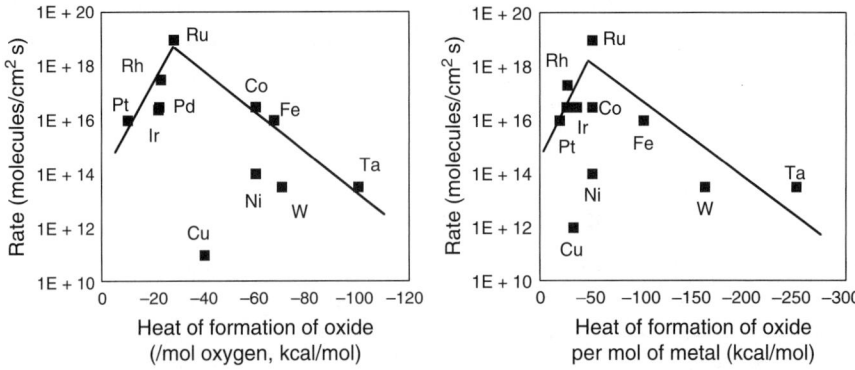

Figure 4.7. Sachtler–Fahrenfort (left) and Tanaka–Tamaru (right) plots for the hydrogenation of ethylene [9,10].

4.2.7 Modeling of Volcano Plots

Several researchers have successfully modeled volcano plots using theoretical equations. Masel has, for example, demonstrated neatly, that a volcano plot can be calculated for the catalytic gas-phase formation of HBr [1]. At low $\Delta H_{adsorption}$, the reaction rate is dominated by the concentration of the intermediate, that is, by the proportion of filled catalytic centers θ. At higher $\Delta H_{adsorption}$, the surface is mostly filled with reactants, so that the reactivity of the intermediate then becomes rate-limiting.

One of the more interesting successes in modeling of a volcano plot is a study by Norskov that predicted reaction rates across the first row of the periodic table for the synthesis of ammonia [16,17]. Using a kinetic model reported by Nielsen, the rate of ammonia formation was calculated as a function of the number of d-electrons in the first-row transition metals. The d-electron contribution to the surface chemical bond depends on the degree of filling of the anti-bonding states. Thus, early transition metals with fewer d-electrons form stronger chemical bonds. The experimental data and the calculated energies for this effect are in good agreement.

When graphed in this way, first-row transition metal catalysts were predicted to display activities that fitted the volcano plot shown in Figure 4.8. This plot suggests that the rate of ammonia formation should rise in the order Cr < Mn < Fe and then fall in the order Fe > Co > Ni. It essentially matches Ozaki and Aika's experimentally measured activity [18] of various transition metals as a function of the degree of filling of the d-bond.

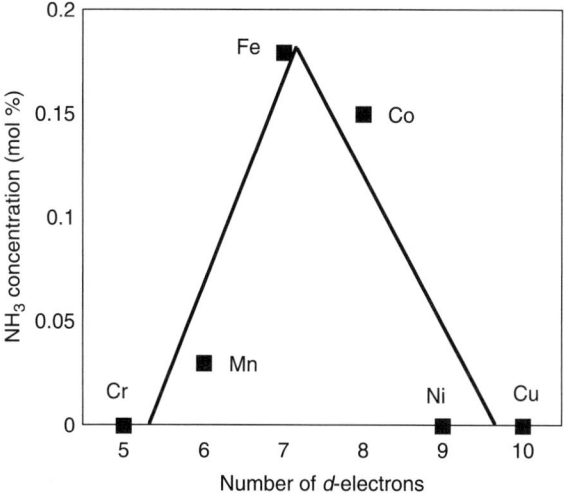

Figure 4.8. Predicted trend for the synthesis of ammonia by first-row transition metals (1 atm, 400 °C, 1 m^2 of catalyst) [15].

4.2.8 Reaction Pathway as a Function of the Most-Favored Transition State

A key difference that is implied to exist between a time-dependent and an energy-dependent catalyst is the way in which they select their most favored transition state. Time-dependent catalysts choose the first-encountered transition state, whereas energy-dependent catalysts select the transition state with the lowest energy. Given that the energetically most stable transition state may, in many cases, also be the first-encountered transition state, one has to wonder whether this selection process can be confirmed experimentally?

Because of the complexity of heterogeneous catalytic systems, it is generally not possible to demonstrate selectivity according to time- or energy-dependence. However, one can demonstrate that the residence time of an intermediate may determine the pathway of their reaction.

An example in this respect is the decomposition of formaldehyde in electroless metal plating [15]. In acid media, formaldehyde is oxidized by transition metals to form CO_2 and H_2, as follows:

$$HCHO + H_2O \rightarrow CO_2 + 2H_2 \tag{4.6}$$

In basic media, however, two types of behavior are observed. On Group 1B metals, such as Cu, Ag, and Au, formaldehyde decomposition is accompanied by hydrogen evolution as shown in Equation (4.6). But, on Group VIII metals, such as Pt and Pd, formaldehyde oxidation is accompanied by water formation according to the general (unbalanced) reaction:

$$HCHO \rightarrow CO_2 + H_2O \tag{4.7}$$

An electrochemical study by Bindra, Roldan, and Arbach [18] demonstrated that, for all catalysts studied, the reaction proceeded according to the mechanism:

$$H_2C(OH)O^- \xrightarrow{K} [H_2C(OH)O^-]_{adsorbed} \tag{4.8}$$

$$[H_2C(OH)O^-]_{adsorbed} + OH^- \xrightarrow{k_2} HCOO^- + M-H^\bullet + H_2O + e^- \tag{4.9}$$

A volcano plot for this reaction indicated that the rate-determining step involved Equation (4.8) (for time-dependent catalysts) *or* Equation (4.9) (for energy-dependent catalysts).

Although both time-dependent and energy-dependent catalysts, therefore, generate the reactive intermediate $M-H^\bullet$ (M = catalyst), the final product of the reaction is determined by the way in which this intermediate reacts. On the group IB metals, which have positive free energies for the hydrogen ionization reaction at the potential applied, the $M-H^\bullet$ species is long-lived and therefore capable of forming H_2 by a hydrogen radical recombination on the metal catalyst:

$$M-H^\bullet + M-H^\bullet \rightarrow 2M + H_2 \tag{4.10}$$

By contrast, on the group VIII metals, which have *negative* free energies for the hydrogen ionization reaction at the applied potential, the M–H$^\bullet$ species is short-lived, so that it simply does not have time to form H_2 on the metal catalyst. Instead, the hydrogen radical dissociates as a hydrogen ion (H^+) and reacts with OH^- in solution to form H_2O according to

$$M-H^\bullet \longrightarrow M + H^+ + e^- \tag{4.11}$$

$$H^+ + OH^- \longrightarrow H_2O \tag{4.12}$$

Thus, the *residence time* of the hydrogen radical on the metal catalyst determines the reaction product. A long residence time allows for the formation of the transition state leading to H_2, which is clearly the most favorable process on the metal catalyst. But a short residence time renders this transition state unavailable and eliminates it as a possibility. Instead, the radical dissociates from the metal and forms H_2O in solution.

4.3 EXCEPTIONS TO SABATIER'S PRINCIPLE

To the best of our knowledge, only one exception to Sabatier's Principle has ever been noted in the scientific literature: polymerization catalysis involving free-radical initiators [1]. Species, like peroxides, initiate polymerization according to mechanisms such as

$$ROOR \longrightarrow 2\ RO^\bullet$$
$$RO^\bullet + CH_2{=}CH_2 \longrightarrow ROCH_2CH_2^\bullet \tag{4.13}$$
$$ROCH_2CH_2^\bullet + CH_2{=}CH_2 \longrightarrow ROCH_2CH_2CH_2CH_2^\bullet$$

Such processes do not obey Sabatier's Principle because the catalyst (the initiator) becomes incorporated into the product. Thus, no need exists for kinetic dynamism on the part of the catalyst and no juxtaposition of energetic efficiency with kinetic dynamism. The stronger the binding between the initiator and the reactant, the faster will be the reaction and the better will be the catalysis.

4.4 SABATIER'S PRINCIPLE IN HOMOGENEOUS CATALYSIS

Sabatier's Principle has not been widely studied in homogeneous catalysis because it is not generally possible to construct volcano plots using a range of homogeneous catalysts. For example, the milieu of the catalytic site in an equivalent range of metal complexes is usually not fixed, being subject to structural vagaries in the bonding about the metal ion.

As noted in Section 3.2.6 of Chapter 3, most uncatalyzed liquid-phase reactions carried out at, or near, ambient temperature are energy-dependent and not time-dependent. The assumption seems, therefore, to have been made that catalyzed homogeneous reactions must also be largely energy-dependent. Indeed, the time-dependent realm has not even been formally recognized in homogeneous catalysis to date.

Sabatier's Principle must, however, apply wherever a catalyst involves both kinetic dynamism and transition state stabilization. It will, necessarily, also occur in almost every homogeneous catalyst.

As noted, a defining property of time-dependent heterogeneous catalysis is the presence of a kinetically detectable, rapidly forming, and dissociating intermediate. Such intermediates are observed in certain homogeneous catalysts. They are, for example, common in the biological catalysts known as enzymes, where they form the basis for *Michaelis–Menten kinetics*. By contrast, they are rare in abiological homogeneous catalysis, which suggests that energy-dependence is the more prevalent form of catalysis in this particular field. Moreover, as noted, several authors have described enzymatic catalysis using analogies that describe time-dependent systems. Based on this evidence, it is not unreasonable to conclude that the realms of time-dependent and energy-dependent catalysis seem to also be represented in homogeneous catalysis.

In the following chapter, we will examine the incidence and character of time-dependence and energy-dependence in homogeneous catalysis.

4.5 CONCLUSIONS. SABATIER'S PRINCIPLE DESCRIBES TWO INDEPENDENT CATALYTIC DOMAINS: ENERGY- AND TIME-DEPENDENT CATALYSIS

In Chapter 3, we reviewed the theory of uncatalyzed and catalyzed reaction kinetics and anticipated the existence of two distinct realms of catalysis: *time-* and *energy-dependent* catalysis. We further expected that these domains would be characterized as follows:

Energy-dependent catalysts are controlled by the entity of the catalyst-bound transition state and its *activation energy* E_a. Energy-dependent reactions involve a thermodynamic equilibrium among the reactants, the transition state, and the products. As such, they are described by *transition state theory*.

Time-dependent reactions are, however, controlled by the *collision frequency* of the catalyst-bound reactants and are therefore independent of the nature of the activated complex, whose energy would necessarily have to be too low to influence the course of the reaction. Such processes are fundamentally nonequilibrium in character and are therefore not subject to *transition state theory*.

In Chapter 4, we sought empirical confirmation of these theorized domains within the field of heterogeneous catalysis. We examined Sabatier's Principle and found that it describes two distinct catalytic domains that correspond to time- and energy-dependent catalysis.

The energy-dependent domain was found to be characterized by a reliance on E_a and the structure and energy of the catalyst-mediated transition state.

The time-dependent domain was found to involve a dependence on the catalyst-mediated collision frequency. This dependence was physically manifested by the presence of a rapidly forming and dissociating intermediate in the rate expression.

The catalytic domains depicted by Sabatier's Principle originate in the need for catalysts to employ simultaneously the most dynamic catalyst–reactant binding AND the most efficient energetics possible. These two fundamental processes are, however, contrarily influenced by the strength of catalyst–reactant binding. Thus, an increase in the one is accompanied by a decrease in the other. When the two processes are unbalanced and one is slower than the other, the catalytic properties, including the overall rate and pathway of the reaction, are necessarily dominated by the slower process.

The two catalytic domains differ in the manner with which the most favored transition state is selected. In the energy-dependent domain, catalysts necessarily select the *lowest energy transition state*. This transition state has the smallest activation energy E_a. In the time-dependent domain, however, catalysts select the first transition state that is formed, that is, the *first-encountered transition state*.

Energy-dependent catalysis currently provides the dominant paradigm of the catalytic process in chemistry and is therefore well known. Numerous chemistry textbooks teach this model of catalysis.

However, time-dependent catalysis is also possible and has been shown to exist in heterogeneous catalysis. Time-dependence of this type has, to date, not been properly recognized or systematically studied in homogeneous catalysis. Indeed, the field of homogeneous catalysis has simply assumed that all catalysts must be energy-dependent. As such, time-dependent homogeneous catalysis remains an unexplored area of research.

ACKNOWLEDGMENTS

The author gratefully acknowledges David Trimm (CSIRO) for his insightful comments.

REFERENCES

1. Masel, R. I. *Chemical Kinetics and Catalysis*, Wiley-Interscience, 2001, p. 717–725 (Sabatier's Principle and Volcano Plots), p. 879–881 (Volcano Plots), and p. 689–793 (Kinetics of Heterogeneous Catalysis), and references therein.
2. Balandin, A. A. *Advances in Catalysis*. **1958**, *10*, 120, and references therein.
3. Bond, G. C. *Heterogeneous Catalysis. Principles and Applications (2nd ed.)*, Oxford Science Publications, 1987, p. 62–64 and references therein; see also: Bond, G. C. *Catalysis by Metals*, Academic Press, 1962, p. 476–478.
4. Fahrenfort, J.; van Reyen, L. L.; Sachtler, W. M. H. In *The Mechanism of Heterogeneous Catalysis, Proc. of the Symposium, Amsterdam, November* 12–13, 1959, Ed. De Boer, J. H., Elsevier, 1960, p. 23–48, and references therein; see also: Sachtler, W. M. H. *Z.*

Phys. Chem. **1960**, *26*, 16; Fahrenfort, J.; van Riegen, L. L.; Sachtler, W. H. M. *Z. Electrochem.* **1960**, *64*, 216.

5. Roberts, M. W.; McKee, C. S. *Chemistry of the Metal–Gas Interface*, Clarendon Press, 1978, p. 30.
6. Thomson, S. J.; Webb, G. *Heterogeneous Catalysis*, Oliver & Boyd, 1968, p. 184–188.
7. Thomas, J. M.; Thomas, W. J. *Principles and Practice of Heterogeneous Catalysis*, Wiley VCH, 1997, p. 29–30 (volcano plots), p. 85–87 (residence time), and references therein.
8. See, for example: Vayenas, C. G.; Bebelis, S.; Pliangos, C.; Brosda, S.; Tsiplakides, D. *Electrochemical Activation of Catalysis; Promotion, Electrochemical Promotion, and Metal-Supported Interactions*, Kluwer Academic, 2001, p. 152–179 and p. 281–298.
9. Sachtler, W. H. M.; Fahrenfort, J. *Proc. Int. Congr. Catal., 1st*, 1958.
10. Tanaka, K.; Tamaru, K. *J. Catal.* **1963**, *2*, 366.
11. For example: Smith, J. M. *Chemical Engineering Kinetics (2nd ed.)*, McGraw-Hill Kogakusha Ltd. 1970, p. 273–281.
12. (a) Campbell, C. T.; Ertl, G.; Signer, J. *Surf. Sci.* **1982**, 115, 309; (b) Ertl, G. *Science* **1991**, *254*, 1750.
13. See: McNaught, A. D.; Wilkinson, A. *IUPAC Compendium of Chemical Terminology, (2nd ed.)*, 1997 (see http://www.iupac.org/publications/compendium/index.html).
14. See, for example: Naidja, A.; Huang, P. M. *Surf. Sci.* **2002**, *506*, L243.
15. Norskov, J. K.; Stolze, P. *Surf. Sci.* **1987**, *189/190*, 91.
16. Somoraj, G. A. In *Elementary Reaction Steps in Heterogeneous Catalysis*, Eds. Joyner, R. W.; van Santen, R. A. Kluwer Academic Publishers, 1993, p. 3–38, and references therein.
17. Ozaki, A.; Aika, K. *Catalysis, Vol. 1*, Eds. Anderson, J.; Boudart, M.; Springer-Verlag, 1981, p. 87.
18. Bindra, P.; Roldan, J. M.; Arbach, G. V. *IBM J. Res. Develop.* **1984**, *28*, 679.

5

TIME-DEPENDENCE IN HOMOGENEOUS CATALYSIS. 1. MANY ENZYMES DISPLAY THE HALLMARKS OF TIME-DEPENDENT ("MECHANICAL") CATALYSIS. NONBIOLOGICAL HOMOGENEOUS CATALYSTS ARE TYPICALLY ENERGY-DEPENDENT ("THERMODYNAMIC") CATALYSTS

ROBIN BRIMBLECOMBE, JUN CHEN, JUNHUA HUANG, ULRICH T. MUELLER-WESTERHOFF, AND GERHARD F. SWIEGERS

5.1 INTRODUCTION

In previous chapters, we have examined and considered *time-dependent* and *energy-dependent* reactions, in both catalyzed and uncatalyzed form.

Arrhenius's theory indicates that the rate of a reaction k is determined by two components according to [1]

$$k = A \exp\left(\frac{-E_a}{RT}\right) \quad (5.1)$$

where the pre-exponential component A describes the average frequency of collisions between reactant molecules (the *collision frequency* of the reaction). The exponential

---

Mechanical Catalysis: Methods of Enzymatic, Homogeneous, and Heterogeneous Catalysis, Edited by Gerhard F. Swiegers
Copyright © 2008 John Wiley & Sons, Inc.

component ($-E_a/RT$) describes the proportion of those collisions that are sufficiently energetic to result in product formation. The *activation energy* E_a is defined as the *average* threshold energy that must be overcome in randomly oriented collisions between the reactants.

The key properties of a catalyzed reaction, including rate and pathway, depend on which of the above components provide for the slower overall reaction rate. Two distinct realms were theorized to exist in this respect:

(1) *Energy-dependent catalytic actions*, which are governed by the threshold energy (E_a) to be overcome in the *transition state*. Such reactions are described by *transition state theory* or the *high-pressure limit* in Hinshelwood–RRK theory.

(2) *Time-dependent catalytic actions*, which are governed by the frequency with which the catalyst-bound reactants are brought into *mechanical* collision with each other (the *catalyst-mediated collision frequency*). Such reactions are best described by the *low pressure limit* in Hinshelwood–RRK theory. They typically involve *non-equilibrium* conditions.

In Chapter 4, we confirmed that time- and energy-dependence unambiguously exist in heterogeneous catalysis. These realms can be conveniently discerned using a so-called "volcano" plot.

In this chapter, we will address the question as to whether time- and energy-dependent realms exist in homogeneous catalysis? If so, how do we recognize and distinguish them? Volcano plots can, unfortunately, not be employed in homogeneous catalysis.

What physical attributes can then be used to establish the existence of these realms in homogeneous catalysis?

One clue noted in Section 4.2.3.3 in Chapter 4 was that time-dependent heterogeneous catalysts necessarily display a rapidly forming and dissociating intermediate in their rate expression. Such intermediates are characteristic of time-dependent catalysis because they indicate that the dynamism of the catalyst–reactant interactions are rate-limiting. Rapidly equilibrating intermediates are not observed in the rate expressions of energy-dependent catalysts because their formation does not constitute the rate-determining step.

The question then becomes, are there homogeneous catalysts that contain a rapidly equilibrating intermediate in their rate expression? There certainly are. The natural catalysts of biology, known as enzymes, generally display so-called *Michaelis–Menten kinetics* [2,3], which is distinguished by a rapidly forming and dissociating intermediate, termed the *enzyme–substrate* or *Michaelis complex*. Such intermediates are otherwise rare in homogeneous catalysis. Nonbiological homogeneous catalysts essentially never display kinetics of this type.

This observation raises the possibility that many enzymes may employ a time-dependent catalytic action, whereas their nonbiological counterparts are energy-dependent.

In this chapter we will examine the common properties of enzymes relative to their equivalent nonbiological homogeneous catalysts. We will show that the general properties of catalysis by enzymes are consistent with, and predictive of, a *time-dependent* (mechanical) catalytic action. Those of equivalent nonbiological catalysts are consistent with *energy-dependent* (thermodynamic) catalysis. In Chapter 6, we will consider and describe in detail the catalytic action of time- and energy-dependent homogeneous catalysts.

5.2 HISTORICAL BACKGROUND: ARE ENZYMES GENERALLY *ENERGY-DEPENDENT* OR *TIME-DEPENDENT* CATALYSTS?

The underlying origin of the remarkable catalytic properties of enzymes has been a subject of intense debate for many years. Numerous theories have been proposed. For example, in accord with Koshland's "induced-fit" process [4], enzymes have been suggested to bring substrate functional groups close together during enzyme binding, thereby creating a "proximity effect" that may speed catalysis [5]. Haldane [6] has proposed that substrates become sterically strained upon binding, with the subsequent release of this strain accelerating their reaction.

Many more such hypotheses have been proposed; at least 21 distinct theories relating to enzymatic catalysis have been described in the chemical and biological literature [5]. In Chapter 7, we will examine these theories in detail.

The most widely accepted explanation for the efficiency of enzymatic catalysis was proposed about 60 years ago by Linus Pauling, who suggested that enzyme active sites are complementary in structure to the transition states of the reactions that they catalyze [7]. This complementarity, he argued, makes them bind their transition states in preference to their substrates, thereby lowering the minimum threshold energy needed in the transition state, namely, the *activation energy* E_a. According to Pauling, the lower the E_a of an enzyme, the greater will be its reaction rate [7].

Numerous studies have since confirmed that enzyme active sites are, indeed, highly complementary to, and therefore strongly bound to, their transition states [8–10]. This feature is consequently considered to be the fundamental origin of the high catalytic activity and specificity of enzymes. Its replication is the central aim of the field of *biomimetic chemistry* [11].

The generality of Pauling's proposal has also been demonstrated in nonbiological systems. Empirical comparisons of acid- or base-catalyzed intramolecular reaction rates within nonbiological organic molecules have shown large rate accelerations when the reactants are optimally arrayed for transition state stabilization [9,12–14]. Despite the detailed nature of these studies, Frederic Menger has identified an anomaly in that *compilations* of data from intramolecular rate studies reported in the chemical literature cannot be rationalized [15,16]. In an attempt to resolve this apparent contradiction, he proposed in 1985 a *spatiotemporal hypothesis* in which it was argued that the rate of reaction between two functionalities is proportional to the *time* that they are within van der Waals contact distance of each other [15].

Menger's spatiotemporal hypothesis is different and, indeed, mutually exclusive to Pauling's proposal in that it invokes the key determinant of *the time available for reaction* rather than *the activation energy* of the system.

As noted in Section 1.5.4 of Chapter 1, various other researchers have invoked time-dependence, in various guises, to explain enzymatic catalysis. For example, Moss (1968) and Williams (1993), as well as Benkovic, Hammes-Schiffer, and others (1970s–2000s) have described enzymatic action in terms of machine-like or mechanical actions involving "coupled" protein motions [17]. McElheny and Dobson (2006) have described catalysis by dihydrofolate reductase as being *stochastic* in character [17]. All of these descriptions employ time- and not energy-dependence.

This raises the question: are enzymes generally *energy-dependent catalysts* (as, effectively, proposed by Pauling [7] and others), or *time-dependent catalysts* (as alluded to by Menger, Moss, Williams, Benkovic, Hammes-Schiffer, and others) [15,17]?

This question has, to the best of our knowledge, never been considered. Because the energy-dependent domain currently constitutes the only intellectual paradigm of homogeneous catalysis articulated in numerous textbooks, it has simply been *assumed* that enzymes *must be* energy-dependent catalysts. But, is this a valid assumption? In this chapter, we will systematically examine the general properties of enzymatic catalysis relative to their nonbiological analogs with the aim of elucidating the character of their catalytic action.

5.3 THE METHODOLOGY OF THIS CHAPTER: IDENTIFY, CONTRAST, AND RATIONALIZE THE COMMON PROCESSES PRESENT IN BIOLOGICAL AND NONBIOLOGICAL HOMOGENEOUS CATALYSTS

The methodology we will employ in this chapter is to review the observable, general properties of catalysis by enzymes and to compare these with analogous generalizations in similar abiological homogeneous catalytic systems. Within these generalizations lie the common underlying processes that define the character of the catalysis. A systematic rationalization of these properties should yield the fundamental nature of the underlying processes. The general properties of time- and energy-dependent homogeneous catalysts will thereby be revealed.

We must emphasize that this methodology is generalist in character. It does not seek to elaborate on the detail of mechanistic enzymology, only to elucidate its most fundamental nature. As such, it focuses on attributes that exist in *many* enzymes. Such attributes will clearly not exist in *every* enzyme.

For the purposes of the above approach, the most appropriate nonbiological comparison with enzymes is arguably provided by homogeneous catalysts that contain multiple catalytic groups, so-called *multicentered* homogeneous catalysts [18]. The binding contacts employed by such species (e.g., coordination bonds) are similar in character to the hydrophobic–hydrophilic, ion-pairing, and hydrogen bonding interactions employed by most enzymes, albeit somewhat stronger. Some multicentered

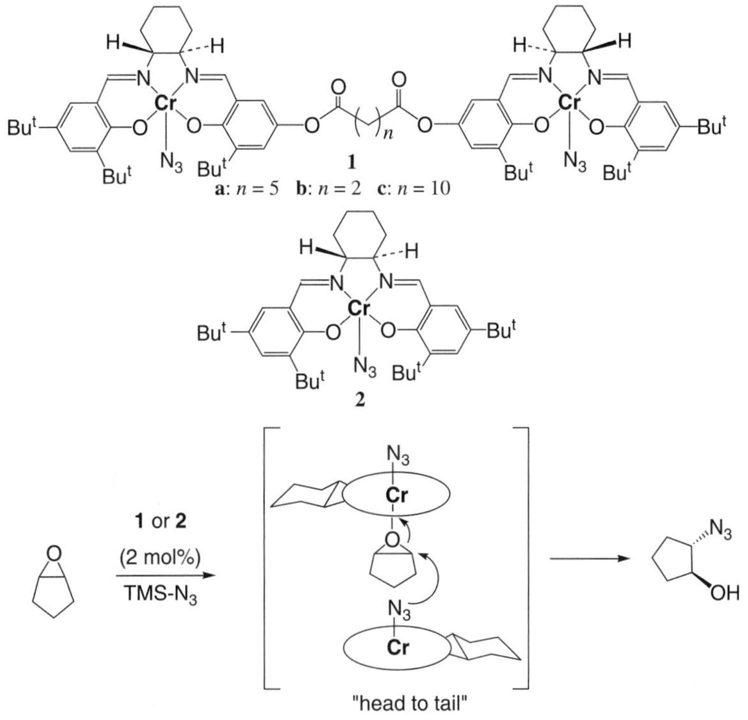

Scheme 5.1.

homogeneous catalysts also undertake covalent and acid–base catalysis in common with enzymes. Several enzymes employ metal-centered coordination bonds during catalysis.

Representative examples of a comparable abiological multicentered catalyst are depicted as **1a–c** in Scheme 5.1. [19] These species contain two Cr-ion catalytic groups, which facilitate epoxide transformations via a two-centered transition state of the type shown at the bottom of the scheme.

5.4 DOES MICHAELIS–MENTEN KINETICS IN ENZYMES INDICATE THAT THEY ARE *TIME-DEPENDENT* CATALYSTS?

5.4.1 Michaelis–Menten Kinetics

Most enzymes display some variant of Michaelis–Menten or so-called *saturation* kinetics [20]. This characteristic originates from the intermediacy of the rapidly forming and dissociating *enzyme–substrate* or *Michaelis complex*, *ES*, shown in Equation (5.2) (E = enzyme, S = substrate, P = product, $K = k_1/k_{-1}$):

$$E + S \xrightleftharpoons{K} [ES] \xrightarrow{k_2} E + P \tag{5.2}$$

Figure 5.1(a) depicts the results of a typical kinetic experiment on an enzyme. At low substrate concentrations [S_o], the initial rate of catalysis V_o is limited by the rate at which ES forms. That is, catalysis is first order with respect to the substrate concentration.

However, at high substrate concentrations, *saturation* occurs, which means that the equilibrium K comes to lie strongly in favor of ES, so that no free enzyme is effectively present. The initial rate of reaction then becomes dependent only on the amount of ES present and is unaffected by the addition of more substrate. Zero-order kinetics therefore prevails, with the overall initial rate tending incrementally toward the maximum velocity V_{max}.

5.4.2 Kinetics in Most Nonbiological Catalysts

Although common in enzymes, saturation kinetics is rare in nonbiological catalysts. Fewer than 50 manmade homogeneous catalysts, among the many hundreds of thousands known, have been reported to display saturation kinetics [21,22]. Most of these catalysts are, moreover, single- and not multicentered.

Nonbiological homogeneous catalysts generally display conventional kinetics of the type depicted in Figure 5.1(b) at all available reactant concentrations [R_o].

For example, the *intramolecular*, multicentered epoxidation catalysts **1a–c** (Scheme 5.1) exhibit simple first-order kinetics with respect to substrate concentration. Second-order kinetics prevails for their monomeric, multicentered counterpart **2**.

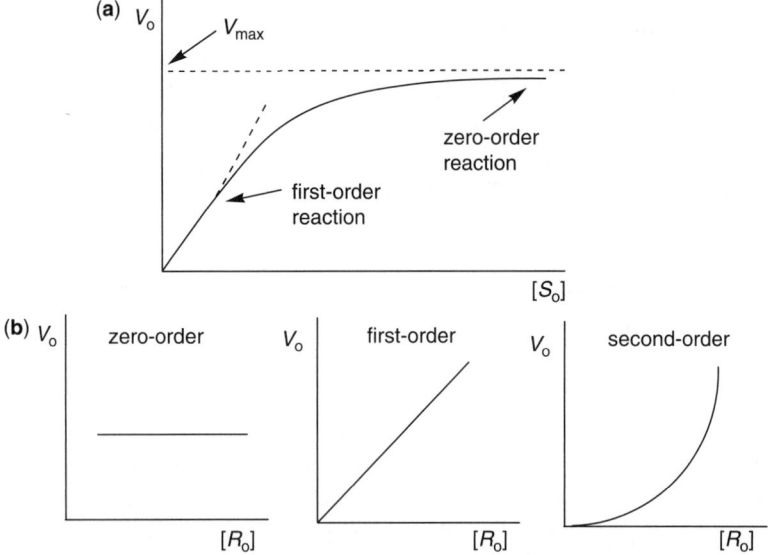

Figure 5.1. Initial reaction rate as a function of concentration in (a) Michaelis–Menten kinetics and (b) conventional kinetics.

5.4 MICHAELIS–MENTEN KINETICS IN ENZYMES

Kinetics of this type is generally not consistent with the presence of a rapidly forming and dissociating intermediate. Instead, as noted in the previous chapter, it fits the reaction scheme in Equation (5.3), where k_2 is the rate-determining step (C = catalyst, R = reactant, P = product):

$$C + R \xrightarrow{k_1} [CR] \xrightarrow{k_2} C + P \tag{5.3}$$

Note that the binding between the catalyst and the reactant in the first step of this scheme is, effectively, a *one-way* process as far as kinetics is concerned. That is, the catalyst essentially binds the reactant and *holds it* until the product is formed. If it does not do that, then this is not discernable using kinetics because k_1 does not appear in the rate expression.

In Appendices A and B (Supplementary Information), we discuss and graphically illustrate why conventional kinetics does not readily result in saturation, whereas Michaelis–Menten kinetics does.

5.4.3 The Contradiction of Saturation Kinetics in Enzymes

The prevalence of saturation kinetics in enzymes throws up an enigma. Enzymes bind their substrates strongly ($1/K_M$ $10^{4 \pm 3}$ M^{-1}) [3]. Why, then, do they involve a rapidly forming and dissociating intermediate? Why do they not bind and hold their substrates until the reaction is complete?

The absence of saturation kinetics in most nonbiological, multicentered catalysts indicates that, from a kinetic point of view, such catalysts bind and hold their reactants until reaction occurs. Even catalysts that bind their reactants more weakly than most enzymes display such one-way binding. Why do enzymes not do this?

5.4.4 Saturation in Time- and Energy-Dependent Catalysts. Saturation Kinetics Is Necessarily an Indication of Time-Dependence

In Sections 4.2.3 and 4.2.4 of Chapter 4, we showed that time-dependent and energy-dependent heterogeneous catalysts interact with their reactants in a very different way. Such catalysts undertake the general processes (C = catalyst, R = reactant, P = product):

$$C + R \xleftrightarrow{K} [CR] \tag{5.4}$$

$$[CR] \xrightarrow{k_2} C + P \tag{5.5}$$

For a time-dependent catalyst, the slowest, rate-determining step is the equilibrium depicted in Equation (5.4). For an energy-dependent catalyst, the slowest, rate-determining step is the formation of products depicted in Equation (5.5).

In a typical *time-dependent* heterogeneous catalyst, the key impediment to reaction is the low rate of catalyst-mediated collisions between the reactants. Thus, reactants rapidly bind and dissociate from the catalyst before they are able to collide and react with each other. On each occasion of binding, the reactants are, in effect, attached to the catalyst for only a very short period of time, known as the *residence time* τ. Because of the brevity of the residence time, there is a low likelihood of the reactants colliding with each other while bound to the catalyst and, therefore, a low *catalyst-mediated collision frequency*.

How, then, does such a time-dependent catalyst become saturated? And what is the result of saturation in such a catalyst?

As noted in Sections 4.2.3.3 and 4.2.3.4 of Chapter 4, the overall rate of a time-dependent catalyst is determined by two factors:

(1) The equilibrium constant K in Equation (5.4)
(2) The average residence time τ, during which the reactant is attached and available for catalysis on each occasion of binding

The equilibrium constant K is directly affected by the reactant concentration relative to the catalyst concentration. Thus, an increase in the reactant concentration drives the equilibrium K in Equation (5.4) to the right and thereby increases the catalyst-mediated collision frequency. This result increases the overall rate.

Time-dependent catalysts, consequently, become saturated once the *relative* reactant concentration is maximized; that is, when the equilibrium shown in Equation (5.4) lies strongly to the right. At that point, all of the catalytic centers will be busy receiving, binding, or releasing reactants/products. However, only a proportion of the catalytic centers will be formally occupied because, at any one time, most catalytic centers are involved in receiving or releasing reactants, not in holding them. The overall rate of the reaction will then be dependent only on the average residence time τ, which is an intrinsic and unchanging feature of the system. The longer this residence time, the greater will be the overall rate of catalysis. That is, the residence time is the *sole and invariant determinant* of the reaction rate upon saturation. Zero-order kinetics therefore prevail.

In a typical *energy-dependent* catalyst, however, the rate-determining step is Equation (5.5). In such a system, the catalyst-mediated collision frequency is high, so that the likelihood of overcoming E_a is the key limiting step. The probability of overcoming E_a is, however, independent of the reactant concentration. Thus, an increase in concentration merely shortens the time between a product leaving the catalyst and the next reactant attaching to it. Such systems become saturated when that time approaches zero, that is, at a high, *absolute* concentration of reactant. At this point, all catalytic centers will be occupied.

In effect, the reactants bind *and remain* attached to the catalytic centers in an energy-dependent catalyst until they are transformed into products. That is, for an energy-dependent catalyst, the arrow in Equation (5.4) is generally one way, and if it is not on occasion, then this does not matter because catalyst–reactant binding is not the slowest step and it does not affect the rate expression.

In summary, time-dependent catalysts become saturated at high reactant concentrations *relative* to the catalyst concentration. That is, they become saturated at a

5.4 MICHAELIS–MENTEN KINETICS IN ENZYMES

high *relative* concentration. Energy-dependent catalysts become saturated at a high *absolute* concentration of reactants. This *absolute* concentration may often not be accessible in liquid phase because the solution becomes saturated with the reactant before the catalytic process achieves saturation. Catalytic saturation is therefore rarely observed in liquid-phase, energy-dependent catalysis.

In Appendices A and B (Supplementary Information), we discuss and graphically illustrate the processes involved in the saturation of a time- and an energy-dependent catalyst.

If these parallels also apply to enzymatic catalysis, and there is no reason to think that they should not, then it is clear that the existence of Michaelis–Menten kinetics indicates that enzymes are generally time-dependent and not energy-dependent catalysts.

Key Point. Indeed, the existence of Michaelis–Menten kinetics is explicitly telling us that enzymes have a low catalyst-mediated collision frequency. It originates because, according to the rate expression, enzymes generally bind and release their substrates rapidly on multiple occasions before reaction, so that there is insufficient time for the reactant functionalities to collide, reactively, with each other on most occasions of binding.

It does not, of course, tell us why this is the case or how it takes place, only that it does happen. In particular, it does not explain why there is a rapid association and dissociation of substrate in the rate expression when enzymes are known to bind their substrates strongly.

Strong enzyme-substrate binding seems to stand in contradiction to the kinetically observed association and dissociation of the substrate. If the substrate is strongly bound, why would it rapidly associate and dissociate? In an archetypal time-dependent *heterogeneous* catalyst, it is *weak catalyst–reactant binding* that causes the observed rapid association and dissociation of the reactants. The binding and dissociation of substrates observed in enzymatic catalysis occurs, however, in the presence of *strong substrate binding*.

By contrast, nonbiological homogeneous catalysts do not exhibit kinetically limiting binding and release of their reactants before reaction. It indicates that they are energy-dependent in their catalytic action. That is, as far as kinetics is concerned, they, effectively, bind their reactants and hold them until reaction occurs. The key impediment to their reaction is then overcoming the activation energy of the reactant, not achieving reactant collision.

5.4.5 Physical Studies of the Rate Processes in Enzymes Are Consistent with a Time-Dependent Action

Numerous physical studies of the processes involved in enzyme-catalyzed reactions have been undertaken. As noted by Cleland [23], such studies demonstrate overwhelmingly that the step of transition state formation leading to products is essentially *never* rate-determining in enzymes. Instead, the slowest step generally involves organizational processes, such as conformational changes within the

Michaelis complex [23]. For example, in the simplest possible case, it has been suggested [8,24] that Equation (5.2) should read as follows:

$$E + S \xrightleftharpoons{K'} [ES] \xrightleftharpoons{K} [ES^*] \xrightarrow{k_2} E + P \tag{5.6}$$

where ES^* represents a modified form of ES and its formation constitutes the slowest step. That is, the slowest step involves the equilibrium constant K''. Recent studies suggest that such steps may involve "coupled" conformational changes in which protein loops or domains close about the catalytic site [9(b)–(d),17(c),25,26].

In evaluating this evidence, one would have to say that, if enzymes were energy-dependent catalysts, it would be truly *extraordinary* for the step involving product formation to *never* be rate-determining. Indeed, it would be, frankly, impossible; the very definition of energy-dependence is that transition state formation leading to products is the slowest of the common steps.

By contrast, if enzymes were time-dependent catalysts, then it would be *entirely conceivable* that a step or steps lying between substrate binding and transition state formation would be rate-determining. Such steps would entail the catalyst bringing the bound and activated reactants into physical contact with each other, that is, mediating their collision.

It may, indeed, involve conformational changes on the part of the enzyme. In general, however, such conformational changes would include only those that participate in bringing about reactant collisions.

Physical studies of enzyme mechanisms, therefore, also point, *unambiguously*, to time- and not energy-dependence as the dominant catalytic realm in enzymes.

Incidentally, it is important to note that other processes, such as product release, may well be rate-determining in many enzymes [23]. However, the slowest process that is *common* to the catalytic action seems to involve mediating reactant collisions. In the absence of other, slower processes, it is this step that sets the *upper limit* for enzymatic activity.

5.4.6 A Time-Dependent Catalyst Cannot Become an Energy-Dependent Catalyst, or vice versa, Without Changing the Temperature or Chemically Altering the Reactivity of the Reactants

It is important to understand that the only way an *energy-dependent catalyst* can be transformed into a *time-dependent* one, or vice versa, is by a significant change in temperature or by a chemical modification to the reactivity of a reactant (see Sections 3.2.7 and 3.3.5 of Chapter 3).

The time- or energy-dependent realm of a catalyst cannot be changed by a variation in the reactant concentration. The reason for this is because an increase in the reactant concentration about a time-dependent catalyst does *not* induce the reactants to remain bound to the catalyst for *longer periods of time* on each occasion of binding. It only increases the frequency of the binding events. Similarly, a decrease in the reactant concentration about an energy-dependent catalyst does not induce the reactants to *dissociate more rapidly* from the catalyst and therefore become more dynamic in their binding. All it does is lengthen the time between product release and binding of the next reactant.

The two regimes of time- and energy-dependence are therefore entirely separate and are not changed by variations in the reactant concentration.

Having said that, there may, of course, be some isolated examples of systems where the rate caused by the dynamism of catalyst–reactant binding, activation, and collision is exactly equal to that of overcoming the E_a threshold. That is, there may be systems where the catalyst sits at the precise nexus of the time- and energy-dependent realms. Concentration effects could then, conceivably, tip the catalyst one way or the other. But this would be rare and certainly not representative of the general case. *Ideal* catalysts of this type are uncommon in heterogeneous catalysis, and there is no reason to doubt that they would also be rare in homogeneous catalysis.

5.4.7 The Current View of Michaelis–Menten Kinetics Is Flawed by an Unwarranted Assumption

How, then, do we rationalize the above results? The existence of Michaelis–Menten kinetics as well as physical studies of the rate processes seem to indicate that many enzymes are time-dependent catalysts. At the moderate temperatures of homeostasis under which most enzymes operate, the catalytic realm is, moreover, fixed; they cannot spontaneously transform themselves into energy-dependent catalysts. However, Pauling and most biochemists believe that enzymes are energy-dependent catalysts. How does modern biochemistry deal with and reconcile this apparent contradiction?

According to the currently accepted view, overall enzymatic rates are determined by K and k_2 in Equation (5.2) at low substrate concentrations but only by k_2 at the high substrate concentrations required to achieve saturation [2,3]. Thus, biochemists conclude that k_2 in Equation (5.2) is ultimately rate-limiting and this seems to involve the step of forming the products from the transition state [2,3]. The Michaelis–Menten scheme is therefore rationalized as being perfectly consistent with energy-dependence.

Although this interpretation is widely cited in textbooks, it is deeply flawed. Indeed, it is a misconception that has *had* to be employed in order to make enzyme kinetics *appear* to fit the assumption of energetic-control.

The scenario it presents involves an *internal contradiction*. At low substrate concentrations, the dependence on K indicates that collision frequency is rate-limiting which is, and can only be, consistent with *time-dependent catalysis*. However, the dependence on k_2 at both high and low concentrations is cited as indicating that the activation energy is *also* rate-limiting, which is consistent with *energy-dependent catalysis*.

It is impossible for both collision frequency *and* activation energy to be *simultaneously* rate-determining over an entire class of catalysts. As noted, there may be rare cases where these two processes are perfectly balanced. But it is not viable for this to be the case in virtually *every* enzyme. It is still less viable to argue that at saturation, every single enzyme catalyst then uniformly slews over to become energy-dependent.

The false nature of this proposal is further demonstrated by the fact that a rapidly forming and dissociating intermediate is *not required* in the kinetic schemes of energy-dependent catalysts. So, why is it a characteristic of so many enzymes? If the slowest step is product formation, there should be no need to hypothesize a rapidly forming and dissociating intermediate at all.

However, the collision frequency is *directly dependent* on the rate and duration of catalyst–reactant binding, which is exactly why a rapidly forming and dissociating intermediate *must* be included in the kinetic schemes of time-dependent catalysts.

In effect, the current explanation of Michaelis–Menten kinetics tries to rationalize away the presence of a rapidly equilibrating intermediate in enzyme kinetics.

What is the flaw in this view of Michaelis–Menten kinetics? And how does one resolve the contradiction it creates?

The answer is actually simple. The Michaelis–Menten scheme was originally developed as a means of theoretically reproducing the saturation curve depicted in Figure 5.1(a) [2,3]. However, the only catalytically relevant property that varies with the substrate concentration in Figure 5.1(a) is the frequency of the catalyst-substrate binding events. The time needed to bring the reactants into physical contact (collision) with each other after binding to the catalyst is an intrinsic, invariant property of the system and does not change. This time was designated t_{TS} in Chapter 3 and termed "the time required for transition state formation."

Thus, the empirically observed equilibrium constant K in Equation (5.2) incorporates *only* the frequency of catalyst–reactant binding, namely, K' in Equation (5.6). It does not incorporate the process of bringing the reactants into collision with each other on the catalyst, which is designated K'' in Equation (5.6).

The K in the Michaelis–Menten scheme shown in Equation (5.2) is, consequently, only the equilibrium constant K' in Equation (5.6).

Key Point. **By default, the process of catalyst-mediated reactant collision (K'') is included in the Michaelis–Menten constant k_2, as shown in Equation (5.7). This is a consequence of the purely empirical basis of the model.**

$$E + S \overset{K'}{\longleftrightarrow} [ES] \underset{k_2 \text{ (in Michaelis–Menten scheme)}}{\underbrace{\overset{K''}{\longleftrightarrow} [ES^*] \overset{k_2}{\longrightarrow} E + P}} \tag{5.7}$$

The empirically observed Michaelis–Menten k_2 therefore incorporates both of the following from Equation (5.6):

- K'' (rate-limiting in time-dependent catalysts)
- k_2 (rate-limiting in energy-dependent catalysts)

As such, the fact that the Michaelis–Menten k_2 becomes rate-limiting at saturation does *not* mean that the catalysis is energy-dependent.

Instead, the fact that at low substrate concentration both the Michaelis–Menten K and k_2 are rate-limiting, but at saturation only k_2 is rate-limiting indicates, *unequivocally*, that the reaction is *time-dependent*. The slowest, common process within the empirically observed Michaelis–Menten k_2 must, necessarily, be that involving K'' in Equation (5.6). It is the only way that both K (indicative of time-dependence) and k_2 (indicative of time-dependence if K'' is rate-limiting) can govern the rate at a low reactant concentration.

This approach is wholly consistent with processes (1) and (2) in Section 5.4.4, which determine the rate of time-dependent catalysts. At a low concentration, K' and K'' govern the overall rate. But at saturation, K' is maximized and K'' becomes the sole determinant of rate. K'' is, in turn, dependent on the residence time.

This interpretation gels perfectly with the physical studies described in Section 5.4.5.

5.4.8 Summary: Michaelis–Menten Kinetics Is Characteristic of Time-Dependent Catalysis. Time-Dependent Catalysis Provides an Explanation for Michaelis–Menten Kinetics in Enzymes

Key Point. Michaelis–Menten or saturation kinetics is therefore entirely consistent with, and predictive of, time-dependent catalysis in enzymes. It is not consistent with energy-dependence.

Perhaps more importantly, time-dependent catalysis provides an explanation for the existence of Michaelis–Menten and related forms of saturation kinetics in enzymes. No such explanation has previously been proposed and none would be possible if enzymatic catalysis were generally energy-dependent.

As noted, however, the Michaelis–Menten scheme does not explain how or why time-dependent catalysis originates in enzymes. In particular, it does not explain how a time-dependent catalytic action comes into being in the face of strong enzyme–substrate binding. To understand this aspect, we need to examine the catalytic action in greater detail. Section 5.6 of this chapter is dedicated to this task.

5.5 OTHER GENERAL CHARACTERISTICS OF CATALYSIS BY ENZYMES AND COMPARABLE NONBIOLOGICAL HOMOGENEOUS CATALYSTS

To clarify and enlarge on the preceding discussions, we will now examine other general characteristics of catalysis by enzymes and comparable nonbiological homogeneous catalysts. We will then rationalize these generalizations so as to try to understand what they are telling us.

As a class, enzymes display several common catalytic properties that are distinctively different to comparable nonbiological homogeneous catalysts. These include the following:

5.5.1 Enzymes Employ Weak and Dynamic Individual Binding Interactions with Their Substrates. Nonbiological Catalysts Do Not

The most obvious physical difference between enzymes and nonbiological homogeneous catalysts is that enzymes typically employ the "weak" forces of biology to bind their substrates, whereas their abiological counterparts generally employ much stronger individual interactions. The weak interactions include van der Waals interactions, hydrogen bonds, hydrophobic–hydrophilic interactions, and ion-pairing.

The significance of this is two-fold. First, the individual binding interactions of enzymes are exceedingly weak, being far weaker than most manmade catalysts. Second, they are highly transient.

As noted in Section 3.3.2 of Chapter 3, the process of catalyst–reactant binding causes the attached reactants to become *activated*. *Activation* involves modification of the electronic structure of the substrate so as to "stretch" substrate bonds in order to make them conducive to reaction, and thereby to initiate the catalytic process [20]. In effect, by polarizing the bound reactant, the catalyst increases the potential energy of the relevant bonds and thereby decreases the threshold energy barrier that must be overcome during collision.

Enzymes must also activate their substrates. Unlike manmade catalysts, however, enzymes typically create this activation using the weak interactions of biology. This means that they are often limited to weak and transient activation.

To illustrate this issue, consider that biological interactions like hydrogen bonds are typically highly dynamic, forming and breaking rapidly. For example, the average lifetime of the $NH_3 \cdots H_2O$ hydrogen bond in aqueous ammonia is only 2×10^{-12} s [27]. During these periods, it seems, additionally, to consist of a mixture of polar interactions, true hydrogen-bonds and ion-pairs [Equation (5.8)]:

$$^{\delta-}R-H^{\delta+} + {}^{\delta-}R'-H^{\delta+} \leftrightarrow {}^{\delta-}R-H \cdots R'-H^{\delta+} \leftrightarrow R^{-\,+}H-R'-H \quad (5.8)$$

polar interaction   hydrogen bond   ion-pair

In only one of these states are the relevant bonds of the substrate likely to be correctly activated and suitable for reaction.

An individual, hydrogen-bonding catalytic group within an enzyme is therefore capable of triggering the reaction of its attached substrate only very briefly and infrequently. Moreover, it is an exceedingly weak trigger. "Strong" hydrogen bonds of the type seen in certain biological structures (15–40 kcal mol^{-1}) may allow for more sustained and extreme activation [28,29].

By contrast, nonbiological multicentered catalysts typically employ much stronger catalyst–reactant binding interactions, such as coordination bonds. Nonbiological catalytic groups therefore have a correspondingly greater likelihood of binding and activating their attached reactants. Moreover, they are capable of activating them more strongly. That is, they polarize the attached reactant more strongly and thereby make it more reactive.

It is important to note that this generalization involves only the strength of the *individual* catalyst–reactant binding interactions and not the strength of *overall* binding. Despite the weakness of their individual binding interactions, enzymes typically have many such interactions and therefore bind their substrates relatively strongly overall ($1/K_M$ $10^{4\pm3}$ M^{-1}) [3].

Many nonbiological homogeneous catalysts have a similarly high, or even higher, overall affinity for their reactants, usually because they involve coordinatively unsaturated metal complexes, sometimes with, effectively, an unpaired charge [30]. No clear distinction therefore exists between enzymes and nonbiological multicentered catalysts in respect of the *overall strength* of catalyst–reactant binding. But a definite distinction exists in the strength and transience of the *individual* binding interactions.

5.5.2 Enzymes Display Transition State Complementarity. Nonbiological Catalysts Do Not

As noted, numerous studies have shown that enzyme active sites are generally structurally complementary to their corresponding reaction transition states and, therefore, strongly bound in this respect (with $1/K_M = 10^{16\pm4}$ M^{-1}) [10]. This is almost never the case for nonbiological catalysts that involve the action of more than one catalytic group [30,31].

For example, **1a–c** and **2** catalytically ring-open epoxides via the transition state shown in Scheme 5.1 regardless of structural variations in the tether linking the two Cr moieties [19]. These ions need not be prearranged in any particular spatial arrangement to achieve catalytic turnover, provided they have freedom of movement. Indeed, they need not even be attached to each other. Transition state complementarity clearly plays little role in the operation of these catalysts. A variety of other bimetallic catalysts also need not be complementary to their transition states. They include Pd-catalyzed aryl couplings with transition states that are simultaneously stabilized by two Pd ions [25], and many Grignard catalysts [30].

A key question in understanding enzymatic catalysis is the following: Why are enzymes almost invariably complementary to their transition states when this is clearly not needed to achieve a catalytic effect? Why has evolution not produced a distribution of catalysts, some of which are complementary to their transition state and others that are not? Instead, it seems that only proteins whose shapes are complementary to a transition state are capable of catalyzing its transformation. Understanding this observation will likely explain many fundamental aspects of enzymatic catalysis.

5.5.3 Enzymatic Catalysis Is "Structure-Sensitive." Nonbiological Catalysis Is "Structure-Insensitive"

The rate of catalysis by enzymes is generally highly sensitive to changes in their structure. For example, mutating residues near to the catalytic groups in enzymes typically results in a complete loss of activity, as classically illustrated by the mutation of residues in catalytic triads [25,32]. Even modifications to physically remote residues often dramatically influences the rate. Thus, enzymes display catalytic activities that are generally $>10^8$-fold larger than those of closely related proteins [25,32]. Enzymatic catalysis consequently has the characteristic of being a highly *structure-sensitive* form of catalysis.

By contrast, multicentered, nonbiological homogeneous catalysts are typically *structure-insensitive* in their action. That is, their rate is relatively unaffected by changes in the catalyst structure. For example, modification of the tether in **1a–c** generates only moderate changes in the reaction rate (Scheme 5.1). This is also true for other multifunctional homogeneous catalysts and explains why they are catalysts even as monomers (e.g., **2**).

With only a few exceptions [33], mixtures in open solution of monomeric or oligomeric amino acid catalytic groups of the type employed by most enzymes produce no catalytic activity whatsoever. This may be, in some cases, because such oligomers contain an incorrect sequence of amino acids to create a particular required electronic effect. But the relative lack of examples suggests a more profound explanation, which is most likely associated with the absence of the necessary complementary spatial arrangement required for catalysis.

It is, indeed, tempting to relate the structure-sensitivity of catalysis by enzymes to their apparent need for complementarity with their transition states. Thus, if enzymes *have* to be complementary to their transition states in order to display a catalytic effect, then mutations that diminish this complementarity will necessarily lead to reduced activity.

5.5.4 Enzymes Transform Catalytically Unconventional Groups into Potent Catalysts. Nonbiological Catalysts Use Only Conventional Catalytic Groups

A curious, but very distinctive, aspect of enzymatic catalysis is that enzymes are often able to transform amino acid residues into extraordinarily powerful catalysts. Amino acids are not known to display dramatic catalytic activity outside of biology. Indeed, only one major system, involving *L*-proline, and a scattering of amino acid oligomers has ever been reported to produce significant catalytic effects [33]. Amino acid residues are therefore very much unconventional catalytic groups outside of biology.

In nonbiological homogeneous systems, the catalytic groups are, by contrast, typically metal ions, like Cr, Pd, and Mg [34]. Such ions are well known as catalysts of a wide and diverse array of reactions because of their intrinsic, and often unique, binding and electronic properties. For example, metal ions of this type often employ vacant coordination sites, *d*-electrons and orbitals, strong polarization, and other electronic or steric effects that are simply not available to amino acids [30]. How, then, are amino acids able to become such remarkably effective catalytic groups in enzymes?

This feature of enzymatic catalysis may well be linked to the activation proferred by enzymes to their bound reactants. As noted, enzymes use the weak interactions of biology to bind their reactants. These interactions can only very poorly "stretch" or polarize bonds in the attached substrate functionalities to thereby make them more conducive to reaction. In manmade systems, such activation is simply too weak to initiate catalysis, so that amino acid groups are not generally recognized as catalytically active. However, enzymes are somehow able to harness this weak activation. If this is the reason that enzymes can use amino acids as catalytic groups, then one must ask: How do they do it?

5.5.5 Enzymes Catalyze Forward and Reverse Reactions. Nonbiological Catalysts Do Not

The direction of catalysis by enzymes can often not be defined in biology [2,3]. Alternatively, it is often context specific, because many enzymes are capable of catalyzing both the forward and the reverse direction of a reaction. Thus, an enzyme that is an *alcohol dehydrogenase* under one set of conditions, *in vitro*, can become a *carbonyl reductase* in another.

This ability for bidirectional catalysis is essential to maintaining homeostasis. It is also, to the best of our knowledge, unique to enzymes, at least in the field of homogeneous catalysis. As far as we know, no nonbiological homogeneous catalysts has been shown to display bidirectional catalysis [30]. Because of its rarity in nonbiological systems, bidirectional catalysis potentially provides a valuable clue as to the catalytic realm of enzymes.

Several explanations have been proposed for this phenomenon. For example, it has been suggested that reactants and products become briefly trapped at the catalytic site in enzymes by the closing of protein loops or domains during catalysis [9(b)–(d),26]. Many forward and reverse transits of the transition state may then conceivably occur before product release, thereby explaining the incidence of bidirectional catalysis in enzymes [26].

In an energy-dependent catalyst, however, such a process can only be readily reversible if the overall activation energy of the forward reaction (E_a^F) is approximately the same as it is for the reverse reaction (E_a^R) (Fig. 5.2). For example, a reverse reaction

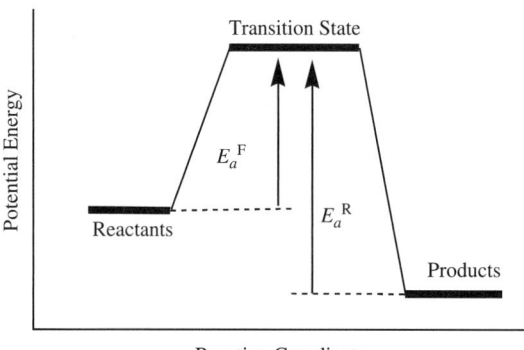

Figure 5.2. Schematic depiction of the activation energy for a forward reaction (E_a^F) and its reverse reaction (E_a^R).

with an E_a that is double that of the forward reaction would occur 100 times more slowly than the forward reaction according to Equation (5.1). It is true regardless of whether the reactants and products are temporarily trapped at the catalytic site.

Under physiological conditions, thermodynamics dictates that every 1.36-kcal/mol free energy difference between the reactant and the product should result in a 10-fold difference in rate [3].

It is very unlikely that E_a^F would be nearly identical to E_a^R for all of the vast array of reactions catalyzed by enzymes. This scenario must therefore be considered implausible.

The only viable alternative explanation is that both E_a^F and E_a^R do not influence the rate of the reaction at all. That is, the rate-determining step in both the forward and the reverse direction does not involve E_a^F or E_a^R, but it is, instead, determined by the rate at which the reactants physically contact each other, namely, the "collision frequency." Because the catalyst-mediated collision frequency is strongly dependent on the reactant concentration, this would also explain how a simple change in the substrate or product concentration could reverse the pathway of the catalysis.

As described in Chapter 3, a dependence on collision frequency is characteristic of time-dependent ("mechanical") catalysis.

Key Point. In Chapter 1 we noted that mechanical processes are often able to proceed in both the forward or the reverse direction with equal efficiency, rather like movies that may be played forward or backward. That is, they follow preordained courses that can be tracked either forward or backward. The bidirectional capacity of enzymatic catalysis seems to fit that expectation.

5.5.6 Enzymes Display High Selectivity and Activity. Nonbiological Catalysts Do Not

Perhaps the most remarkable features of enzymatic catalysis are the extraordinary selectivities and activities they achieve. A classic illustration of enzymatic selectivity is given by glycerol kinase, which converts glycerol to exclusively α-glycerolphosphate [35]. Catalysts displaying such extraordinary fidelity are effectively unknown in nonbiological homogeneous systems.

The catalytic activity of most enzymes is also high, lying typically between 1 and 1000 turnovers s^{-1} [3]. The most active catalyst known is the enzyme carbonic anhydrase, which can turn over 600,000 CO_2 molecules per second [3]. Several enzymes also achieve kinetic perfection, with their catalytic velocity restricted only by the rate at which they encounter substrate in solution [3]. Catalysis of this type is extremely rare in nonbiological homogeneous systems, although instances have been reported [21].

5.5.7 Enzymes Display Convergent Synergies. Nonbiological Catalysts Display Complementary Synergies

When the whole of a system exceeds the sum of its parts, it is said to be *synergistic*. Thus, when the properties of a catalyst exceed the sum of the properties of its

5.5 OTHER GENERAL CHARACTERISTICS OF CATALYSIS BY ENZYMES

individual catalytic groups, the species displays *synergy*. As noted in Section 1.4.3 of Chapter 1, *synergy* is a key ingredient in mechanics; all mechanical devices rely on, or harness, synergy to some extent or another. As such, mechanical processes (time-dependent) invariably display higher levels of synergy than thermodynamic ones (energy-dependent).

All multicentered catalysis is necessarily synergistic since such catalysis cannot, by definition, involve only one catalytic group. However, the synergies in enzymes are very different, and of immensely higher order than those in their nonbiological counterparts. Can the different levels of synergy be distinguished and, if so, how?

Several different types of synergy have been identified in *Complex Systems Science* [36]. To illustrate two of these types, consider the example of two people building a cabinet, with one doing the sawing and the other doing the nailing. If these two tasks can be performed independently, then the tasks are complementary and building the cabinet involves only the scheduling of a combination of these activities. That is, it involves the simple sum of the two tasks. Synergies of this type are known as *functional complementarity* [36].

But if the one person depends on the other to hold up the cabinet or otherwise assist them when they are sawing or nailing, then each task cannot be done on its own. Instead, the tasks are interdependent and can only be completed simultaneously with a coordinated effort. In effect, the activities of the builders must *converge*. In completing each of their respective tasks, each builder also enables the other to complete their task. Synergies of this type are known as *functional convergence* [36].

In most nonbiological catalysts, such as **1a–c** (Scheme 5.1), each catalytic group carries out a separate task that complements the tasks of the other catalytic groups. As long as each group can carry out its task, and these can be combined at some point in time, catalysis is possible. Thus, the groups do not depend on each other to carry out their respective functions. Monomeric catalytic groups in open solution like **2** are, therefore, perfectly capable of producing multicentered catalysis. Catalysts of this type display *functionally complementary* synergies [36].

*Key Point. **In enzymes, however, each catalytic group seems to depend on the other catalytic groups to carry out its task successfully. That is, the different tasks must all be performed correctly, at the same time, in unison. If any one group acts incorrectly, all of the other groups also fail in their tasks. Thus, there seems to be a high degree of interdependence in which all of the catalytic groups either succeed together or fail together. In effect, the actions of the individual catalytic groups must be coordinated in order to achieve a catalytic effect. This conclusion is reflected in the fact that enzymes rely on rather precise spatial arrangements in their catalytic groups to bring about catalysis. It also potentially explains the high catalytic selectivities of enzymes; all of the catalytic groups work together to achieve great fidelity. Thus, monomeric catalytic groups, such as monomeric amino acids, do not generate a catalytic effect. However, when many such amino acids act convergently within an enzyme active sites, they are potent catalysts. Enzymes consequently display functionally convergent synergies.***

Much of the uniqueness of enzymatic catalysis is arguably summarized in the functionally convergent synergies that they display. The requirement for high-order synergies in enzymes is very typical and distinctive of time-dependent, mechanical processes. As noted in Section 1.4.3 of Chapter 1, mechanical processes often depend on synergy to create an effect. It appears to also be the case in enzymes.

By contrast, their nonbiological counterparts do not need extensive synergies of action in order to create a catalytic effect. Their lower synergies are consistent with energy-dependence, which relies on a large input of energy to achieve its effect (see Section 1.4.3 of Chapter 1). This energy input is, arguably, provided by the strong and sustained activation provided to bound reactants by the metal ions employed as catalytic groups in most nonbiological homogeneous catalysis. Because of their synergies, enzymes do not need such extreme activation and can, instead, bring about catalysis using weakly activating amino acid catalytic groups.

In Chapter 8 we will discuss the issue of catalytic synergy in greater detail.

5.5.8 Summary

We have described several general properties of enzymes and their comparable non-biological catalysts. Of these properties, we have shown that Michaelis–Menten kinetics, the bidirectionality of enzyme catalysis, and the synergies of enzymes are, arguably, directly consistent with and predictive of *time-dependent catalysis*. They are not consistent with *energy-dependent catalysis*. Moreover, other properties of enzymes, such as their ability to use weakly activating amino acids as catalytic groups, their structure sensitivity, and their complementarity to their transition states, are likely involved in creating the necessary synergies of action that characterize enzymatic catalysis. As such, they must be considered indirectly consistent with *time-dependent catalysis*.

But, if enzymes are time-dependent, from where does this derive? In heterogeneous catalysts, time-dependence originates in weak and short-lived reactant binding. Enzymes, however, bind their substrates strongly overall, albeit with the use of weak and transient individual binding interactions. In the following sections, we will examine and try to explain the common processes at work in multi-centered homogeneous catalysts such as enzymes.

5.6 RATIONALIZATION OF THE UNDERLYING PROCESSES. THE MECHANISM OF ACTION IN TIME-DEPENDENT AND ENERGY-DEPENDENT CATALYSTS

5.6.1 Common Processes in Multicentered Homogeneous Catalysts

As noted in Section 3.3.2 of Chapter 3, all multicentered catalysts, whether they are biological or nonbiological, must necessarily perform the same basic processes. These processes involve:

(1) Reactant binding
(2) Reactant activation

5.6 RATIONALIZATION OF THE UNDERLYING PROCESSES

(3) Transition state formation
(4) Product formation and release, leading to regeneration of the catalyst

For illustrative purposes, Figure 5.3 schematically depicts the common processes that must occur during catalysis by a representative three-centered species. The moieties C, C′, and C″ represent catalytic groups. Each moiety is free to move within a

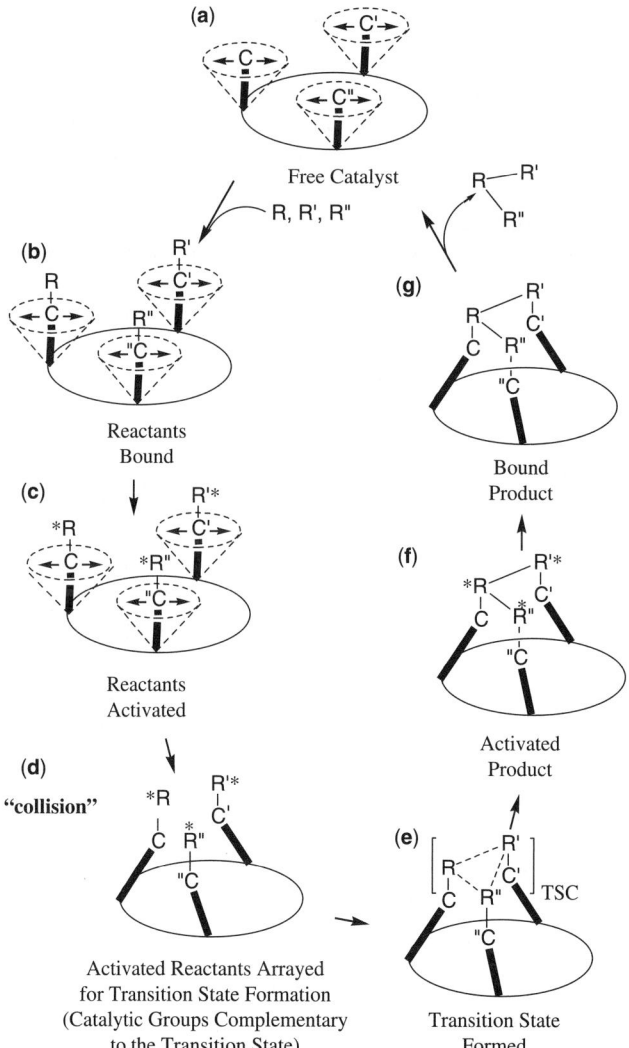

Figure 5.3. Schematic depiction of the common processes that occur in multisubstrate catalysis by a representative three-center catalyst (C, C′, C″ = catalytic groups; R, R′, R″ = reactants).

specified volume during conformational flexing. These volumes are depicted as the cones about each catalytic group.

The arrows show the sequence of events during a reaction involving three reactants: R, R', and R''. These reactants may be covalently attached to one another; in which case, the reaction involves a single reactant with three binding groups R, R', and R''.

In the first step, each catalytic group must bind a reactant [Fig. 5.3(a) to (b)]. The reactant must then be activated by its respective catalytic group [20]; that is, bonds within the reactants must be polarized or "stretched" for initiation of the catalytic process. The extent of this activation will vary from reactant to reactant and from functionality to functionality. Where a catalytic group simply acts to bind a reactant group in a particular location, such as a remote hydrogen bonding interaction in enzyme-substrate binding, little activation will be required. However, where a catalytic group must actually induce and bring about a reaction, a more extreme activation will be required.

For convenience we will consider all of the reactants in Figure 5.3 to need some sort of activation; the activated reactants are depicted as R^*, R'^*, and R''^* in Figure 5.3(c).

The activated reactants may individually come into physical contact with each other during conformational flexing of the catalyst, but these contacts do not generally result in formation of the product, $R''-R-R'$, unless the reactants are arrayed to form a transition state at the precise instant of the collision. This will occur when the catalytic groups are in a conformation that complements this transition state [Fig. 5.3(d)]. When that happens, the activated reactants can successfully form the transition state, shown in Figure 5.3(e), thereby generating the product, $R''-R-R'$, still in an activated form [Fig. 5.3(f)]. After deactivation [Fig. 5.3(g)], $R''-R-R'$ dissociates, regenerating the free catalyst [Fig. 5.3(a)].

For convenience, all of the arrows in Figure 5.3 have been shown to be unidirectional. However, in practice, all steps up to the point of transition state formation may be in equilibrium, involving double-headed arrows.

If this process is time-dependent, it will be controlled by the catalyst-mediated collision frequency. That is, steps (a)–(d) in Figure 5.3 will be rate-determining. If, however, the process is energy-dependent, it will be controlled by the need to overcome the threshold energy barrier of the transition state. Steps (d)–(f) in Figure 5.3 will then be rate-determining.

5.6.2 The Influence of the Strength of the Individual Catalyst–Reactant Binding Interactions

We will now try to explain the observed properties of enzymes and nonbiological homogeneous catalysts by considering the catalytic action of two hypothetical homogeneous catalysts, A and B. The two catalysts both have the general structure shown in Figure 5.3(a). They also catalyze the same reaction at the same rate using identical mechanisms. They differ only in the strength of their individual catalyst–reactant binding interactions. Catalyst A has strong individual binding interactions with the reactants, such as the coordinate bonds employed by many abiological multicentered catalysts. Catalyst B has weak individual binding interactions with the reactants, such as the weak interactions of biology typically employed by enzymes. Catalyst A therefore binds and activates the reactants for longer and more sustained periods than catalyst B.

Figure 5.4(a) schematically depicts the two catalysts. If we start with catalyst A and arbitrarily decide that each catalytic group, C, C′, and C″, binds its corresponding activated reactant, R, R′, or R″, for 90% of the time, then the probability that any one group will bear an activated reactant at any one instant will be $K = K' = K'' = 0.9$ [Fig. 5.4(b)–(d)].

The likelihood that all of the catalytic groups will simultaneously bear activated reactants is consequently $L = K \times K' \times K'' = 0.9 \times 0.9 \times 0.9 = 0.73$ or 73% [Fig. 5.4(e)]. Thus, just less than three quarters of the catalysts in solution will contain concurrently activated reactants on all of the catalytic groups.

We will then decide that the flexibility of catalyst A is such that the catalytic groups in catalyst A are perfectly complementary to the transition state for 3% of the time. Thus, the catalyst can mediate a successful collision of the reactants leading to product formation for 3% of the time (probability: $P = 0.03$) [Fig. 5.4(f)].

The overall probability Q that all of the necessary components in catalyst A will then be properly synchronized for successful transition state formation becomes accordingly: $Q = P \times L = 0.03 \times 0.73 = 2.2 \times 10^{-2}$ or 2.2% [Fig. 5.4(g)].

Thus, 2.2% of the molecules of catalyst A in solution will form the transition state leading to the products, at any one instant. This will be reflected in the overall catalytic rate.

Now consider hypothetical catalyst B in Figure 5.4, which is identical in form to catalyst A except that it employs shorter lived individual binding interactions with the reactants.

The probability of reactant binding and activation, K, K′, and K″, in catalyst B must necessarily be less than in catalyst A. If we decrease K, K′, and K″ to, say, one third of that in catalyst A, namely, to 30% (0.3) [Fig. 5.4(b)–(d)], then the probability of concurrent reactant activation, L, in catalyst B declines to $L = K \times K' \times K'' = 2.7\%$, which is a fall of more than 27-fold relative to catalyst A [Fig. 5.4(e)]. It reflects the cumulative effect of the shorter lived individual binding interactions.

To maintain the same overall probability of forming a successful transition state [$Q = 2.2\%$; Fig. 5.4(g)], the likelihood of complementarity with the transition state P must then be increased to 81% [Fig. 5.4(f)].

In other words, the catalytic groups in catalyst B must spend 81% of their time in the spatial arrangement that complements the transition state and not 3%, in order to realize the same likelihood of successful transition state formation that exists in catalyst A.

To sustain an identical reaction rate, the catalytic groups in B must therefore be *more consistently complementary to the transition state* than is necessary in catalyst A. That is, the decline in probability L caused by the weaker individual binding interactions employed must be compensated by an increase in the complementarity for the transition state (probability P).

The above arguments are true regardlesss of the specific numbers used in the above example, provided that the values of K, K′, and K″ are less for catalyst B

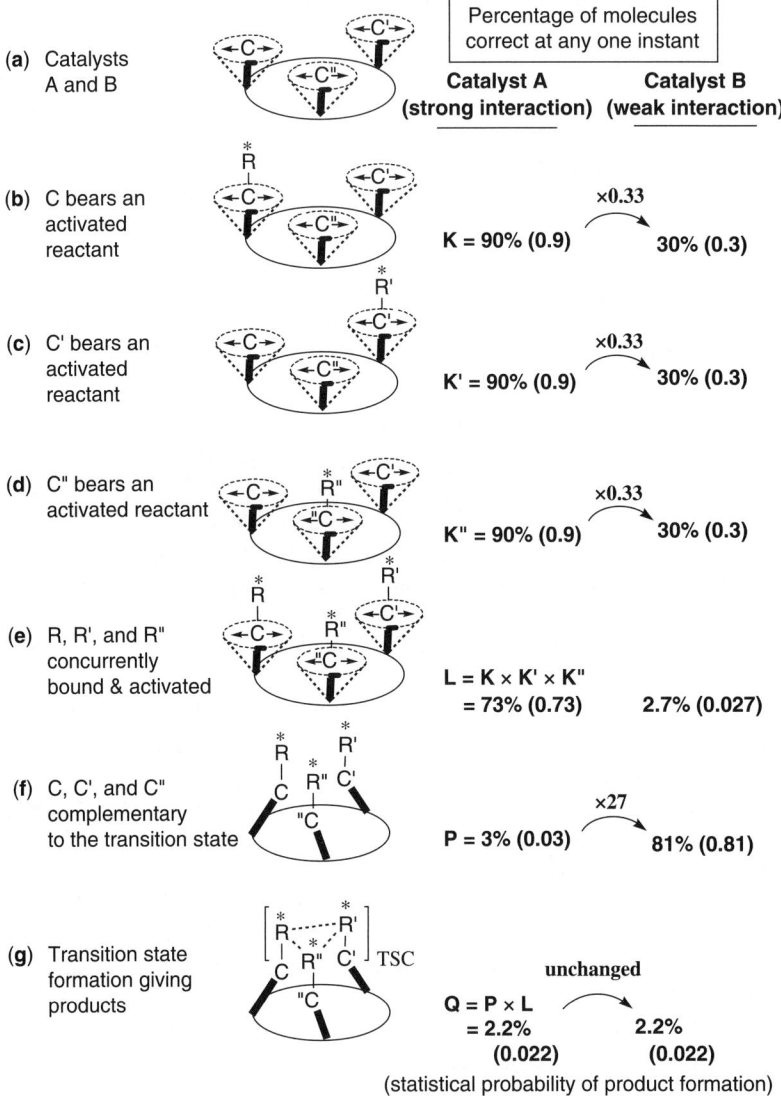

Figure 5.4. Schematic determination of the statistical probability of transition state formation in the three-center catalysts A and B (C, C', C″ = catalytic groups; R, R', R″ = reactants; K–Q = the percentage of molecules that are catalytically optimum).

than for catalyst A, and Q is kept the same for both catalysts. It is also true for any type of multicentered homogeneous catalyst that one may care to consider.

The point of this exercise is to illustrate the very important relationship that exists between collision frequency and energetic efficiency in multicentered homogeneous catalysts.

5.6 RATIONALIZATION OF THE UNDERLYING PROCESSES

In the above example, collision frequency is, effectively, represented by the quantity L, which describes the steps involved in bringing the reactants together. L represents the steps depicted in Figure 5.3(a)–(d).

The extent of transition state stabilization is represented by the quantity P. The higher the value of P, the more the catalyst is constrained to flex about a structure that is complementary to the transition state of the reaction. As P increases, the ability of the catalyst to drive its bound reactants into collision is increasingly limited to pathways that are optimum for successful transition state formation. A high P consequently corresponds to a low *average* energy barrier for reaction E_a. In effect, the more complementary the catalyst is to its transition state, the greater is its energetic efficiency. As such, P represents the steps depicted in Figure 5.3(d)–(f).

The overall reaction rate Q is the mathematical combination of L and P in the same way that is seen in Arrhenius's Equation (5.1).

In an energy-dependent catalyst, L must be greater than P, because the greatest impediment to reaction involves forming the transition state leading to the products. Catalyst A is therefore energy-dependent because in its case, L > P.

In a time-dependent catalyst, however, P must be more than L, because the greatest impediment to reaction involves bringing the activated reactants into collision with each other. Thus, catalyst B is time-dependent because P > L.

5.6.3 The Coexistence of Transition State Complementarity, Structure-Sensitive Catalysis, and Unconventional Catalytic Groups in Enzymes Is Caused by their Weak Individual Binding Interactions

Using the above example, we can now better understand the general properties of enzymes and their nonbiological counterparts.

Catalyst B is structurally highly complementary to its transition state. This is necessary to compensate for its weak individual binding interactions. Catalyst B is therefore also *structure sensitive*; that is, small changes in its structure will alter its complementarity and thereby dramatically affect its overall rate. Catalyst B can, moreover, use unconventional catalytic groups like amino acids, because its structure compensates for the fact that they bind and activate their reactants only weakly and transiently.

Key Point. In other words, there is an internal consistency in the coexistence of transition state complementarity, structure-sensitive catalysis, and unconventional catalytic groups in catalyst B. These features originate because of the weak and dynamic individual catalyst–reactant interactions employed. They must be present in catalysts employing such interactions, because this is the only way in which such species can facilitate catalysis.

This finding implies that enzymes are time-dependent catalysts. It also explains many of their general properties as being an internally consistent outcome of that fact.

By contrast, catalyst A employs strong individual binding interactions. For this reason, it does not need to be complementary in structure to its transition state.

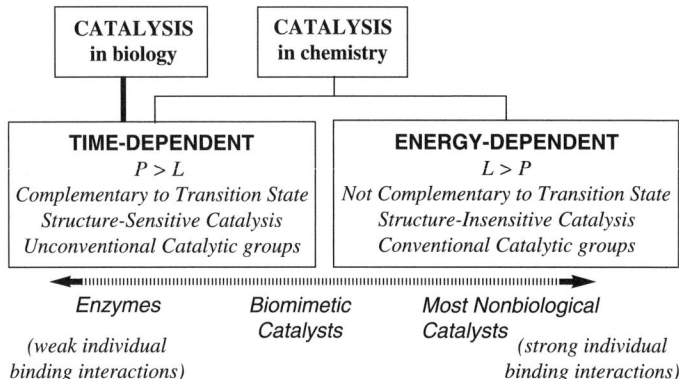

Figure 5.5. The conceptual overlap of multicentered catalysis in chemistry and biology.

Thus, it is structure-insensitive in its catalytic action; small changes to its structure will have little effect on its overall rate; it need only be complementary to the transition state for 3% of the time. However, catalyst A cannot use unconventional catalytic groups such as amino acids, because it depends on sustained reactant-binding interactions to generate a catalytic effect. Amino acids do not provide such interactions.

Thus, multicentered, nonbiological homogeneous catalysts are implied to be energy-dependent, with many of their general properties being an internally consistent outcome of that fact.

To summarize, many enzymes seem to depend on their structure during conformational flexing in order to generate a catalytic effect (quantity P in Fig. 5.4). Their nonbiological counterparts seem to depend on the persistence of their individual binding interactions (quantity L in Fig. 5.4). An influence that alters the structure of an enzyme will therefore dramatically affect its catalysis because it depends strongly on its structure to achieve a catalytic effect. By contrast, an influence that increases the dynamism of binding in a non-biological catalyst will destroy its catalysis (e.g., the use of amino acid catalytic groups), because it depends on sustained and strong individual binding interactions to achieve a catalytic effect.

Enzymes and nonbiological homogeneous catalysts are, consequently, suggested to occupy opposite ends of a continuum of multicentered catalytic action defined by the strength of the individual catalyst–reactant binding interactions employed (Fig. 5.5).

5.6.4 The Origin of the Time-Dependence and the Synergies of Enzymes

The fundamental origin of time-dependence in many enzymes can now also be understood. Figure 5.6 schematically illustrates the transit times of a typical sequence of events during catalysis by a multicentered time-dependent catalyst of the type shown in Figures 5.3 and 5.4. The bottom axis is time. Each catalytic group C, C′,

5.6 RATIONALIZATION OF THE UNDERLYING PROCESSES 123

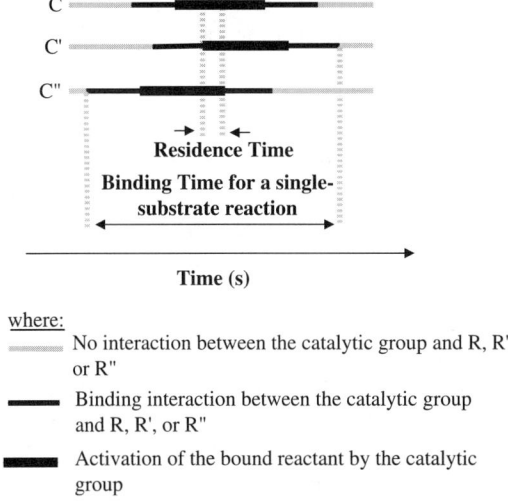

Figure 5.6. Residence time in enzymatic catalysis and its relationship to the binding time. In a single-substrate reaction, the time during which concurrent activation occurs (the "residence time") is shorter than the time of substrate attachment (the "binding" time). Brevity in the residence time is the fundamental origin of time-dependence in enzymatic catalysis.

and C″ starts off on the left of the figure without an interaction with its respective reactant R, R′, and R″. This is indicated by the gray lines that commence on the left-hand side of Figure 5.6 for each catalytic group.

As time passes, the various catalytic groups bind their reactants, each at a different time. The onset of binding is indicated by the appearance of the thin black lines as one goes from the left to the right in the figure.

During periods of binding, the reactants also become activated. It is shown, for each catalytic group, by the thick black lines that emerge as one moves further to the right of Figure 5.6. As noted, the extent of this activation may vary. Some catalytic groups, such as groups that simply facilitate correct positioning of the substrate, may not need to provide significant activation. Others, such as participating groups within the active site, may have to be more strongly activating in order to initiate catalysis.

Eventually activation ceases, leaving the reactants still bound to their catalytic groups, but no longer activated. This is shown by the thin black lines to the right of the figure.

Finally, the reactants disengage from their catalytic groups; this is depicted by the gray lines at the far right of the figure.

This sequence of events may occur independently for each catalytic group. However, if the reactants R, R′, and R″ are all part of a single substrate, then the establishment of one interaction is likely to induce the others to follow in short order.

In such a case, the time during which the substrate is "bound" by the catalyst (namely, the *binding time*) stretches from the start of the first binding interaction (C″) to the end of the last one (C′).

However, the time during which the reactant groups R, R', and R" are all *simultaneously* bound and activated and therefore available to form the transition state is much shorter. As noted in Section 4.2.3.1 of Chapter 4, this time is known as the *residence time* τ.

Key Point. Brevity in the residence time is the cause of time-dependence in heterogeneous catalysts. It is now revealed to also be the cause of time-dependence in many enzymes. But it originates in a very different way. In heterogeneous catalysts, it derives from brevity in reactant binding and activation. In enzymes, however, it is caused by brevity in concurrent reactant binding and activation. That is, the time during which all of the reactant functionalities are simultaneously bound and activated by the enzyme is exceedingly short and this is what causes the action to be time-dependent. The substrate need not formally dissociate during this process.

In other words, the brevity of concurrent reactant binding and activation diminishes the likelihood that the bound reactants will physically and mechanically collide with each other in an arrangement suitable for reaction. This ensures that the catalysis is governed by the *catalyst-mediated collision frequency*.

5.6.5 The Mechanism of Time-Dependence in Enzymes Resolves the Contradiction of a Kinetically Observed Rapidly Forming and Dissociating Intermediate in the Face of Strong Overall Substrate Binding

We can now understand how a rapidly forming and dissociating intermediate can exist in enzyme kinetics despite the fact that enzymes bind their substrates so strongly.

Key Point. What happens is that enzymes actually bind their substrates for sustained periods of time. But, it is not a static binding. During these periods, there is a dynamic engagement and disengagement of individual substrate functionalities. That is, the many interactions involved in the binding of the substrate form and dissociate continuously and individually. When even a few of these interactions are present, the substrate is formally "bound" to the enzyme. But catalytic reaction is only possible when all of them are bound simultaneously. Because of the sheer dynamism of these interactions, this generally occurs only sporadically and for brief periods at a time.

Thus, the rapidly equilibrating [ES] intermediate in the Michaelis–Menten equation (5.2) is identified as, and originates in the dynamic binding of the substrate and enzyme at multiple contact points. It is the dynamism of substrate binding that constitutes the "rapid equilibration." The system spends most of its time in dynamic "equilibration" and this shows up in its kinetics.

Despite the fact that enzymes bind their substrates strongly overall, the residence time during which catalysis is possible is very short. Enzymatic rates are consequently dependent on bringing the individual substrate functionalities into contact during the brief times that they are bound. That is, it depends on the catalyst-mediated collision frequency and not on the underlying energy of the process.

5.6.6 Catalysis in Enzymes Involves Synchronization of Enzyme Binding and Enzyme Flexing

Enzymes are therefore indicated to be catalysts that conformationally flex about a structure that is complementary to their transition states. During this flexing, they drive any bound substrate functionalities into reactive contact ("collision") with each other. A specific pathway is followed in each of these actions. This pathway involves a low-energy barrier for the collision because enzyme flexing is constrained to maintain a high level of complementary to the transition state. In other words, the conformational limitations of the enzyme ensures that near-optimal approach trajectories are followed by the bound reactants immediately before their collision. It does not allow for substantial deviations from these approach pathways; that is, the collision process is *path-specific*. Because of the resulting, low-energy barrier to reaction, the possibility exists for a successful collision, leading to products, during each cycle of conformational flexing. However, the collision can only be successful if all of the participating reactant functionalities are bound to their respective enzyme groups at the precise instant of collision. The individual enzyme-substrate binding interactions are highly dynamic, though, with the functionalities constantly being bound and released. It arises because of the weak interactions of biology employed by enzymes.

*Key Point. **The overall rate of an enzyme-catalyzed reaction therefore depends on the extent to which its conformational flexing and its substrate binding are synchronized***. This feature is schematically illustrated in Figure 5.7.

That is, the success of each collision action depends on the overlap that exists between the time during which the substrate is fully bound and the time during which the substrate functionalities are brought into contact by enzyme flexing. The longer this time on average, the greater will be the catalytic rate.

Enzymatic rates are, of course, formally expressed in terms of the *inverse* of this overlapping time, namely as the average number of successful collisions per second between reactant functionalities (the "collision frequency").

This dependence on *synchronized* action is the origin of the time-dependent character of enzymatic catalysis. It also explains the *mechanical* character of enzymatic catalysis; enzymes literally "put reactants together" in a mechanical way and thereby turn them into products. As with any other mechanical device, small changes in the structure of the enzyme or the conditions of the process can have an unpredictable effect on the required synchronization with accompanying drastic changes in its overall efficiency.

5.6.7 Summary: The Origin of the General Properties of Enzymes

We can now summarize how most of the general properties of enzymes originate.

*Key Point. **Enzymes are structurally complementary to their transition states because this is the only way to get any sort of synchronization between enzyme binding and flexing***.

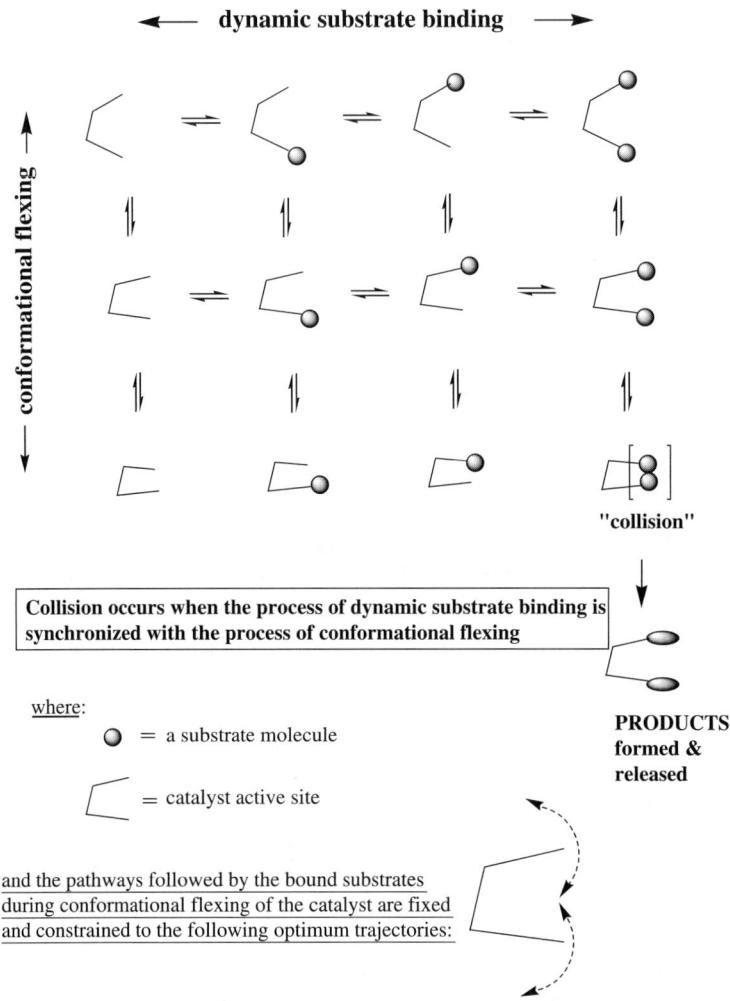

Figure 5.7. *Time-dependent catalysis*: Schematic representation of the two key processes present in many enzymes. Shown on the left, from top to bottom, is the process of *conformational flexing* of the enzyme active site. During conformational flexing, the groups in the enzyme that bind the substrate are repeatedly moved back and forth over the same, optimum pathways in space, relative to each other. Shown at the top, from left to right, is the process of *dynamic substrate binding*. In this process, the substrate repeatedly binds and releases the relevant enzyme groups. A successful "collision" occurs when these processes of *conformational flexing* and *dynamic substrate binding* are synchronized; that is, when the substrates are fully bound at the instant that they are brought into physical contact with each other.

For the same reason, enzymes are structure-sensitive catalysts. The large reaction rates displayed by enzymes relative to structurally similar but inactive proteins derive from the fundamental path- and time-dependent nature of enzymatic catalysis. Even minor changes that alter the approach pathways of the reactants before collision, or that change the synchronization of binding and flexing, will destroy the effect. Enzymic structure-sensitivity therefore originates because there is a nonlinear relationship between enzyme structure and catalytic rate.

Enzymes can employ nonconventional catalytic groups, like amino acid residues, because they can synchronize their actions to compensate for the weak and dynamic binding that they provide.

Michaelis–Menten kinetics derives from the fact that the reaction cannot go any faster than the rate at which enzyme flexing brings the bound reactants into collision with each other. Once the reactant concentration is enough to ensure that essentially every conformational cycle involves bound reactants and results in a successful collision, the catalyst will be saturated. This will occur at a certain relative concentration of the reactant with respect to the catalyst. This does not happen in catalysts controlled by E_a because the rate is not limited by the collision frequency. In such cases, saturation occurs when the time between the release of products and the uptake of new reactants approaches zero. This will take place at a high absolute reactant concentration, not a relative reactant concentration.

5.6.8 Catalysis in Nonbiological Analogues Depends on the Activation Energy E_a

By contrast, almost all known nonbiological homogeneous catalysts are implied to be *energy-dependent* in their mode of catalytic action.

In such species, the catalyst also creates collisions between bound reactant functionalities during flexing. But the approach pathways involved in these collisions are essentially random, or close thereto, because of greater conformational freedom in the catalyst. This conformational freedom does not negatively affect the catalytic process because such catalysts bind and activate their reactants for sustained periods of time. The bound reactants can therefore simply wait for the infrequent occasions that the catalyst happens to becomes complementary to the transition state. Thus, such catalysts need not display a predilection for structurally complementarity to the transition state. The reactant functionalities remain bound to the catalyst more or less continuously until a successful transition state is obtained. There is, consequently, no requirement for synchronization, or, certainly, a far less demanding one than that which exists in many enzymes.

The result is that each collision between the bound reactants in such a catalyst involves a different energy barrier, with only those collisions that have a low barrier being successful. The catalytic rate is therefore dependent on the *activation energy* (E_a) of the reaction. E_a is formally defined as the *average* energy barrier when the approach trajectories of the bound reactants immediately before the collision are random.

5.6.9 Enzymatic Selectivity and Synergies Derive from Time-Dependence

The selectivity and *convergent synergies* of time-dependent catalysts originate in the great importance of correct component arrangement during the very brief periods available for transition state formation. If even one element is incorrectly arrayed at the critical juncture of synchronous binding at the point of collision, successful reaction is not possible. Catalysis is therefore achievable only very transiently and for a very particular structural arrangement. This arrangement, along with the brevity of the opportunity, ensures excellent selectivity in the catalytic process.

It also ensures *convergent* actions on the part of the various elements involved. If any of the reactant functionalities or catalytic groups do not act synchronously at the point of collision, no reaction can occur. In other words, if the coordination of the participating elements is not optimum or near-optimum, a product is not formed.

The resulting fidelity in the reaction process not only enhances its selectivity but also prevents the formation of nonfunctional intermediates, making such catalysts long-lived. It potentially explains why enzymes can operate without being poisoned in the extraordinarily mixed reactant streams of biology.

Energy-dependent homogeneous catalysts do not operate under the same constraints. Because the likelihood of simultaneous reactant activation is higher, bound reactants are free to participate in any and all of the different transition states that may be formed in the many collisions that occur during flexing of the catalyst. Thus, all possible transition states are populated. Although the product will largely derive from the energetically most stable transition state, other products from other transition states are also possible. In effect, the opportunity for transition state formation is not transient and this destroys the selectivity of the catalysis.

*Key Point. **The high selectivities and the convergent synergies displayed by enzymes are therefore entirely consistent with time-dependent catalysis. The observed durability of enzymes in the extraordinarily mixed reactant streams that exist in biology supports this view. The poor selectivity of nonbiological, multi-centered catalysts and their often modest durability even in highly purified reactant streams is generally more consistent with energy-dependent catalysis.***

5.6.10 Enzymatic Activity Is Consistent with Time-Dependence

The higher activities achieved by enzymes relative to nonbiological catalysts are caused by the extreme rapidity and dynamism of enzyme-substrate binding. They are also consistent with the greater intrinsic efficiency of functionally convergent synergies relative to functional complementarity [36]. Convergent systems require coordinated actions and are therefore inherently more disposed to achieving cooperative amplifications than are complementary ones [36]. As noted in Section 1.4.4 of Chapter 1, examples of such amplifications are seen in all manner of convergent, time-dependent systems including evolution, DNA replication, and even in laissez-faire economic systems [36]. In Chapter 8, we will discuss the issue of synergy in catalysis in greater detail.

5.7 ALL GENERALIZATIONS SUPPORT TIME-DEPENDENCE IN ENZYMES

The following evidence therefore favors the assertion that many enzymes are time-dependent catalysts:

(1) Physical studies of the rate-determining process in enzymes are consistent with time, and not energy-dependent catalysis.
(2) Michaelis–Menten kinetics is consistent with, and predictive of, time-dependent, not energy-dependent catalysis.
(3) Time-dependent catalysis provides a conceptually consistent explanation for Michaelis–Menten kinetics in enzymes where no such explanation has previously existed. Energy-dependent catalysis cannot provide such an explanation.
(4) The prevalence of bidirectional catalysis by enzymes is only generally possible in time-dependent, not energy-dependent, systems.
(5) The coexistence of (a) transition state complementarity, (b) structure-sensitive catalysis, (c) unconventional catalytic groups, and (d) weak individual binding interactions can only be explained in an internally consistent way if enzymes are generally time-dependent catalysts.
(6) Time-dependence in enzymes can be explained as originating from brevity in the residence time, in complete analogy with time-dependent heterogeneous catalysts.
(7) Enzymatic synergies are explained by time-dependent catalysis but not by energy-dependent catalysis.
(8) The propensity of enzymes to be structurally complementary to their transition states is necessary to ensure synchronization between enzyme binding and flexing. Such synchronization is needed to generate a catalytic effect in time-dependent homogeneous catalysts but not in energy-dependent ones.
(9) The large reaction rates displayed by enzymes relative to structurally similar but inactive proteins are explained by a fundamental path- and time-dependence, which is intrinsic only to time-dependent processes, not to energy-dependent ones.
(10) Enzymatic specificity and activity are more consistent with time-dependent than with energy-dependent catalysis. The limited time available to bring about a successful collision and its dependence on correct binding at the instant of collision explains enzymatic specificity.

The distinctive features of enzymatic catalysis are therefore in all respects consistent with, and indicative of, time-dependent catalysis. They are inconsistent with energy-dependent catalysis.

5.8 TIME-DEPENDENCE IN A NONBIOLOGICAL CATALYST GENERATES THE DISTINCTIVE PROPERTIES OF ENZYMES

Deductively derived understandings of natural systems can only be considered valid if the presence of the theorized attribute can be shown to replicate the phenomenon in a nonbiological system. Only one example of such a system is needed, since this provides the exception that unequivocally proves the rule.

With this in mind, we reviewed the scientific literature in search of nonbiological, multicentered, homogeneous catalysts whose use of dynamic, weak binding interactions leads to time-dependent catalysis. Because of the rarity of such catalysts, only a few examples could be found. In Chapter 10 we will discuss all of these examples in detail.

An unambiguous example of an abiological time-dependent catalysis exists, however, in the proton reduction catalyst [1.1]ferrocenophane **3** (Scheme 5.2).

This species converts strong acids (H^+) to dihydrogen (H_2) in the presence of a sacrificial reductant [37,38]. The mechanism of this catalyst involves simultaneous homolytic cleavage of transient Fe–H species as shown in Scheme 5.2 [38]. Theoretical calculations indicate that the bound H^+ ions in intermediate **4** are effectively atomic hydrogen ($H^{\bullet}$) with most of their positive charge resident on the metal [38–40].

The key feature of catalysis by **3** is weak and dynamic proton binding. The ferrocenes in **3** are very weakly basic (pK_{a1} −6.5, pK_{a2} −7.1) [40] so that strong acids are required to protonate them. During the infrequent periods of ferrocene–H^+ binding, the protons participate in a rapid, dynamic equilibrium involving Fe–H and agostic Cp-ring C–H isomers as shown in Scheme 5.3 [38,39]. Density functional theory (DFT) calculations indicate that the Fe–H state is of higher energy and that its lifetime is exceedingly short [39]. As such, it must be considered to be the activated form of the bound reactant. Recently, **4** could be directly observed by ^{1}H NMR at −122 °C [39]. This confirmed that both iron atoms became protonated and that the exchange processes were too rapid to be resolved by NMR.

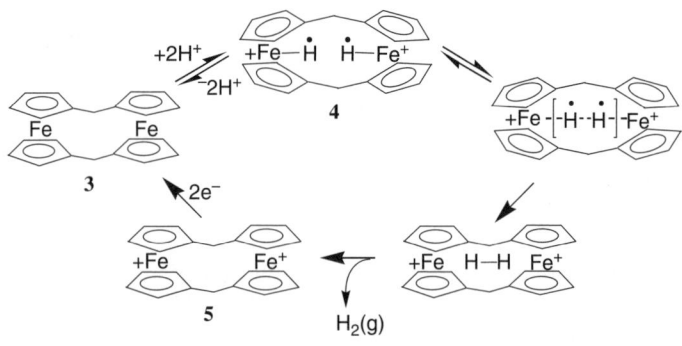

Scheme 5.2.

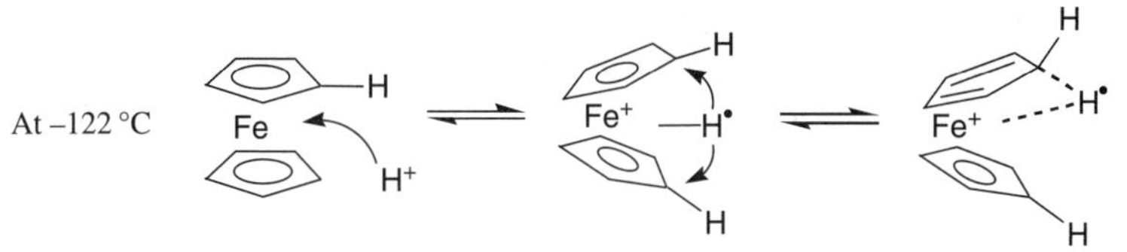

5.8 TIME-DEPENDENCE IN A NONBIOLOGICAL CATALYST

Scheme 5.3.

The effect of this rapid and dynamic proton binding is that the ferrocene groups can only act catalytically if they have a high probability of being optimally arranged to form the transition state during conformational flexing. That is, the arrangement of the ferrocene catalytic groups must be such that flexing sees the activated H• substrates brought into reactive contact with each other along the optimum trajectory to form the transition state; this is the case in **3** and **6** [38]. Structural modifications that decrease the rate of flexing (as in **7** in Scheme 5.4) or that alter this optimum arrangement during flexing (as in **8–9** in Scheme 5.4) destroy the catalytic effect.

Thus, although **3** and **6** are active catalysts of proton reduction, **7–9** are entirely inactive (Scheme 5.4), as are diferrocenylethane and free ferrocene [38]. A complete loss of catalytic activity upon minor structural modification is also a feature of enzymatic catalysis, which is similarly *structure-sensitive*. Ferrocene, which is not otherwise known to be a catalytic species, is, moreover, transformed into a potent catalyst in **3**, in further analogy with enzymes.

Catalysis by **3** consequently involves a low likelihood of binding protons, a still lower likelihood of activating them, and a yet lower likelihood of simultaneously activating two bound protons. This system clearly has a low collision frequency. In compensation, the structure of **3** is, effectively, highly complementary to the H⋯H transition state during flexing, so that the activation energy will be overcome when a collision occurs. Thus, **3** must be a time-dependent and not an energy-dependent catalyst.

The time-dependent nature of the catalytic action in **3** becomes clearer when one considers that this catalyst depends on physically and mechanically driving the activated H• species in **4** into contact with each other during the brief and infrequent times that both ferrocenes happen to bear an activated proton. This must happen along trajectories that correspond to, and lead to, formation of the transition state. Thus, **3** relies on synchronized binding and flexing to create a catalytic effect.

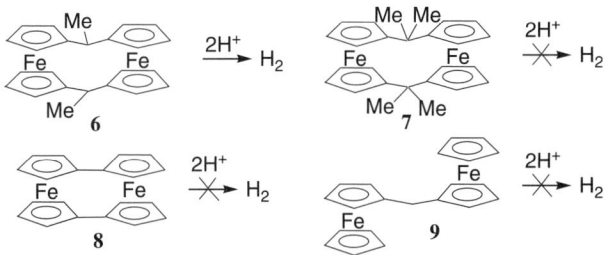

Scheme 5.4.

Key Point. Catalyst 3 can be aptly termed a mechanical catalyst since it exhibits numerous characteristics common to mechanical processes in general. For example, like essentially all mechanical devices, the movements and actions of the components of 3 must be synchronized in order to realize an effect. That is, the catalytic groups must act in a concerted, coordinated manner or, more correctly, a convergent manner. Moreover, catalysis by 3 is driven by a mechanical impulse, namely conformational flexing. The catalytic action of 3 is, in fact, intrinsically machine-like. That is, when the processes of catalyst flexing and binding are synchronized, protons are dynamically bound, activated, and carried by the mechanical impulse of conformational flexing along specific, highly repeatable and near-optimum pathways into reactive collisions with each other. These collisions occur within a structure that complements the desired outcome, namely the transition state. The products are then dynamically ejected and new reactants taken up for the next cycle. Thus, just as, say, a bottle-making machine dynamically takes up input materials and mechanically drives them into each other within a structure that complements the shape of a bottle, so 3 dynamically takes up protons and mechanically drives them into a reactive collision with each other within a structure that complements the transition state.

The action of 3 is an intrinsically non-equilibrium event. That is, it takes place not as a response to an energy gradient, but rather because the impulses and the mechanics of the situation are such that there is no alternative but for it to occur. In other words, 3 generates products because its mechanical action and spatial arrangement at the point of collision is such that H_2 is inevitably formed.

[1.1]Ferrocenophane is very active and long-lived as a homogeneous catalyst. Individual molecules of **3** have been shown to turn over 1 million H_2 molecules, on average, without any noticeable loss of activity.[1] When bound to polystyrene and coated on a p-type silicon photocathode, **3** turns over an estimated five molecules of H_2 per second, over five days of continuous operation [41].[2] The sheer durability of this catalyst can only be caused by its excellent selectivity, which prevents the formation of nonfunctional intermediates and hinders deactivation. Kinetic studies have verified the mechanism of catalysis by **3** [40], but the possibility of saturation kinetics has not been investigated, probably because of the high acid concentrations needed to achieve a catalytic effect. Nor has the possibility of bidirectional H_2-2H^+ catalysis been examined.

The effect of time-dependence is therefore to impart **3** with the following distinctive properties:

(1) Weak and dynamic individual catalyst–reactant binding interactions
(2) A high incidence of structural complementarity with the transition state during conformation flexing
(3) Highly structure-sensitive catalysis

---

[1] 10 mg of **3** will dissolve at least 5 g of lead as sacrificial reductant during H_2 generation; see Reference 38.
[2] A coating containing 2 M [**3**] over an area of 0.16 cm^2 and a response thickness estimated to be 1 μm generates a saturation current of 37 mA during catalysis.

(4) A transformation of nonconventional catalytic groups (ferreocenes) into potent catalysts
(5) High selectivity and activity
(6) Functionally convergent synergy

It should be added that intermediate **4** also displays induced strain. The ferrocene cyclopentadienyl rings cannot accommodate a proton without loss of coplanarity (Scheme 5.2), so that dihydrogen formation is favored by the release of steric strain [38]. The release of steric strain has been proposed to be important in enzymes (see Section 7.2.5 in Chapter 7 *vide infra*) [6].

Except for the presence of saturation kinetics and bidirectional catalysis, time-dependence in **3** consequently creates all of the same distinctive catalytic features displayed by enzymes. This can surely not be coincidental.

These conclusions imply that the above properties are characteristic of *all* time-dependent homogeneous catalysts. In other words, they are not so much the properties of enzymes as they are of time-dependent catalysts in general.

5.9 CONCLUSION: MANY ENZYMES ARE TIME-DEPENDENT CATALYSTS

Key Point. The only reasonable conclusion that can be drawn from these results is that many enzymes are time-dependent catalysts. All of the general properties of enzymes are consistent with, or predictive of, time-dependence. They are inconsistent with energy-dependent catalysis. Moreover, a rare example of a time-dependent, nonbiological, homogeneous catalyst displays the same general properties as enzymes. In acting as the exception that proves the rule, this catalyst verifies that enzymatic catalysis is often time-dependent. According to the principles of logic, only one example is needed for such a proof and the [1.1] ferrocenophane catalyst **3** arguably provides that example.

The assumption of energy-dependence in enzymes must therefore be considered generally inaccurate. Enzymatic catalysis does not appear to be energy-dependent, by and large. Most enzymes, in fact, seem to stabilize their transition states to such an extent that their *activation energies* E_a become irrelevant to the catalytic process, which is then, instead, controlled by the *collision frequency* of the bound reactants.

Although catalysis by enzymes does not seem to be generally controlled by the activation energy of the reaction, it is certainly dependent on the conformational flexing of the protein and its binding of the substrate, as well as the synchronicity of these two processes. These processes are, of course, intrinsically connected to the steric and other energies involved in protein flexing and binding. But this is different to the energies involved in the reaction itself. It is different in a profound and fundamental way: It involves a time-dependent, not an energy-dependent reaction process.

Key Point. The key intellectual error made in enzymatic catalysis to date seems to have been to accept the dominant paradigm of catalysis, namely, energy-dependence. Although energy-dependence certainly does hold for most nonbiological homogeneous catalysts, it does not seem to be generally true for enzymes.

Given the academic ascendancy of the energy-dependent description of catalysis [42], and the relative obscurity of the time-dependent version [43], one can hardly blame chemists and biochemists for making this mistake. Nevertheless, a recent issue of *Chemical Reviews* dedicated to enzymatic catalysis contained not a single paper that even considered the possibility that enzymatic catalysis may be limited by the collision frequency of the bound substrates [44]. A clarification of the central role played by catalyst–reactant dynamism in this respect is clearly required.

In the following chapter, we will discuss and describe in greater detail the processes that occur in time- and energy-dependent multicentered homogeneous catalysts.

ACKNOWLEDGMENTS

The authors thank for their insightful comments: Chuck Dismukes and Damian Carrieri (Princeton University), Wolf Sasse (CSIRO), Bob Williams (Oxford University), and Craig Hutton (University of Melbourne).

REFERENCES

1. Atkins, P. W. *Physical Chemistry*, Oxford University Press, 1978, p. 864–866, p. 899–925.
2. Kuchel, P. W.; Ralston, G. B. *Theory and Problems of Biochemistry*, Schaum's Outline Series, McGraw Hill Inc., 1988, p. 225.
3. Stryer, L. *Biochemistry (3rd ed.)*, W. H. Freeman and Company, 1988, p. 191, p. 182–183. Kinetic "perfection" in an enzymatic system is described in: Knowles, J. R.; Albery, W. J. *Acc. Chem. Res.*, **1977**, *10*, 105.
4. Storm, D. R.; Koshland, E. E. *Proc. Natl. Acad. Sci. USA* **1970**, *66*, 445.
5. Page, M. I. Chap 1 in *Enzyme Mechanisms*, Eds. Page, M. I., Williams, A., *Royal Soc. Chem.*, 1989, p. 1, and references therein.
6. (a) Haldane, J. B. S. *Enzymes*, Longmans, Green and Co., 1930, p. 182; (b) Fersht, A. *Enzyme Structure and Mechanism (2nd ed.)*, W. H. Freeman and Company, 1977, p. 331, and references therein.
7. (a) Pauling, L. *Chem. Eng. News* **1946**, *24*, 1375; (b) Pauling, L. *Nature* **1948**, *161*, 707.
8. Wolfenden, R.; Frick, L. Chap 7 in *Enzyme Mechanisms*, Eds. Page, M. I.; Williams, A., *Royal Soc. Chem.*, 1989, p. 97, and references therein.
9. (a) Bruice, T. C. *Acc. Chem. Res.* **2002**, *35*, 139, and references therein; see also: (b) Bruice, T. C.; Lightstone, F. C. *Acc. Chem. Res.* **1999**, *32*, 127; (c) Lau, E.; Bruice, T. C. *J. Am. Chem. Soc.* **2000**, *122*, 7165; (d) Bruice, T. C.; Benkovic, S. J. *Biochemistry* **2000**, *39*, 6267.

10. Williams, D. H.; Stephens, E.; Zhou, M. *Chem. Commun.* **2003**, 1973, and references therein.
11. Breslow, R. *Acc. Chem. Res.* **1995**, *28*, 146, and references therein.
12. Koshland, D. E. *Proc. Natl. Acad. Sci. USA* **1958**, *44*, 98.
13. Hillery, P. S.; Cohen, L. A. *J. Org. Chem.* **1983**, *48* 3465.
14. Jencks, W. P.; Page, M. I. in *Enzymes: Structure and Function, Vol. 29*, North-Holland/American Elsevier, 1972, p. 45.
15. (a) Menger, F. M. *Acc. Chem. Rev.* **1993**, *26*, 206, and references therein; (b) Menger, F. M. *Acc. Chem. Rev.* **1985**, *18*, 128, and references therein.
16. Kirby, A. J. *Adv. Phys. Org. Chem.* **1980**, *17*, 183.
17. For example: (a) Moss, D. W. *Enzymes*, Oliver and Boyd, 1968, p. 68, and references therein; (b) Agarwal, P. K. *Microb. Cell Fact.* **2006**, *15*, 2, and references therein; (c) Williams, R. J. P. *Trends Biochemi. Sci.* **1993**, *18*, 115; (d) Hammes-Schiffer, S.; Benkovic, S. J. *Annu. Rev. Biochem.* **2006**, *75*, 519; (e) Benkovic, S. J.; Hammes-Schiffer, S. *Science* **2003**, *301*, 1196; (f) Boehr, D. D.; McElheny, D.; Dyson, H. J.; Wright, P. E. *Science* **2006**, *313*, 1638; (g) Vendruscolo, M.; Dobson, C. *Science* **2006**, *313*, 1586.
18. Steinhangen, H.; Helmchen, G. *Angew. Chem. Int. Ed. Engl.* **1996**, *35*, 2339.
19. Konsler, R. G.; Karl, J.; Jacobsen, E. N. *J. Am. Chem. Soc.* **1998**, *120*, 10780.
20. Masel, R. I. *Chemical Kinetics and Catalysis*, Wiley-Interscience, 2001, p. 717–725, p. 879–888, p. 689–793, and references therein.
21. Pirrung, M. C.; Liu, H.; Morehead, A. T. *J. Am. Chem. Soc.* **2002**, *124*, 1014, and references therein.
22. For example: Corey, E. J.; Noe, M. C. *J. Am. Chem. Soc.* **1996**, *118*, 319.
23. Cleland, W. W. *Acc. Chem. Res.* **1975**, *8*, 145, and references therein.
24. Palmer, T. *Understanding Enzymes (4th ed.)*, Prentice Hall Ellis Horwood, 1991, p. 103.
25. (a) Tousignant, A.; Pelletier, J. N. *Chem. Biol.* **2004**, *11*, 1037; (b) Hammes-Shiffer, S. *Biochem.* **2002**, *41*, 13335.
26. Workshop on future directions of catalysis science, *Catal. Lett.* **2001**, *76*, 111, and references therein.
27. Emerson, M. T.; Kaplan, M. L. *J. Am. Chem. Soc.* **1960**, 82, 6307.
28. Desiraju, G. R. *Encyclopaedia of Supramolecular Chemistry*, Eds. Atwood, J. L.; Steed, J. W. Marcel Dekker Inc., 2004, p. 658, and references therein.
29. Aakeröy, C. B. *Encyclopaedia of Supramolecular Chemistry*, Eds. Atwood, J. L.; Steed, J. W. Marcel Dekker Inc., 2004, p. 1379, and references therein.
30. Stern, E. W. Chap 4 in *Transition Metals In Homogenous Catalysis*, Ed. Schrauzer, G. N. Marcel-Dekker Inc, 1971, p. 93, and references therein. For a description of homogeneous catalysis in general, see the above title and: Parshall, G. W.; Ittel, S. D. *Homogeneous Catalysis: The Applications and Chemistry of Catalysis by Soluble Transition Metal Complexes*, Wiley, 1992, and references therein.
31. Tietze, L. F.; Ila, H.; Bell, H. P. *Chem. Rev.* **2004**, *104*, 3435, and references therein.
32. Lipscombe, W. N. In *Structural and Functional Aspects of Enzymatic Catalysis*, Eds. Eggerer, H.; Huber, R. Springer-Verlag, 1981, p. 17, and references therein.

33. (a) Cobb, A. J. A.; Shaw, D. M.; Longbottom, D. A.; Gold, J.; Ley, S. V. *Org. Biomol. Chem.* **2005**, *3*, 84, and references therein; (b) Gröger, H.; Wilken, J. *Angew. Chem. Int. Ed. Engl.* **2001**, *40*, 529; (c) Kelly, D. R.; Roberts, S. M. *Chem. Commun.* **2004**, 2018; (d) Kelley, D. R.; Meek, A.; Roverts, S. M. *Chem. Commun.* **2004**, 2021; (e) Cordova, A.; Zou, W.; Ibrahem, I.; Reyes, E.; Engqvist, M.; Liao, W.-W. *Chem. Commun.* **2005**, 3586; (f) Davies, S. G.; Sheppard, R. L.; Smith, A. D.; Thomson, J. E. *Chem. Commun.* **2005**, 3802.
34. For example: (a) Shibasaki, M.; Sasai, H.; Arai, T. *Angew. Chem. Int. Ed. Engl.* **1997**, *36*, 1237; (b) van Beuken, E. K.; Feringa, B. L. *Tetrahedron* **1998**, *54*, 12985; (c) Chin, J. *Curr. Op. Chem. Biol.* **1997**, *1*, 514; (d) Liu, S.; Luo, Z.; Hamilton, A. D. *Angew. Chem. Int. Ed. Engl.* **1997**, *36*, 2678; (e) Williams, N. H.; Takasaki, B.; Wall, M.; Chin, J. *Acc. Chem. Res.* **1999**, *32*, 485.
35. Moss, D. W. *Enzymes*, liver and Boyd, 1968, p. 68, and references therein.
36. Corning, P. A. *Systems Research* **1995**, *12*, 89 (published on the web at http://www.complexsystems.org/publications/pdf/synselforg.pdf).
37. Bitterwolf, T. E.; Ling, A. C. *J. Organomet. Chem.* **1973**, *57*, C15.
38. Mueller-Westerhoff, U. T. *Angew. Chem. Int. Ed. Engl.* **1986**, *25* 702, and references therein. See also: Mueller-Westerhoff, U. T.; Haas, T. J.; Swiegers, G. F.; Leipert, T. K. *J. Organomet. Chem.* **1994**, *472*, 229.
39. (a) Karlsson, A.; Broo, A.; Ahlberg, P. *Can. J. Chem.* **1999**, *77*, 628; (b) Karlsson, A.; Himersson, G.; Ahlberg, P. *J. Phys. Org. Chem.* **1997**, *10*, 590.
40. Hillman, M.; Michalle, S.; Felberg, S. *Organometallics* **1985**, *4*, 1258.
41. see Mueller-Westerhoff, U. T.; Nazzal, A. *J. Am. Chem. Soc.* **1984**, *106*, 5381.
42. Laidler, K. J.; King, M. C. *J. Phys. Chem.* **1983**, *87*, 2657.
43. Pilling, M. J.; Seakins, P. W. *Reaction Kinetics*, Oxford University Press, 1996, p. 121.
44. Various authors. *Chem. Rev.* **2006**, *106*, issue 8.

6

TIME-DEPENDENCE IN HOMOGENEOUS CATALYSIS. 2. THE GENERAL ACTIONS OF TIME-DEPENDENT ("MECHANICAL") AND ENERGY-DEPENDENT ("THERMODYNAMIC") CATALYSTS

Robin Brimblecombe, Jun Chen, Junhua Huang,
Ulrich T. Mueller-Westerhoff, and Gerhard F. Swiegers

6.1 INTRODUCTION

In Chapter 1, we described two fundamental types of action that are observed in physical science. The first is created by the laws of *thermodynamics*, which describe, in essence, how *changes in energy create actions* or set processes in train. Because the field of thermodynamics today extends far beyond this most elementary description and to avoid any possible confusion, we have termed actions or processes of this type *energy-dependent* in character. An example of an energy-dependent action is that of a ball falling to Earth under the influence of gravity. The gravitational field is invisible and intangible. However, it induces movement. We can often only perceive energy gradients in the physical effect that they create.

The second type of action is *mechanical* in character and derives from an action–reaction sequence (see Chapter 1). Movement or action is therefore created, not by an invisible energy gradient, but by another, first, movement or action. An example is a billiard ball that is struck by another, moving billiard ball. In its turn,

---

Mechanical Catalysis: Methods of Enzymatic, Homogeneous, and Heterogeneous Catalysis,
Edited by Gerhard F. Swiegers
Copyright © 2008 John Wiley & Sons, Inc.

it may strike a third billiard ball, causing it to move. Actions or processes of this type are governed by the laws of *mechanics*. A *mechanical process* is defined as the *physical* evolution of a system as it proceeds from an initial state *over time*. We have therefore termed such movements or actions as *time-dependent* to avoid any confusion with other aspects of the field of mechanics. This terminology reflects the fact that the key determinant of such actions is their dependence on the *time that has elapsed* and on the *time required for each step*, rather than on the relative energies of each step, or overall.

When viewed at the most fundamental level, time-dependent processes have the quality that they seem to be preordained and perfectly predictable. An example in this respect is a line of dominoes. When the first domino is knocked over, it is preordained by virtue of its spatial position to knock over the next one and thereby set off or propagate the collapse sequence.

Systems of this type are said to be *deterministic* (see Chapter 1); that is, their outcome is predetermined by the spatial positioning and timing of the events. This preordination extends in both the forward or the reverse direction; that is, the dominoes can be made to fall in the same, predictable way in either one direction or its reverse. This is true of all mechanical processes, which typically resemble a movie that may be played forward or backward in time.

A key distinction between time-dependent and energy-dependent processes is therefore that the former are *spatiotemporal* in character (see Chapter 1), which means that they have to do with spatial positioning and time. They are also path-dependent. That is, their outcome depends on the precise pathways that are followed and on the exact times at which they occur. Even slight variations in either of these can drastically change the outcome of the process. For this reason, sequences of time-dependent actions may display *chaotic* or *complex* behavior. That is, their final outcome can be dramatically influenced in an unpredictable, *stochastic* way by extraordinarily tiny changes in the initial conditions. This is not true of energy-dependent processes.

To illustrate this concept, consider a machine in which one cog has been fabricated slightly imperfectly. The behavior of that machine will be completely unpredictable. It may jam immediately on starting, or it may take 10 years to finally jam. What is clear, however, is that it *will* eventually jam because all mechanical devices rely on *synchronized, coordinated actions* in all of their parts. When such synchronization is present, mechanics can achieve very significant feats. It does this by *finessing* problems with clever *synergies* of action. That is, each component interacts with the other components in a way that drastically amplifies both of their contributions to, thereby, bring about a solution.

Energy-dependent processes do not employ such synergies of action. They must instead resort to overwhelming the problem with a large input of energy (see Chapter 1).

In Chapter 5, we examined the issue of time- and energy-dependence in homogeneous catalysis. We showed that many of the catalysts of biology, known as enzymes, seem to employ a time-dependent mode of action, whereas most nonbiological homogeneous catalysts seem to be energy-dependent.

We did not, however, elaborate on our findings to describe, in detail, the actions of time- and energy-dependent homogeneous catalysts. Nor did we show how these fitted into the general characteristics of time- and energy-dependence, as described above.

In this chapter, we will address these issues. We will draw on the previous chapter to discuss and describe the general operation of time- and energy-dependent catalysts. We will particularly examine why they are so different in their modes of action. We will also show how their properties relate to the general characteristics of mechanical ("time-dependent") and thermodynamic ("energy-dependent") processes. In so doing, we will seek to explain why enzymes are such extraordinarily efficient catalysts. In the final sections, we will examine the implications of these findings for biology and the field of homogeneous catalysis.

The current chapter builds on the previous chapter and is, therefore, intended to be read in close conjunction with it.

6.2 *TIME-* AND *ENERGY-DEPENDENT,* MULTICENTERED HOMOGENEOUS CATALYSTS

As noted in Section 3.3.2 of Chapter 3, all catalysts, whether they are biological or nonbiological, must necessarily perform the same basic processes. These processes involve:

(1) Reactant binding.
(2) Reactant activation. Activation involves the process whereby the catalyst polarizes the electronic structure of the reactant to thereby make it conducive to reaction.
(3) Transition state formation.
(4) Product formation and release, leading to regeneration of the catalyst.

These steps are schematically depicted in Figure 6.1(a). As noted in Figure 6.1(a), the slowest and least likely step in a *time-dependent* catalyst involves mechanically bringing the bound reactants into physical contact ("collision") with each other. That is, the catalytic rate is limited by a low *collision frequency*. This property originates because the reactants tend to bind and release the catalyst rapidly. Because of the brief residence time of the reactants on the catalyst, there is a low likelihood that they will come into physical contact with each other while they are simultaneously bound to the catalyst.

To compensate for a low catalyst-mediated collision frequency, such species can only generate a significant catalytic effect if they are energetically (thermodynamically) efficient. That is, they only act as catalysts if the few collisions that do occur have a high likelihood of yielding products.

In an *energy-dependent* catalyst by contrast, the key impediment to reaction is the need to overcome the threshold energy in the transition state known as the *Activation*

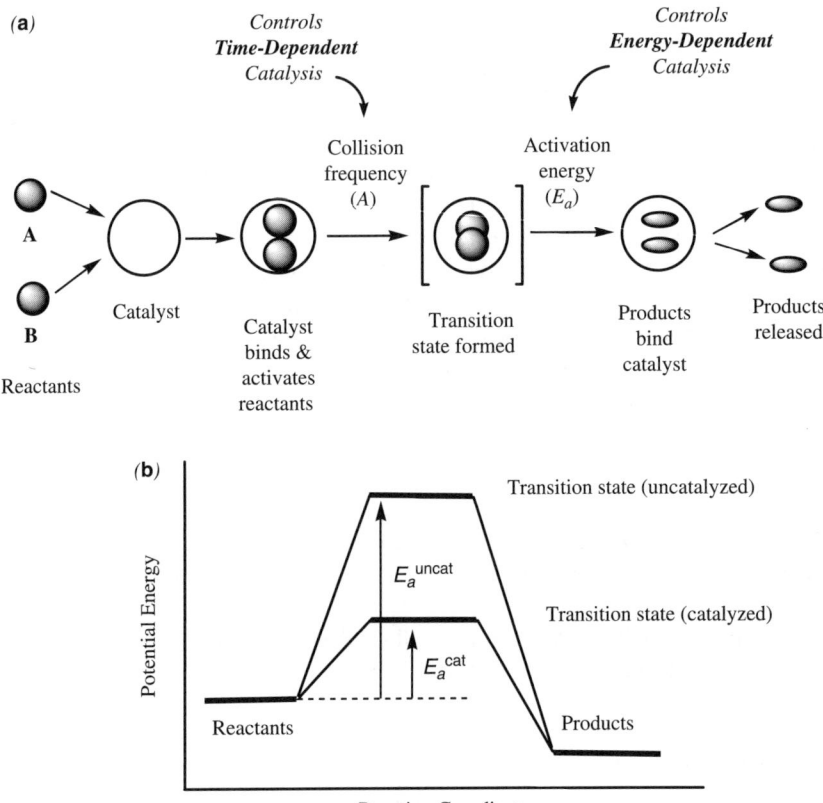

Figure 6.1. Schematic depiction of (a) a catalyzed, reactive encounter between two molecules, A and B, leading to the formation of products, and (b) the energy profile followed during the catalyzed reaction, showing the minimum threshold energy (the activation energy E_a^{cat}) needed for reaction, relative to the same threshold for the uncatalyzed reaction (E_a^{uncat}). In an *energy-dependent* catalyzed reaction, the step that comprises the greatest impediment to reaction involves overcoming the activation energy, E_a^{cat}. In a *time-dependent* catalyzed reaction, the greatest impediment involves bringing the bound reactants into collision with each other before they disengage from the catalyst. In this case, the key determinant is the *collision frequency*.

Energy or E_a [see Figure 6.1(a)]. In such catalysts, the reactants tend to bind and remain bound to the catalyst, undergoing many collisions with each other before one is successful and products are generated. The catalytic rate is therefore determined by the thermodynamic efficiency of the catalyst, not by the collision frequency, which is high.

We must make an important note here about the quantity known as the *activation energy* (E_a). The concept of an *activation energy* originated in collision theory [1]. It is, in fact, a theoretical construct that describes the proportion of molecules in a gas, which collide with sufficient energy to react in a noncatalyzed reaction. Thus, the activation energy E_a is calculated from the Arrhenius equation (6.1), where the pre-exponential term A describes the frequency of collisions [1].

$$k = A \exp\left(\frac{-E_a}{kT}\right) \quad (6.1)$$

In gases, molecules collide with random orientations relative to each other. The activation energy therefore provides an *average* measure of the energy required to bring about reaction during a collision involving *randomly oriented* reactants. This average incorporates both high energies, where the reactants collide with inopportune relative orientations, and low energies, where the reactants happen to be ideally oriented for reaction.

6.3 THE ACTION OF *ENERGY-DEPENDENT*, MULTICENTERED HOMOGENEOUS CATALYSTS

How, then, do the above processes play out in the catalytic action of a multicentered, energy-dependent, homogeneous catalyst? To answer this question, we will consider, by way of example, the catalysts in Scheme 6.1, which we noted in the previous chapter to likely be energy-dependent in their action.

In these species, the Cr ion catalytic groups in **1** or **2** bind to an epoxide or an azide (N_3) as depicted at the bottom of Scheme 6.1 [2]. In so doing, they *activate* these reactants, which means that the electronic arrangement of the reactants becomes polarized in such a way that they will react with each other if they are brought into physical contact in suitable orientations and dispositions. The extent of the activation provided depends on the nature of the catalytic group. A transition metal catalytic group such as a Cr ion will typically create a more extreme polarization of the reactant electronic structure than, say, an amino acid catalytic group.

Having bound and polarized the reactants, the catalyst then brings them into physical contact ("collision") with each other. To examine how this contact occurs, let us consider, first, the case of the monomeric catalyst **2**, which is substantially less active overall than **1a–c** at the same, equivalent concentration.

For catalyst **2**, collisions between the Cr-bound azide and epoxide reactants will occur with entirely *random orientations* because they will take place in open solution between free, monomeric species.

Most of these orientations will be *incorrect* for product formation, which means that the activation provided by the Cr-ion catalytic groups will be insufficient to bring about reaction in that particular orientation. In other words, in most collision orientations, the reactant *will not be polarized enough* to react. In such collisions, the catalyst will have provided *insufficient energy* to bring about the reaction.

In a smaller proportion of the collisions, however, the reactants *will* happen to be *correctly* oriented and disposed for reaction. In those cases, the activation provided by the catalyst *will* be sufficient to bring about reaction. That is, in such collision

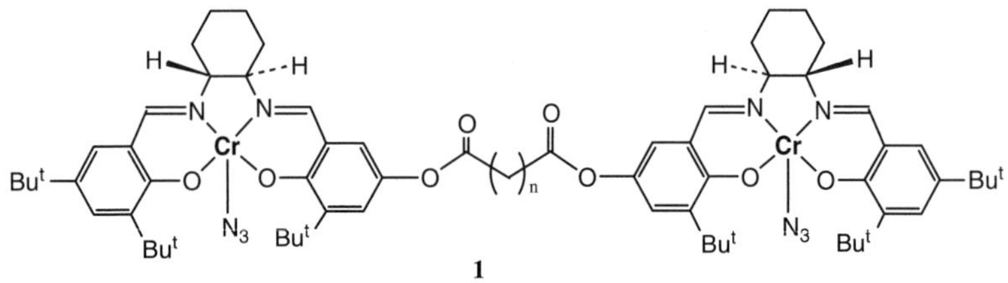

Scheme 6.1.

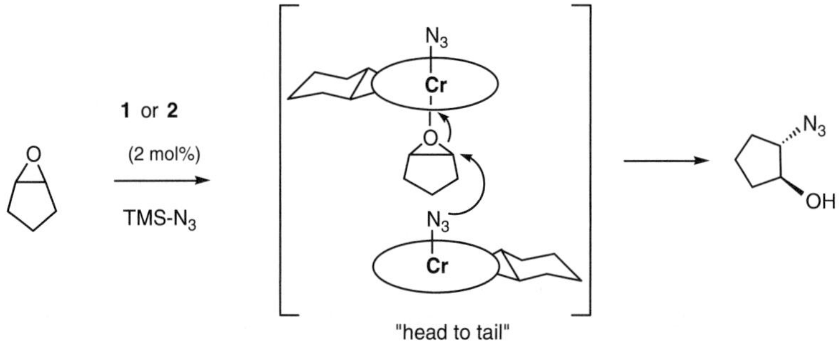

Intramolecular reaction rate constant (k_{intra}):

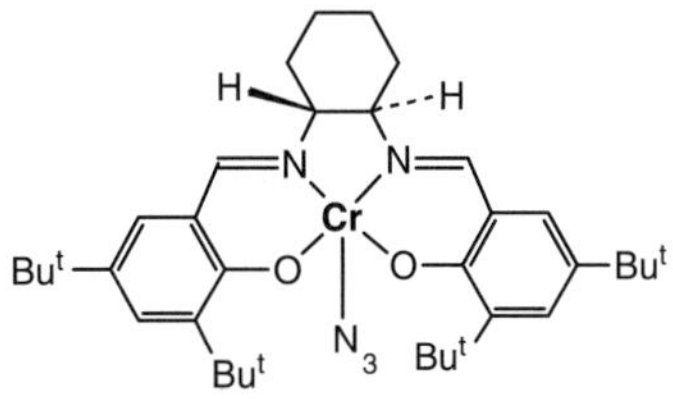

1a: 42.9×10^{-2} min^{-1}
1b: 4.4×10^{-2} min^{-1}
1c: 3.8×10^{-2} min^{-1}

orientations, the reactants *will be polarized enough to react with each other*. The catalyst will then have provided sufficient energy for reaction.

Thus, catalyst **2** will facilitate, on average, many collisions of its attached reactants before achieving a collision that leads to products. The successful collisions will have involved the bound reactants approaching each other along certain, optimum, or near-optimum trajectories and pathways. When these trajectories and pathways are followed, the energy provided by the catalyst during activation will exceed the minimum threshold required in the transition state and reaction will take place. When they are not followed, the transition state will be of insufficient energy for reaction.

Precisely which trajectories and pathways are suitable for reaction will depend entirely on the nature and the extent of the activation provided by the catalytic

6.3 THE ACTION OF *ENERGY-DEPENDENT*

groups. The stronger and more intense the activation provided by the catalytic groups, the wider will generally be the range of approach trajectories and pathways that will lead to reaction.

How do we know that this happens? How, for example, do we know that the reactants will only form products when they collide with each other in certain orientations? How do we know that the epoxide and azide reactants bind to the Cr-ions in **2** and remain attached during the catalysis?

Well, numerous studies over many years have confirmed that the approach orientations and trajectories of catalyst-bound reactants during collision have a significant effect on whether they react [3]. In Chapter 7 we will describe some of these studies and the theories that have resulted. Although a clear relationship is known to exist between the approach trajectories/pathways of the reactants and the activation provided by a catalytic group, the precise parameters and nature of the relationship has not, as yet, been quantified.

We can practically demonstrate the role of the approach pathways in catalysis by comparing **2** with the catalysts **1a–c** (Scheme 6.1). These species differ from **2** in having the catalytic groups physically connected to each other. They differ among themselves in the length of the tethers that link the two Cr-ion catalytic groups. The tethers will, obviously, control the way in which any reactants bound to the two Cr-ions approach each other during collision, as well as the frequency with which they approach each other. Their catalytic activity originates from a combination of these two factors.

Listed at the bottom of Scheme 6.1 are the comparable intramolecular reaction rate constants for **1a–c** [2]. As can be seen, a fairly substantial difference can be found in these constants for **1a** compared with **1b** and **1c**. Whereas **1a** has the highest rate constant, the tether in **1b** is clearly too short, whereas that in **1c** is too long.

If we compare **1b** with **1a**, then it is clear that, because it has three fewer atoms in its linker, the two Cr-ions in **1b** must spend a greater proportion of their time close to each other than do their counterpart Cr ions in **1a**. That is, the Cr-ion catalytic groups in **1b** must physically approach each other more frequently than is the case in **1a**.

The fact that **1a** has the higher rate constant can therefore only be because of the *manner* in which the Cr-ions approach each other in this molecule; this is governed by the steric conformations that are present and populated. In other words, a set of key conformational interconversions that bring about reactant collisions must be more populated in **1a** than in **1b**. These conformational changes must drive the attached and activated reactants along particular, suitable approach pathways into collisions that lead to reaction. Because these pathways are more likely in **1a**, it presumably has a higher rate constant than **1b**.

Although **1c** has, likely, the same set of key conformational interconversions available, these are presumably less populated than in **1a** because the longer tether diminishes the frequency with which the reactant-bound Cr groups approach each other.

Thus, the comparative example of **1a–c** indicates and supports the critical role played by the approach trajectories and pathways followed by the bound reactants at collision. The efficiency of a homogeneous catalyst depends on *how well and how often* it can bring its attached reactants into collision along these pathways.

The more frequently and the more reproducibly it can do so, the more thermodynamically efficient it will be.

As depicted schematically in Figure 6.2, the conformational freedom of catalysts like **1a–c** means that they cannot restrict the approach pathways of the reactants to this optimum. A variety of approach pathways are, instead, followed, resulting in collisions involving many different relative orientations between the reactants. Most of these collisions involve non-optimum approach trajectories, and they will therefore be ineffective in generating products.

However, a few collisions will happen to occur along the optimum or near-optimum pathways (in accordance with the prevalence of the relevant conformational changes as a proportion of all possible conformational changes). These collisions will result in product formation. To get to this point, it is necessary for the reactants to be bound to the catalyst for a sustained period of time.

To maximize the number of approach pathways that will result in reaction, it is also necessary for the catalyst to employ strongly activating catalytic groups. The more strongly the catalytic groups can polarize the reactants, the greater the variety of dispositions at collision that will result in reaction. For this reason, strongly binding and activating catalytic groups, like Cr-ions, are required. Only by binding and holding the epoxide and azide reactants for the duration of the many collisions that occur, can their reaction ultimately be facilitated.

In effect, therefore, **1a–c** and **2** rely on powerful catalytic groups that produce large inputs of energy to achieve a catalytic effect. With such groups, the catalyst binds the reactants in a sustained way and simply keeps knocking them together until they finally react with each other. The problem of facilitating the reaction is therefore *overwhelmed* through the application of a large input of energy.

How, though, do we know that the reactants bind the catalytic groups and largely remain attached during the catalytic process?

The fact that the monomer **2** generates a catalytic effect in open solution indicates, conclusively, that the reactants do not rapidly bind and release the Cr-ions. More accurately, it indicates that the Cr-ions have reactants attached to them for a significant proportion of their time. They may certainly release these reactants, but there can only be short gaps in the time during which they do not hold a reactant. How do we know that?

As noted in Section 5.6.2 of Chapter 5, in order to maintain a particular activity, the effect of weak binding interactions in a multicentered catalytic system must be counteracted by making the catalyst more structurally complementary to its transition state. Thus, whereas ***catalyst A*** in Section 5.6.2 need not have any particular structure because it enjoys strong binding interactions with its reactants, ***catalyst B***, which has much weaker binding interactions, must spend 81% of its time in a structure that complements its transition state.

This process occurs because a catalyst having strong binding interactions can hold its reactants through many different, unsuccessful collisions before a transition state leading to products is formed. Thus, such a catalyst can employ even monomeric catalytic groups in open solution. Such monomers will constantly bump into each

6.3 THE ACTION OF *ENERGY-DEPENDENT*

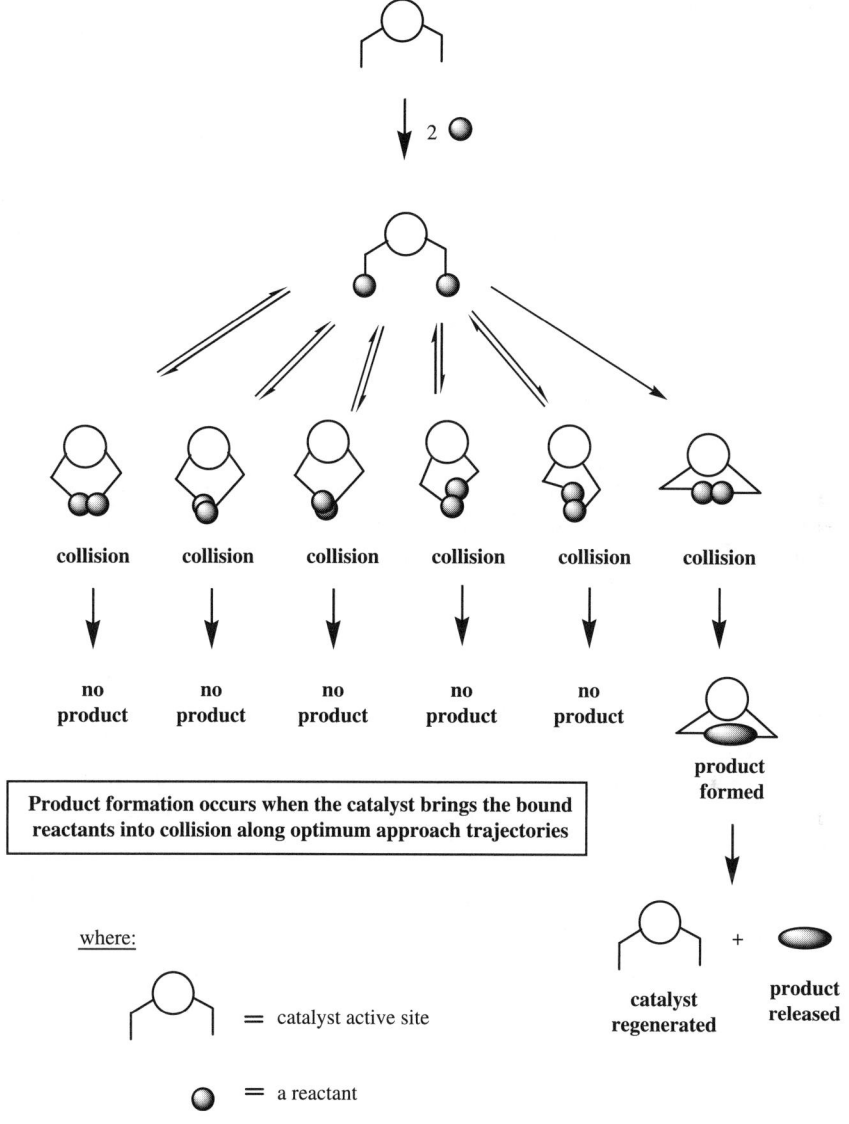

Figure 6.2. *Energy-dependent catalysis*: Schematic representation of the catalytic action in a conformationally unconstrained dicentered homogeneous catalyst. The catalyst first binds and holds the two reactants. Then it brings them into collision with each other multiple times. Because it is conformationally unconstrained, these collisions involve a wide variety of relative orientations. In only one set of these orientations do the reactants approach each other along trajectories and pathways that are suitable for reaction. That is, the *activation* provided to the reactants by the catalyst is such that reaction will only occur if the reactants collide along these particular pathways and trajectories. When this happens, the activation provided by the catalyst exceeds the minimum threshold energy and products are formed.

other until eventually their attached reactants collide in an orientation that is suitable for reaction.

By contrast, a catalyst having weak and dynamic binding interactions with its reactants does not have this luxury of time. It cannot hold its reactants for extended periods of time. At best, it holds its reactants for very brief slivers of time. If it is to generate a catalytic effect, it is critical that, at that juncture in time, it is set up to drive the bound reactants along optimum pathways into collision with each other.

In open solution, two monomeric species like **2** can clearly never spend more than a minuscule fraction of their time in a spatial arrangement that complements their transition state. Thus, if a monomeric catalytic group generates a catalytic effect, it *must* employ strong and sustained binding interactions with its reactants.

This requirement is confirmed by the fact that the intramolecular rate constants of **1a–c** are of similar magnitude to each other; that is, they are relatively *structure-insensitive* catalysts, which could not be the case *unless* the catalyst–reactant binding was strong and sustained.

Consider, by contrast, the structure sensitivity of the time-dependent catalyst **3** in Chapter 5. Although **3** generates a substantial catalytic effect, the modified species **9** (in Chapter 5) does not, nor does free ferrocene. This result can only be because of the brief residence time of the reactant on the catalytic group.

6.4 THE ACTION OF *TIME-DEPENDENT*, MULTICENTERED HOMOGENEOUS CATALYSTS

What about the catalytic action in a time-dependent, multicentered homogeneous catalyst? How does this action differ from that of an equivalent energy-dependent catalyst?

The key distinguishing feature of time-dependent catalysis is *dynamic reactant binding*. That is, the reactants constantly bind and release the catalyst. Their residence time, when they are fully attached to the catalyst, is typically short for each occasion of binding. As a result, the catalyst cannot simply hold the reactants and bump them together over and over again in many different orientations until they react. The reactant does not remain bound to the catalyst long enough for this strategy to be successful.

Instead, a new tactic must be employed. The catalytic groups must be arranged in space so that they can rapidly carry any attached reactants along optimum pathways and trajectories into collisions leading to product formation. This process must occur during the brief time that the reactant is fully bound to the catalyst. In effect, the catalytic groups must rapidly oscillate, back and forth, about a structure that is complementary to the transition state of the reaction.

Time-dependent catalysts therefore involve two separate and interdependent dynamic processes: (1) reactant binding and (2) catalyst flexing. As schematically depicted in Figure 6.3, successful collisions only occur when these two processes are synchronized. That is, they only occur when the reactants are fully bound to the catalyst at the precise instant that they are also brought into collision along the optimum or near-optimum approach trajectories.

6.4 THE ACTION OF *TIME-DEPENDENT*

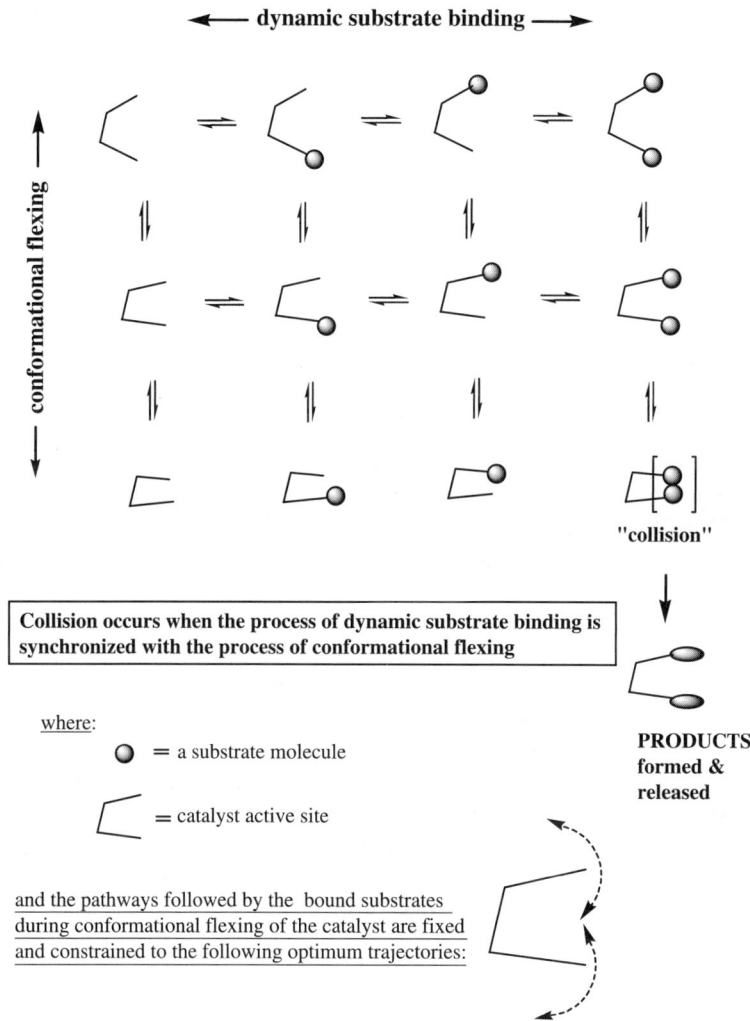

Figure 6.3. *Time-dependent catalysis*: Schematic representation of the two key processes present in many enzymes. Shown on the left, from top to bottom, is the process of *conformational flexing* of the enzyme active site. During conformational flexing, the groups in the enzyme that bind the substrate are repeatedly moved back and forth over the same, optimum pathways in space, relative to each other. Shown at the top, from left to right, is the process of *dynamic substrate binding*. In this process, the substrate repeatedly binds and releases the relevant enzyme groups. A successful "collision" occurs when these processes of *conformational flexing* and *dynamic substrate binding* are synchronized, that is, when the substrates are fully bound at the instant that they are brought into physical contact with each other.

This characteristic is very different from the energy-dependent analog described in the previous section. In energy-dependent catalysis, only one dynamic variable exists: catalyst flexing. Consequently, there is little need for synchronization with the process of reactant binding. As a result, the catalyst can have virtually any shape and oscillate through more or less any set of conformations, so long as *one* of those conformational changes brings the attached reactants into an optimally oriented collision with each other. The catalyst can even comprise monomeric catalytic groups that must find each other in open solution.

The addition of a new dynamic process—catalyst binding—drastically complicates the situation for a time-dependent catalyst. It no longer has an almost unlimited period of time to bring the bound reactants into successful collision with each other. Instead, it has only a very limited time available. To exploit that brief slice of time, the conformational change leading to reactive collision must be very highly populated. In other words, the catalyst must flex rapidly and exclusively to drive the attached reactants along the optimum or near-optimum pathways for collision leading to product formation. This is the only way in which it can create successful collisions during the time it has available.

Time-dependent, multicentered homogeneous catalysts must therefore necessarily be constrained to undergo rapid conformational flexing about a structure that is complementary to the reaction transition state. Only when they oscillate about such a structure will they carry any attached reactants along the required trajectories for reactive collision.

It is for this reason that the active site of many enzymes are complementary to their transition states, as was first noted by Pauling [4]. Without such complementarity, the substrates will not be driven into successful collisions leading to products in the brief periods that they are fully bound.

It is also the reason that enzymatic catalysis is a very *structure-sensitive* form of catalysis. That is, it explains why small changes in the structure of the active site tend to have disproportionately large effects on the catalytic activity. Such changes alter the approach trajectories of the bound reactants, so that they are no longer near optimum, with resulting sharp declines in catalytic activity.

What becomes clear, then, is that the key impediment in a time-dependent catalyst does not involve overcoming the threshold energy of the reaction. Rather, it involves binding the reactants and then bringing them into a successful collision with each other.

In effect, spatial and temporal optimization in time-dependent catalysts diminishes the energy barrier of the reaction to the point that it becomes irrelevant to the catalysis. Instead, achieving the spatial and temporal optimization becomes the key limiting process.

We will now discuss several implications that originate from this mode of action.

6.4.1 The Activation Energy E_a Does Not Provide a True Measure of the Threshold Energy in Time-Dependent Catalysts

As noted, the approach of the reactants in catalysts that are complementary to their transition states (including many enzymes) is not random. Instead, it is carefully controlled with the participating orbitals on the bound and polarized ("activated") reactants favorably oriented for reaction.

The energy barrier that must be overcome during the reaction is therefore not formally comparable with the *activation energy* E_a, which is a measure of the *average* energy barrier that must be traversed during *randomly oriented* collisions between the reactants.

Instead, it is best compared with the energy barrier involved in an uncatalyzed gas-phase collision where the reactants are *optimally oriented* for reaction. Although this energy cannot be practically measured, it must, necessarily, be substantially less than the average represented by E_a. Indeed, it is likely to be at or near the *lowest possible* energy for reaction, depending on how perfectly complementary to its transition state the catalyst is during flexing.

6.4.2 Weak and Dynamic Binding and Activation Is Sufficient to Fulfill the Threshold Energy in Time-Dependent Catalysts

We can now also understand why time-dependent catalysts, like many enzymes, can employ weakly and dynamically binding and activating catalytic groups, like amino acid residues.

In energy-dependent catalysts, like **1a–c** and **2**, strong activation (polarization) of the reactant electronic structure is needed. The stronger and more extreme this activation is, the greater will be the variety of approach trajectories that will result in reaction. Because a catalyst like **2** depends on statistical chance to achieve a suitable collision, the more approach pathways that will result in reaction, the more efficient will be the catalysis. There is, therefore, a significant advantage to maximizing the number of approach trajectories that will result in product formation. This is true also for **1a–c**.

However, in a comparable time-dependent catalyst, where a single collision-inducing conformational change occurs repeatedly and rapidly, there is no need to maximize the diversity of viable approach trajectories. Moreover, there is no need for sustained binding and activation, because the applicable conformational change occurs extremely often. In short, there is no need for extreme activation of the type required in energy-dependent catalysts.

Thus, weakly binding and activating groups, like amino acids, are eminently suitable as catalytic groups. Such amino acids provide sufficient activation to bring about reaction by dynamically attached reactants.

Amino acids of this type do not act as catalytic groups in manmade catalysts simply because manmade homogeneous systems are generally insufficiently complementary to their reaction transition states. That is, the key collision-inducing conformational changes occur too infrequently in these catalysts to allow for the use of amino acids as catalytic groups. If, however, one could achieve the necessary complementarity in a manmade homogeneous catalyst, amino acids could, certainly, be used.

To put this another way, because they are highly complementary to their transition state, time-dependent catalysts like many enzymes enjoy a *much lower* energy requirement for reaction than do comparable manmade catalysts. This is because they are conformationally limited to driving their attached reactants into collision along only one, optimum approach pathway. For this reason, they can use catalytic groups that provide only very weak *activation*, such as amino acid residues.

Thus, it is not only the extent of activation provided by an amino acid that is relevant. The *way* in which the catalytic action is set up also influences how much activation is required to initiate the catalysis. We will discuss and explain this concept in the next section.

6.4.3 Transition State Formation in a Time-Dependent Catalyst Can Be Thought of as a Coordinated Mechanical Process

As noted, time-dependent, mechanical action–reaction sequences exist in many different forms in chemistry and in human experience in general. They typically rely on a coordinated progression of events in which the first event causes the next one, which causes the next one, and so forth (see Chapter 1).

Perhaps the best example of such a process is a series of dominoes stacked in a line. As noted in Chapter 1, each step in a sequence of dominoes falling is individually favored by the spatial positioning of the dominos and the timing of their fall. The overall process need not be thermodynamically favorable. The dominos may, for example, be stacked up the side of a hill, resulting in the initial impulse being ultimately transferred to a point of higher potential energy. Coordinated mechanical processes of this type are therefore considered to operate independently of the underlying energy landscape.

Time-dependent homogeneous catalysts, like many enzymes, display a similar type of effect in their catalytic action. The energy barrier to reaction in time-dependent catalysts is sufficiently low that the collision frequency controls the process, which manifests itself in the form of an extreme structure sensitivity. In effect, the catalytic action in such cases is entirely dependent on the bound and activated reactants being in the right place at the right time, which is conceptually equivalent to the example of the dominoes. Everything must be in the right place at the right time, for reaction. And, when everything *is* in the right place at the right time, a very small impetus—for example, weak activation by amino acid groups—is sufficient to set off the process. Indeed, the more optimum is the setup, the smaller need be the initial impulse that sets it off.

The extent of the activation that must be provided by the catalytic groups is therefore dependent *entirely* on how optimum are their spatial arrangements, approach pathways, and timing. The more perfectly these features are synchronized, the smaller need be the activation provided by the catalytic groups.

In effect, time-dependent catalysts, like many enzymes, depend on *coordinated, synchronized* actions on the part of their catalytic groups. In this respect, their mode of action closely resembles that of a machine.

6.4.4 Time-Dependent Catalysts Are Machine-Like (Mechanical) in Their Catalytic Action

Machines operate by repeatedly and reproducibly carrying out the same actions over and over again. In so doing, they follow the same pathways, at the same time intervals, and generate the same outputs. Materials are fed into the machine and products are ejected from the machine in a synchronized and dynamic way.

Key Point. Time-dependent homogeneous catalysis is machine-like in its action. It involves catalytic groups (and their attached reactant functionalities) being repeatedly impelled by the effects of conformational flexing along the same, optimum pathways in space and time. In parallel with that, the catalytic groups dynamically bind and release the reactants.

When these parallel processes of catalyst flexing and binding are synchronized, then the catalyst is, effectively, a machine. Reactants are fed in, bound, activated, and carried along optimum pathways into collisions with each other that almost inevitably produce products. The resulting products are then dynamically ejected, and new reactants are taken up for the next cycle.

On some occasions, the reactant functionalities will not be fully bound to their respective catalytic groups at the instant of collision, which will result in a failed collision attempt. Before the next collision cycle, however, that functionality will have left and another will have taken its place.

This "collision" and "ejection" process is perfectly analogous to a *mechanical device*. It is, arguably, also the true origin of Michaelis–Menten kinetics.

6.4.5 The Origin of Michaelis–Menten Kinetics in Time-Dependent Catalysts

We can now better understand the processes that take place in the Michaelis–Menten kinetics that characterizes many enzyme catalysts. In the most general case, the sequence of events is that shown in Equation (6.2), where ES^* represents the point at which the bound and activated substrate functionalities collide.

$$E + S \xrightleftharpoons{K'} [ES] \xrightleftharpoons{K''} [ES^*] \xrightarrow{k_2} E + P \tag{6.2}$$

where E = enzyme, S = substrate, and P = product.

In the first step, a substrate binds to the enzyme, in a rapid and dynamic equilibrium, described by the constant K'.

In the following step involving K'', the fully bound and activated substrate is brought into collision by the conformational flexing of the enzyme. In the example given in Equation (6.2), "collision" involves pulling apart or rearranging the substrate S.

It is important to note that the point of collision $[ES^*]$ also represents the transition state because virtually every collision must necessarily be successful in order to realize a catalytic effect. Moreover, $[ES^*]$ is reached only if the equilibrium involving K' and the equilibrium involving K'' are synchronized. That is, the formation of $[ES^*]$ involves two dynamic processes (K' and K''). The step involving K'' is, consequently, a *mechanical* equilibrium, not a *chemical* one. That is, the point of collision (namely, $[ES^*]$) is only reached if the catalyst can bring the bound functionalities into reactive contact (K'') during the brief time that they are simultaneously fully bound to the enzyme (K'). This is *wholly dependent* on (1) the conformational flexing of the enzyme, (2) the binding of the substrate functionalities, and (3) their synchronization. It is *not dependent* on the energy profile of the reaction as depicted in Figure 6.1(b).

The slowest and least likely step in this overall process is simply getting to the transition state [ES^*]. Once the transition state is reached, it will form the product P because of the optimized or near-optimized approach trajectories and pathways employed by a time-dependent catalyst.

Contrast this situation with what happens in an analogous energy-dependent catalyst:

$$C + R \xrightarrow{K'} [CR] \xleftrightarrow{K''} [CR^*] \xrightarrow{k_2} C + P \qquad (6.3)$$

where C = catalyst, R = reactant, and P = product.

In the first step, a reactant (R) binds to the catalyst (C) in what is effectively a one-way process (K'). The catalyst then brings about numerous collisions [CR^*] during conformational flexing (K''). Only the final collision generates a transition state that leads to products.

The slowest and least likely step in this case is not achieving a collision *per se*. Many such collisions occur, but only a few of them are successful. The slow step involves achieving a collision that involves forming a successful transition state that yields products. That is, the slowest step involves k_2.

The role of the mechanical equilibrium K'' is quite different in Equations (6.2) and (6.3). In Equation (6.2), it must synchronize with substrate binding to bring about collision. In Equation (6.3), it is the means by which the energy barrier for the reaction is overcome.

A critical difference in these roles can be found. In Equation (6.3), the equilibrium K'' is the means by which a thermodynamic (energetic) equilibrium is established between the substrate and the products. That is, equilibrium K'' is a mechanical process operating under, and carrying out, a thermodynamic imperative. It ensures that the substrate, the transition state, and the products exist in a thermodynamic equilibrium with each other.

In Equation (6.2), however, equilibrium K'' is the means by which collision is achieved. That is, it is a mechanical process that operates simply as a mechanical process, not as a proxy for an underlying thermodynamic one. The thermodynamics of the reaction do not figure in equilibrium K'' in Equation (6.2).

In other words, the bound substrate functionalities involved in equilibrium K'' in Equation (6.2) collide with each other in a purely physical, mechanical way according to external, conformational impulses that are unrelated to the reaction itself. They are brought into collision by the conformational flexing of the enzyme. The overall process of the reaction is governed by the rate at which the enzyme does this. This rate has nothing to do with, and is entirely unrelated to, the energetics of the chemical reaction. In effect, the reaction process involves an equilibrium that is *purely mechanical* in character. It does not mediate a thermodynamic equilibrium between the reactants and the products.

The concept of an equilibrium that is purely mechanical in character has not enjoyed wide recognition in chemistry and catalysis to date. Instead, equilibria have generally and universally been considered to be thermodynamic in character. However, we have in Equation (6.2) an example of a reaction whose rate is determined by an equilibrium that is purely mechanical in character. This equilibrium occurs only because the mechanics and circumstances within the system are such that it has no alternative but to occur.

Key Point. Much of the confusion about Michaelis–Menten kinetics can be traced back to an assumption that the equilibrium described by K'' in Equation (6.2) is thermodynamic in nature. It is not. As in the machine-like example described above, it is wholly physical and mechanical.

This finding leads to another important point. Since a reversible thermodynamic equilibrium does not exist among the reactants, the transition state, and the products, the key assumption on which transition state theory is based does not hold [5]. Thus, a catalyst of the type shown in Equation (6.2) is not subject to transition state theory.

Key Point. Instead, so-called "nonequilibrium" conditions pertain [5]. That is, the process of the chemical reaction does not occur in response to, or as part of, a thermodynamic equilibrium. It occurs simply because the mechanics of the situation cause it to occur.

6.4.6 Time-Dependent Catalysts like Many Enzymes Display All of the Characteristic Hallmarks of Mechanical Processes

In the analogy of the machine, we see numerous common features that time-dependent catalysts share with other mechanical processes.

Key Point. The operation of time-dependent catalysts is, first, path-dependent and spatiotemporal in character (that is, to do with time and space). Moreover, because the pathways involve relatively small energy barriers, the interaction of the participating elements is, effectively, independent of the energies involved in the reaction process. That is, the underlying energy landscape of the process is, effectively, flat (see Chapter 1). The process is therefore governed by the physical and the mechanical interaction of the participating elements.

Viewed in isolation, the whole catalytic process seems to be preordained; that is, it is deterministic in character. But this is, in fact, only because all elements of the system are set up to interact correctly in a coordinated, synchronized manner. If they did not interact in this way, catalysis would not be possible.

Thus, the individual components, such as the amino acid catalytic groups in many enzymes, achieve a remarkable synergy; that is, their effect in combination is substantially larger than the simple sum of their individual capacities. This synergy is harnessed to finesse the problem of facilitating a reaction by the substrate. As a result, the catalytic process can be achieved using truly minimal energy inputs, like those provided by amino acids. These inputs are substantially smaller than those produced by the transition metal ions that must typically be employed in manmade, multicentered, homogeneous catalysts like 1a–c and 2.

Like a machine, any changes in the components or the timing of this system will rapidly degrade its action. Thus, time-dependent catalysts are subject to a nonlinear structure sensitivity; small modifications in their structure have a disproportionately large effect on their catalytic abilities. In the case of many enzymes, they are also temperature-dependent in their efficiency, which is likely caused by temperature-induced alterations to their timing.

Like the cogs and fabrication point in a machine, structural complementarity is a critical feature of the mechanical action. In the case of time-dependent catalysis, the catalyst active site must have a shape that is structurally complementary to its transition state. It is necessary to ensure that, within this site, the reactants are assembled and brought into reactive contact in a manner that will ensure their formation of the new structure. If the active site is not structurally complementary to its transition state, then the new structure cannot be formed (catalysis is impossible). This is also true for the fabrication point in a manufacturing machine.

In short, time-dependent catalysts like many enzymes are truly and in all respects mechanical and not thermodynamic in their action. They are governed by the rate at which their bound reactants physically collide with each other. They involve specific, reproducible pathways in space and time. They depend on synchronization and synergy. They minimize the amount of energy required by finessing the problem of a chemical reaction using coordinated actions. They display all of the characteristic hallmarks of mechanical devices.

6.4.7 Additional Insights into Enzymatic Catalysis: The Bidirectionality of Enzymatic Catalysis Originates from the Mechanical Nature of the Catalytic Action

These findings prospectively also explain the bidirectionality of many enzymatic catalysts. Just as a line of dominoes can, equally easily, be made to fall over in one direction as the other, so enzymes can catalyze both the forward and the reverse of a reaction. This action occurs because mechanical processes are not dependent on the underlying energy landscape.

Key Point. Thus, if the required binding and flexing processes are synchronized, catalysis can proceed in either the forward or the reverse direction, depending only on the initial impulse. This impulse is provided by the nature of the species that the catalyst encounters. If it consistently encounters reactants, it will consistently convert them to products. However, if, under different circumstances, it consistently encounters products, it will, more or less equally consistently, convert them to reactants. The outcome of the catalytic process therefore depends only on the local concentration of the reactants or on the products about the catalyst. This local concentration (along with the binding affinity of the catalyst for the reactants and products) provides the initial impetus that decides the direction of the process.

This explanation is consistent with previous suggestions that, in some cases, forward and reverse transits of the transition state may conceivably occur prior to product release [6(b)–(d)].

6.4.8 Additional Insights into Enzymatic Catalysis: Many Enzymes Select the First-Encountered Transition State, Rather than the Lowest Energy Transition State

In Section 3.3.5 of Chapter 3, we noted that, at the most fundamental level, time- and energy-dependent catalysts are distinguished by the manner in which they select their most-favored transition state. Time-dependent catalysts select the *first-encountered transition state*, whereas energy-dependent catalysts select the *most stable transition state*.

Key Point. This process must also be true for many enzymes, except that, in these cases, the most stable transition state is often also the first-encountered transition state. On a superficial level, it may therefore seem that the above distinction is academic. However, it is arguably important to understand this difference because:

(1) *It cuts to the core of theoretical enzymology and the manner in which experimental results are interpreted*
(2) *It is the only means with which one can develop truly authentic mimics of enzymes in manmade catalytic systems*

The latter consideration is particularly important. An ability to mimic wholeheartedly the catalytic feats of enzymes prospectively opens extraordinary new opportunities in many fields, including alternative energy and transformational chemical technologies.

6.5 THE IMPORTANCE OF RECOGNIZING *TIME-DEPENDENT* CATALYSIS

In many cases it may seem that the distinction between energy-dependent and time-dependent catalysis sounds awfully abstract. However, ignorance of this distinction has had far-reaching and very real consequences. Two examples may be cited out of the many that exist, to illustrate this contention.

In the 1970s, Ronald Breslow at Columbia University established the field of *biomimetic chemistry*, which sought to develop catalysts that mimic the actions of enzymes [7]. The guiding philosophy assumed an energy-dependent catalytic action, in accord with the common wisdom. Thus, catalysts were designed to have structures that were complementary to their transition states, as proposed by Pauling [4]. This was, of course, perfectly accurate and necessary. However, no account was taken in *biomimetic chemistry* of the need for, or the role of, dynamism in binding, activating, and collision. Such dynamism is, in fact, the dominating feature in time-dependent catalysts and therefore key to its authentic mimicry.

A substantial literature of such *biomimetic* catalysts now exists [8]. Although many are certainly impressive catalysts, it is generally acknowledged that none achieve the selectivity of enzymes and few achieve rates comparable with enzymes. A key reason can now be largely understood to be that they do not involve highly dynamic catalyst–reactant interactions.

The second example involves another group of, mainly, organic chemists, who have, since the 1960s, been studying the origin of rate accelerations in *intramolecular* organic reactions [9]. These researchers were also guided by an assumption that all catalysis is energy-dependent. As a result, an entire field of comparative mechanistic organic chemistry was developed that focused exclusively on the connection between the energy barrier of chemical reactions and the structure of their catalysts. No account was taken of the dynamism of the catalyst–reactant interactions involved. At least partly for this reason, this field has become bogged down in trying to rationalize compilations of comparative rate constants that represent, according to one researcher, "one of the largest and most variant bodies of unexplained data in physical organic chemistry" [10].

Key Point. The scientists involved in biomimetic chemistry and organic intramolecular rate studies have not been well served by a one-sided focus on energy-dependent catalysis. A clear and systematic description of the key role played by catalyst–reactant dynamism is urgently required. In particular, chemists need to understand how such dynamism leads to time-dependent catalysis. It is the central intention and impetus of this volume.

6.6 TIME-DEPENDENT CATALYSIS IS VERY DIFFERENT TO ENERGY-DEPENDENT CATALYSIS AND THEREFORE SEEMS UNFAMILIAR

The atypical controlling step in time-dependent catalysis manifests itself in ways that are very different to its energy-dependent counterpart. Given that the latter is extremely familiar to chemists and biochemists, many of these differences will be considered by current-day researchers to be unusual and even peculiar. Indeed, in many respects, the features of time-dependent catalysis flatly contradict the key principles of energy-dependent catalysis that are so deeply ingrained into chemistry, biochemistry, and chemical engineering students at present.

For example, the sort of arithmetic employed in conventional chemical catalysis, in which the bond enthalpies of the reactants and the products are tallied in order to establish whether a reaction is thermodynamically favored, is *irrelevant* in time-dependent catalysis. The catalytic process is not influenced by the underlying thermodynamic landscape. Indeed, the very concept of bond enthalpy itself is of dubious value in time-dependent catalysis because bond enthalpy data are not determined under approach-optimized conditions.

As previously, energy-dependent catalysts employ a brute force approach. They overwhelm energy barriers with an even greater input of energy. Time-dependent catalysts, however, finesse their way around the problem by arranging and choreographing a dance that finds a path through these obstacles.

Key Point. Whereas researchers must focus on the thermodynamic aspects of reactions in energy-dependent catalysis, they must, in time-dependent catalysis, focus on the structure and conformational flexing of the catalyst, as well as on the dynamism of its reactant binding.

Time-dependent catalysis is therefore truly a foreign and a very different entity to the energy-dependent catalysis that is familiar to chemists, biochemists, and engineers at present. Given the subtlety and singularly unusual nature of time-dependent catalysis, it is not surprising that researchers have grappled so intensely and so long to explain the key underlying processes in enzymatic catalysis. In Chapter 7, we describe some of these struggles and interpret their results in light of the conclusions we have drawn regarding time-dependent catalysis. Their fundamental and underlying unity is demonstrated.

6.7 CONCLUSIONS FOR BIOLOGY

The recognition of time-dependent catalysis in many enzymes sheds some light on their biological role. It becomes clear that such enzymes do not act to speed up the attainment of a thermodynamic equilibrium between reactants and products. That is, whereas most catalysts facilitate a thermodynamic, reactant-product equilibrium without changing its position, time-dependent enzymes act without regard to such equilibria. In being machine-like and mechanical in their operation, they simply eliminate local numerical excesses of reactants or products and bring the system into a state that reflects the relative binding affinities of the enzyme for the reactants and products [11]. *Thus, they will catalyze both the forward and the reverse of a reaction depending only on the circumstances present. In effect, enzymes act to achieve a local "kinetic" equality. This may be expected to seldom equate to the thermodynamically expected equilibrium concentrations.*

6.8 CONCLUSIONS FOR HOMOGENEOUS CATALYSIS

These findings also have implications for the rational design of new, nonbiological, homogeneous catalysts.

Key Point. A gap seems to exist in the field of nonbiological homogeneous catalysis. Time-dependent catalysis has not been recognized or seriously explored. The only form of catalysis that has been earnestly studied in manmade homogeneous catalysts is energy-dependent in character. Catalysis of this type requires strongly binding and activating catalytic groups, such as transition metal ions. However, weakly and dynamically binding catalytic groups can be used in time-dependent catalysts. Thus, there is a much wider range of possible catalytic groups than has hitherto been considered.

How, then, does one create a time-dependent catalyst? We will discuss this matter in greater detail in Chapter 10. For now, however, we can note that time-dependent homogeneous catalysts display two key properties that must be properly achieved:

(1) Their catalytic groups must be arrayed so that, during conformational flexing, they rapidly oscillate about a structure that is complementary to the transition state.
(2) The catalytic groups must employ weak and highly dynamic individual catalyst–reactant interactions. The catalytic groups must, nevertheless, provide sufficient activation of their bound reactants to bring about their reaction when brought into contact with suitable dispositions and orientations.

Attribute (1) ensures that any bound reactants are transported along pathways in space that are most likely to result in collisions yielding products. Attribute (2) provides for the required catalyst–reactant dynamism. This dynamism limits the duration of reactant binding and thereby limits the time available for a reactive collision between the reactant functionalities.

The combined effect of these measures is to create a system whose energy efficiency allows for a faster reaction than its kinetic dynamism. The system is, therefore, time-dependent.

6.9 THE "IDEAL" HOMOGENEOUS CATALYST

At this stage, one may ask the following question: Why not develop a catalyst that is complementary to its transition state *and* which binds its reactants strongly and sustainably? Such a catalyst would enjoy the best of both worlds: It would employ the lowest energy approach pathways during collision, without any limitations in the time available to bring about collision. Thus, it would have the advantages of both time- and energy-dependent catalysts, without the disadvantages of either. Such a catalyst would, in effect, be an "ideal" catalyst.

This point, is perfectly valid except insofar as the dynamism of catalyst–reactant binding is likely also to extend to dynamism in catalyst-product binding. That is, products formed by such a catalyst will not be as readily and rapidly ejected as would have been the case if weak and dynamic binding interactions are employed.

We will, nevertheless, examine in greater detail the concept of the "ideal" catalyst in 8.7.2 in Chapter 8, *vide infra*.

6.10 CONCLUSIONS FOR THE CONCEPTUAL UNITY OF THE FIELD OF CATALYSIS

It seems that the prevalence of energy-dependent catalysis in nonbiological systems and the comparative difficulty of studying the catalytic action in biological systems has obscured the fact that many enzymes seem to employ a fundamentally different catalytic action to comparable nonbiological catalysts. This potentially explains why it has been so difficult to unite the fields of enzymatic catalysis and nonbiological homogeneous catalysis; they may involve different modes of action that require different theoretical treatments. We will discuss in Chapter 9 how this recognition may affect our understanding of the conceptual unity of the field of catalysis.

ACKNOWLEDGMENTS

The authors thank for their insightful comments: Chuck Dismukes and Damian Carrieri (Princeton University), Steve Benkovic (Pennsylvania State University), and Bob Williams (Oxford University).

REFERENCES

1. Atkins, P. W. *Physical Chemistry*, Oxford University Press, 1978, pp. 864–866 & p. 899–925.
2. Konsler, R. G.; Karl, J.; Jacobsen, E. N. *J. Am. Chem. Soc.* **1998**, *120*, 10780.
3. Some examples of the studies and theories describing an angular-dependence in the approach pathways of reactants during collisions that generate products, are given in: (a) Bruice, T. C. *Acc. Chem. Res.* **2002**, *35*, 139, and references therein; (b) Storm, D. R.; Koshland, E. E. *Proc. Natl. Acad. Sci. USA* **1970**, *66*, 445; Storm, D. R.; Koshland, D. E. *J. Am. Chem. Soc.* **1972**, *94*, 5805; Dafforn, A.; Koshland, D. E. *Proc. Natl. Acad. Sci. USA* **1971**, *68*, 2463; Hoare, D. G. *Nature* **1972**, *236*, 437; (c) Milstein, S.; Cohen, L. A. *Proc. Natl. Acad. Sci. USA* **1970**, *67*, 1143.
4. (a) Pauling, L. *Chem. Eng. News* **1946**, *24*, 1375; (b) Pauling, L. *Nature* **1948**, *161*, 707.
5. Pilling, M. J.; Seakins, P. W. *Reaction Kinetics*, Oxford University Press, 1996, p. 66–85, p. 106–120.
6. See, for example: (a) Bruice, T. C.; Lightstone, F. C. *Acc. Chem. Res.* **1999**, *32*, 127; (b) Lau, E.; Bruice, T. C. *J. Am. Chem. Soc.* **2000**, *122*, 7165; (c) Bruice, T. C.; Benkovic, S. J. *Biochemistry* **2000**, *39*, 6267; (d) No author stated, *Catal. Lett.* **2001**, *76*, 111, and references therein.
7. Breslow, R. *Acc. Chem. Res.* **1995**, *28*, 146, and references therein.
8. (a) Breslow, R. *Chem. Soc. Rev.* **1972**, *1*, 553; (b) Breslow, R. In *Bioinorganic Chemistry, Advances in Chemistry*, Vol. 100, American Chemical Society, 1971; (c) Breslow, R. *Acc. Chem. Res.* **1980**, *13*, 170; (d) Breslow, R.; Dong, S. D. *Chem. Rev.* **1998**, *98*, 1997; (e) Breslow, R. *Chem. Rec.* **2001**, *1*, 3.
9. See, for example: (a) Kirby, A. J. *Adv. Phys. Org. Chem.* **1980**, *17*, 183; (b) Kirby, A. J. *Angew. Chem. Int. Ed. Engl.* **1996**, *35*, 707, and references therein.
10. Menger, F. M. *Acc. Chem. Res.* **1985**, *18*, 128.
11. See, for example: Mudd, S. H., Mann, J. D. *J. Biol. Chem.* **1963**, *238*, 2164.

7

UNIFYING THE MANY THEORIES OF ENZYMATIC CATALYSIS. THEORIES OF ENZYMATIC CATALYSIS FALL INTO TWO CAMPS: ENERGY-DEPENDENT ("THERMODYNAMIC") AND TIME-DEPENDENT ("MECHANICAL") CATALYSIS

GERHARD F. SWIEGERS

7.1 INTRODUCTION

The most remarkable catalysts known to man are the natural catalysts of biology, enzymes. Enzymes display two very dramatic effects that have captured and focused the attention of researchers: *molecular recognition* and *rate acceleration*. It was Linus Pauling who first observed that enzymes selectively stabilize ("recognize") the transition states that they form during catalysis [1]. Quite naturally, he concluded that this must also be the cause of the rate accelerations displayed by enzymes relative to similar but inactive proteins [1].

However, the extent to which reaction rates are increased by enzymes as opposed to nonenzyme proteins cannot be easily quantified. Indeed, it is extraordinarily difficult to evaluate precisely. The mechanism of an enzyme-catalyzed reaction is often different to that of any comparable nonenzyme. Moreover, the structural

---

Mechanical Catalysis: Methods of Enzymatic, Homogeneous, and Heterogeneous Catalysis,
Edited by Gerhard F. Swiegers
Copyright © 2008 John Wiley & Sons, Inc.

complementarity that Pauling considered so obviously important in enzymes does not exist outside of biology and is not easily duplicated in nonbiological catalysts.

Many biochemists therefore concluded that enzymes could simply not be accurately replicated and that all attempts in this respect were futile [2]. A recent publication recounts a conversation with a prominent biochemist in which the remark was made: "It does not matter how elegant an enzyme model you organic chemists construct, no biochemist will ever pay much attention to it" [2].

Such anti-reductionist sentiments did not deter numerous chemists, mainly organic chemists, from trying to understand and replicate the fundamental effects at work in enzymes. A wide range of proxy and model systems were developed in organic chemistry to study enzymatic rate enhancements [2–5]. These systems drastically simplified the problem. In so doing, however, they could be accused of inaccurately mimicking the problem [2]. Still worse, they were open to suggestions that they described an altogether different problem [2].

This possibility seemed to be confirmed when different proxy systems led to different explanations for the efficiency of enzymatic catalysis [2]. Indeed, a veritable profusion of theories and hypotheses have been developed [3]. At least 21 different theories of enzymatic catalysis were noted to exist in a 1989 publication [3]. More have been added since [2(b),6,7].

Difficulties, therefore, developed in this field. Academic disputes erupted among the champions of the various theories [2(a)]. Some of these disputes could never be entirely resolved.

The problem is that the fundamental, underlying common processes that create the phenomenon of enzymatic catalysis are well hidden from view, because of the complexity and diversity of the natural systems. These common processes manifest themselves in various nonobvious and unintelligible ways in all participants in the phenomenon. However, so do a lot of other processes that may coincidentally also be common in some, or even all, of the examples.

Complex and diverse natural systems derived from evolution have historically proved extremely challenging to understand. For example, the principles of flight by heavier-than-air objects were not properly understood for the entirety of human existence until simple wind tunnel experiments by the Wright brothers clarified the Law of the Aerofoil. At that point, the key principles behind the phenomenon of flight were suddenly revealed in simple, intelligible terms and manmade flying machines could be built.

The natural phenomenon of enzymatic catalysis is clearly even more challenging to disentangle. To unwrap this puzzle, one must—like the Wright brothers—find a means to elevate the key underlying processes to prominence. The best way to do this has, historically, been by comparisons with nonbiological, manmade systems. However, independent studies that do not involve references to manmade systems are also essential.

Depending on the complexity of the natural system, such examinations will typically give snapshots of what is important. The key to understanding the phenomenon is then to assemble the snapshots into a conceptual whole in the same way that one may put together a puzzle.

7.2 THEORIES OF ENZYMATIC CATALYSIS

In the two previous chapters, we have sought to assemble the available snapshots of enzymatic catalysis into a single, coherent, and meaningful picture. We have demonstrated that two classes of homogeneous catalysis exist: *energy-dependent* ('thermodynamic") and *time-dependent* ("mechanical") catalysis. The latter is controlled by the *catalyst-mediated collision frequency*, that is, by the rate at which the catalyst brings the bound reactants into reactive collision with each other. Many enzymes appear to employ a catalytic action of this type. By contrast, most multicentered, manmade homogeneous catalysts are energy-dependent in their action. That is, their catalytic action is governed by the threshold energy that must be overcome in the transition state of the reaction.

The distinction between an energy-dependent and a time-dependent catalytic action has not been explicitly recognized to date. It has simply been assumed in both homogeneous and enzymatic catalysis that all catalysts are energy-dependent. Given that this difference is now clear, it is feasible to reconsider the major theories of enzymatic catalysis and to see whether they can be assembled into a more coherent, conceptual whole.

In this chapter, we will seek to do that. We will consider how the major theories of enzymatic catalysis overlap and knit together in light of the distinction between energy- and time-dependent catalysis. Our intention in doing so is to coalesce the various theories of enzymatic catalysis into a coherent, unified, and accurate whole.

7.2 THEORIES OF ENZYMATIC CATALYSIS

The catalytic properties of enzymes have been used for centuries in roles such as the brewing of beer. The term "enzyme" was first coined in 1867 [27]; it comes from the Greek expression for "in yeast." In this section, we will discuss the major theories and underlying themes that have been proposed for enzymatic catalysis over the years. These theories are presented in rough chronological order. A more comprehensive listing of theories is provided in Table 7.1 [8–26].

7.2.1 Adsorption Theory

The earliest attempts to explain the action of enzymes envisaged that substrate molecules become adsorbed on the surface of enzymes in much the same way as occurs in modern heterogeneous catalysis [28]. In this way, the substrate was thought to become concentrated in and about the enzyme, thereby realizing a greater opportunity for reaction. For this to happen, however, enzymes would have to have many different binding sites on their surface, which did not gel with the specificity of enzymatic catalysis. Moreover, it later became apparent that many enzymes have only one substrate binding site.

7.2.2 "Lock-and-Key" Theory

In 1894, Emil Fischer recognized that the specificity of glycolytic enzymes means that they must have a particular shape into which the substrate fits exactly. He

TABLE 7.1. Theories of Enzymatic Catalysis Not Described in the Text [3]

Theory Title	Ref.
Near Attack Conformers	[4](a)
Covalent Catalysis	[6]
Approximation	[8]
Togetherness	[9]
Rotamer Distribution	[10]
Anchimeric Assistance	[11]
Distance Distribution Function	[12]
Stereopopulation Control	[13]
Substrate Anchoring	[14]
Vibrational Activation	[15]
Vibrational Activation Entropy	[16]
Orbital Perturbation Theory	[17]
Group Transfer Hydration	[18]
Electrostatic Stabilisation	[19]
Electric Field Effect	[20]
Catalytic Configurations	[21]
Directed Proton Transfer	[22]
Coupling between Conformational Fluctuations	[23]
Gas Phase Analogy	[24]
Torsional Strain	[25]
Freezing at the Reactive Centres of Enzymes	[26]

subsequently proposed his "lock-and-key" theory, which argues that enzyme–substrate interactions are similar to a lock and key [27]. That is, the enzyme active site has a structure which is complementary to that of the substrate.

This theory explained why enzymes are so specific in their catalysis; they will only bind and, therefore, transform certain substrates. But it did not explain in a meaningful way their high catalytic rates.

7.2.3 Haldane's Strain Theory

In 1930, J. R. Haldane proposed a modification of Fischer's lock-and-key principle [29]. He suggested that enzymes and their substrates bind in somewhat imperfect, rather than perfect, complementarity with each other. As a result, the substrate becomes somewhat strained upon binding. Haldane put it thus [29]: "Using Fischer's lock and key simile, they key does not fit the lock quite perfectly, but exercises a certain strain on it." The strain that is created in the substrate becomes released during reaction. Thus, according to Haldane, the reaction of a bound substrate is accelerated by the release of the strain that is induced during binding.

This proposal is still widely accepted today. In support of it, many biochemistry textbooks note that the catalytic ring-opening of many strained cyclic monomers in polymer chemistry is drastically faster than that of their comparable unstrained monomers under otherwise identical conditions [27]. This acceleration is unequivocally caused by the strain that is released during reaction.

7.2.4 Pauling's Theory of Transition State Complementarity

In 1946, Pauling noted that many enzyme active sites were, in fact, structurally more complementary to the catalytic transition state than they were to the substrate, *per se* [1]. He proposed that this allowed enzymes to bind their transition states strongly, thereby reducing the activation energy of the catalytic process and speeding up the rate [1]. This proposal was in accord with the dominant intellectual paradigm of catalysis, which suggests that the key impediment to a catalytic reaction is overcoming the threshold *activation energy* (E_a) during the formation of products.

The suggestion that enzymes bind their transition states more strongly than their substrates has been intensively investigated over the years. Studies have shown that enzyme–substrate binding is relatively strong, with $1/K_M$ in the order $10^{4\pm3}$ M^{-1} [27]. However, transition state binding does, indeed, seem to be substantially stronger, with $1/K_M$ calculated to fall in the range $10^{16\pm4}$ M^{-1} [4,6,27,30].

Thus, Pauling's proposal seemed to be vindicated. Tight binding of the transition state also explains why specific substrates are cleaved more rapidly than nonspecific substrates.

However, for Pauling's proposal to be truly verified, it was still necessary to demonstrate its generality. In other words, the same effect must be shown to exist in nonbiological catalysts. Several theories described below were developed with that aim in mind. They seemed to confirm that Pauling's proposal was, indeed, valid in nonbiological catalysts.

7.2.5 Koshland's Induced Fit Theory. Fersht's Concept of Stress and Strain

In 1958, D. E. Koshland proposed that enzyme active sites are not rigid but relatively flexible [31]. Thus, the binding of a substrate induces a conformational change in the three-dimensional structure of the enzyme active site. In effect, the active site distorts in order to accept the substrate, which then itself also becomes physically misshapen.

In light of this proposal, Fersht subsequently offered a modification of Haldane's strain theory [32]. He suggested that the distortion in the enzyme during binding is greater than the distortion in the substrate. Thus, the enzyme is *strained*, but the substrate is only *stressed* [32].

According to Fersht, the formation of the transition state is therefore also, or mainly driven by, the need to reduce strain in the enzyme. That is, the impetus to reduce strain in the enzyme impels the system toward a particular transition state.

7.2.6 Intramolecularity

Starting in the 1960s, several authors noted that when the rates of unstrained intramolecular organic reactions were compared with those of the corresponding bimolecular reaction, large rate enhancements were observed [2–6].

An example in this respect is depicted in Figure 7.1, which compares the rate constants for the unimolecular (k_1) and bimolecular (k_2) formation of anhydride from the succinate half ester **A** and two molecules of the equivalent acetate ester **B** [33]. As can be seen, k_1 is five orders of magnitude larger than k_2. Some reactions show accelerations of 10^8-fold, which is believed to be similar to the rate accelerations brought about by enzymes.

Effects such as these led researchers increasingly to study the reason that intramolecular reactions proceed so much more rapidly than their intermolecular counterparts. Such studies were believed to provide potential insights into the efficiency of enzymatic catalysis.

In effect, intramolecular organic reactions were harnessed to serve as proxy's for mechanistic studies of enzymatic reactions.

The underlying proposal outlined by the concept of intramolecularity is that enzymes effectively sequester their substrates from solution, thereby increasing their "effective concentration" in the enzyme active site [2,27]. The observed rate accelerations are proposed to be caused by this increase in concentration.

An *effective molarity* (EM) parameter was subsequently devised in an attempt to quantify intramolecularity [2]. EM was defined as k_{intra}/k_{inter} for corresponding intra- and intermolecular reactions operating under otherwise identical mechanisms [2].

Although for any one reaction the intramolecular pathway was, essentially, always faster than the equivalent intermolecular pathway, the absolute EM values of a series of different reactions seemed, in general, to display few clear trends.

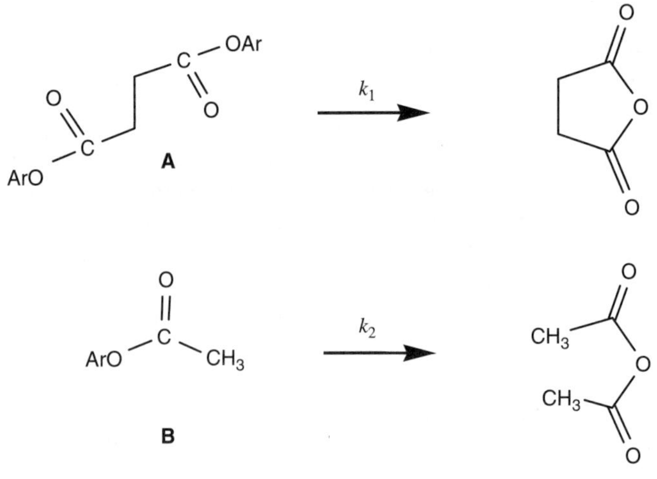

$$\frac{k_1}{k_2} = 10^5 \text{ M}$$

Figure 7.1. Rate acceleration during the intramolecular formation of an anhydride relative to its corresponding monomer.

7.2 THEORIES OF ENZYMATIC CATALYSIS

Thus, detailed compilations of EM values [34] indicated that they fluctuated wildly from reaction to reaction. They could be very small (<0.3 M) or very large (>10^{10} M) or anything in between. EM values were hypothesized to depend on a range of influences, including ring size, substituent, solvent, and reaction type [2]. Clear structure-activity relationships could not be discerned.

No theory could therefore be devised to rationalize EM values. A complaint raised in a 1985 publication [2](a) stated that a recently published and comprehensive list of EM values "represents one of the largest and most variant bodies of unexplained data in physical organic chemistry."

Although EM values could not be rationalized over the entire body of data, explanations and theories based on intramolecular rate accelerations could be derived for specific reactions.

7.2.7 Orbital Steering

In 1971, Koshland suggested that a possible explanation for the high rates of intramolecular reactions may lie in a severe angular dependence in the interaction of the relevant orbitals on the participating substrate functionalities [35]. Thus, a perfectly linear approach of the interacting orbitals creates a much lower energy barrier for reaction than is the case if the orbitals are displaced from linear by even a few degrees. In other words, even a minor optimization of the trajectories with which the substrate functionalities approach each other during collision should generate a much lower energy barrier. This should translate, in turn, into a far more rapid overall reaction.

This theory came to be known as *orbital steering* [35]. Its central thesis is that optimum orientation and disposition in the reactant functionalities at the point of collision provide for a transition state with the very lowest possible energy, which decreases the activation energy E_a and speeds up the reaction rate.

Orbital steering induced an immediate negative outcry upon its publication. Bruice, for example, attacked the formal equations proposed for orbital steering on the grounds that they involve unreasonably large force constants [36]. Support for Koshland came from Hoare, however [37], who showed that the inclusion of solvent effects confirmed the accuracy of orbital steering. Monte Carlo methods were later also used to show that small changes in geometry may, indeed, result in large variations in activity [12].

A wide variety of theories describing variations on this "optimum approach" theme have, moreover, since been proposed. These include theories entitled *rotamer distribution* [10], *stereopopulation control* [13], and others.

Underlying all of these conceptions is the suggestion that the energy barrier to reaction in an enzyme manifests itself in the form of an extreme structure sensitivity. That is, the more structurally complementary the enzyme is for the transition state, the lower is the activation energy E_a of the reaction.

Although the formal basis of the concepts surrounding orbital steering are not as yet settled, there seems little doubt that a strong relationship *does* exist between the way reactants approach each other and the threshold energy needed to bring about their reaction during the subsequent collision.

7.2.8 Entropy Traps

The role of entropy in intramolecular and enzymatic rate accelerations was perhaps first raised by Westheimer in 1962 [38]. It has since been studied with particular thoroughness. Perhaps the definitive explanation involving entropy was provided in 1971 by Page and Jencks [39] who reasoned that the translational and rotational freedom enjoyed by reacting functionalities in an *intermolecular* reaction is frozen out in an equivalent *intramolecular* reaction. Using standard formulas, they showed that the difference in the $\Delta G^{\ddagger}$ of the reaction of two nonpolar molecules in water was capable of generating a 10^8-fold increase in rate [39].

Thus, enzymes, with their many weak binding interactions, may be considered "entropy traps." This entropic contribution potentially explained the large reaction rates displayed by enzymes.

7.2.9 The Proximity (Propinquity) Effect

The first step in an enzyme-catalyzed reaction involves the bringing together of the enzyme and the substrate. This binding process entails localizing the reacting groups in close proximity to each other. Thus, Bruice noted in 1976 [40] that it is common sense for a "proximity" or "propinquity" effect to speed the rate of their reaction.

Unfortunately, although the term "proximity" seems to be self-explanatory, it has proved impossible to define formally in the context of enzymatic catalysis. Quantification of the "proximity effect" has therefore not been possible.

7.2.10 "Coupled" Protein Motions

Since at least the 1970s molecular motions have been proposed to be involved in enzymatic catalysis [7]. Extensive experimental and theoretical studies have addressed the linkage between conformational changes in enzymes and catalysis [7]. Evidence has been collected for a network of coupled motions that facilitate the chemical reaction [7]. This network seems to compose of fast, equilibrium thermal motions that contribute to slower conformational changes that are involved in the reaction [7].

Recent nuclear magnetic resonance (NMR) studies on enzymes such as *cyclophilin A* and *triosephosphate isomerase* have detected conformational fluctuations within the active site that occur on a time scale that correlates strongly with the microscopic rate of substrate turnover [41,42]. These conformational changes therefore seem to occur in tandem with the catalysis. Using the available structural data, it is possible to predict the approach trajectories of bound substrates before collision in such enzymes [41].

Although the connection between conformational changes in enzymes and their catalytic action has been well established, the formal role of these conformational fluctuations in the catalytic process has not been clearly defined. Researchers have, for example, suggested that the conformational changes act to exclude solvent from the active site, thereby accelerating the formation of the transition state and the reaction overall [42,43]. Loops or domains have been proposed to close over the catalytic site and thereby to isolate the reactants from the bulk solution, exclude solvent molecules, gate the entry and release of reagents to the catalytic site, and for a finite time, make the exit of reactants impossible [42,43]. During the period that the reactants are excluded from the bulk solvent, it has been further suggested that many forward and reverse transits of the transition state may occur, thereby explaining the incidence of bidirectional catalysis in enzymes [43].

An alternative or perhaps a complementary view, which was perhaps best summarized by Williams [44] in 1993 (but also espoused by others [7,45]), is that enzymes are "designed" dynamic mechanical devices. The purpose of the protein movement is to guide the substrate continuously through a limited set of motions that are controlled by the active site in much the same way that a glove is guided onto the fingers of a hand. In this way, the substrate is coaxed into the transition state with all of the tensions and stress energies induced in it. The formal role of protein motions is consequently proposed to activate the relevant substrate functionalities and bring them into reactive contact along a trajectory that is optimum for formation of the transition state. The concept of a machine-like action has also been invoked by others [46].

7.2.11 The Spatiotemporal Hypothesis

In trying to make sense of the extraordinary and anomalous variations that exist from system to system in the earlier-mentioned *effective molarity* (EM) parameter, Menger proposed in 1985 a *spatiotemporal hypothesis* [2]. This concept suggests that the rate of a reaction between two functionalities is proportional to the time that they spend in van der Waals contact with each other. Thus, time and distance constitute the key components of reactivity. In effect, intramolecular reaction rates are suggested to be greater than their intermolecular analogs because the reactant functionalities spend more time in physical contact with each other.

7.3 THEORIES EXPLAINING ENZYMATIC CATALYSIS FALL INTO TWO CAMPS: ENERGY-DEPENDENT AND TIME-DEPENDENT CATALYSIS

The first thing that becomes immediately apparent when one considers the above theories is that many of them involve concepts that derive from energy-dependent catalysis. That is, they hypothesize systems that are dominated by the drive for the lowest energy pathway, in which the successful formation of the transition state is the rate-determining step.

Theories of this type include those of Pauling, Haldane, Fersht, the *entropy theories*, as well as a certain vein of thought in *intramolecularity*. All invoke energy-based explanations for enzymatic rate accelerations or use energy-dependent nonbiological examples. Several theories explicitly employ *transition state theory*.

Orbital steering and related theories also make conclusions based on an energy-dependent conceptualization. But these theories are enabling rather than descriptive; they can potentially explain the origin of either energy- or time-dependence in catalysis.

All of the remaining major theories, however, cite notions compatible with time-dependent processes, although this is inevitably expressed in a vague and indirect way. Theories of this type include *coupled protein motions*, the *spatiotemporal hypothesis*, the *proximity effect*, and to some extent, the concept of *effective molarity*, which underlies the studies of intramolecular reaction rates. The common feature of these theories is the idea that the catalytic action of enzymes is fundamentally dependent on time, position, and pathway, not on the energy barrier of the reaction.

For example, *coupled protein motions* proposes that a series of sequential and very specific actions in the enzyme plays the central role in enzymatic catalytic action. These actions are unrelated to the formal activation energy of reactant collision.

Although *coupled protein motions* is fairly direct in its contradiction of the energy-dependent model of catalysis, other theories are less so.

For example, the *effective molarity* parameter essentially implies that the reactant concentration within the active site is the key fundamental determinant of enzymatic rates. We know, however, that the *activation energy* of a transition state is formally unaffected by the reactant concentration. By contrast, *collision frequency* is fundamentally and intrinsically dependent on the reactant concentration. Thus, the theory of *intramolecularity* may be construed to propose, at heart, that *collision frequency* is rate-determining in enzymes. It does not state that directly, presumably because this was not apparent to the proposers and would contradict the prevailing paradigm of catalysis, namely energy-dependence.

It is not only the intellectual dominance of energy-dependent catalysis that has hindered the clear expression of theories associated with time-dependence. Suitable terminology has also been profoundly and fundamentally absent. One cannot properly describe something for which no words or concepts exist.

A good example in this respect is Bruice's *proximity effect*. As noted, the term "proximity" sounds substantial. However, it proved impossible to define formally in the context of enzymatic catalysis. What is absent from this theory is the accompanying notion of time. The concept of proximity implicitly, but not explicitly, expresses the notion of substrates being near to each other at the *same point in time*. The ultimate proximity is being in physical contact with each other, or colliding. Why is proximity important therefore? Because becoming increasingly proximate is what a collision is all about and the frequency of such collisions is what governs enzymatic catalysis.

Menger, in effect, recognized that implication and stated it in his *spatiotemporal hypothesis*. In an inspired intuitive leap, his theory also notes the importance of the time available to achieve contact.

Both the *proximity effect* and the *spatiotemporal hypothesis* can therefore be interpreted to imply a dependence on the collision frequency.

Thus, the major theories of enzymatic catalysis divide into two camps. The theories of Pauling, Haldane, Fersht, the *entropy theories*, and a vein of thought in intramolecular

7.3 THEORIES EXPLAINING ENZYMATIC CATALYSIS FALL INTO TWO CAMPS

studies, describe energy-dependent catalysis and variations thereof. *Coupled protein motions*, the *spatiotemporal hypothesis*, the *proximity effect*, and the *effective molarity* concept in intramolecular rate studies express time-dependent catalysis in various guises.

Given that many enzymes seem to employ a time-dependent catalytic action, we can justifiably reassess the reasoning of the energy-based theories as explanations for the rate accelerations observed in enzymes. We can also explain in greater detail how the time-dependent theories describe enzymatic catalysis.

7.3.1 Haldane's Strain Theory and Fersht's Concept of Stress and Strain Are Valid Explanations for Rate Accelerations but Do Not Seem to be Responsible for the Rate Accelerations of Many Enzymes

Haldane's strain theory and Fersht's concept of stress and strain invoke a thermodynamic driver to explain enzymatic rate accelerations, namely, the release of pent-up energy in the form of conformational strain. As such, they can formally only be relevant for *energy-dependent* catalysts in which transition state formation leading to product formation is the slowest step. In such a case, the release of strain in the reactant will, certainly, lead to an increase in the reaction rate.

This can be practically illustrated by the fact that the nonbiological, catalytic ring-opening of strained cyclic monomers is always dramatically accelerated relative to nonstrained analogs. The more strained the monomer, the greater is the rate acceleration.

The effect of pent-up strain within a reactant is to increase its potential energy. As noted in Chapter 3 (Section 3.2.6 and Fig. 3.6), this causes the activation energy to be decreased, with a resulting acceleration in the overall reaction rate of such a system.

However, such an effect can only be important when the reaction is energy-dependent; that is, where its rate is determined by the activation barrier (E_a) that must be overcome during the reaction. Where this is *not* the slowest step, a purely thermodynamic impetus like strain-release *cannot* influence the catalyst-mediated collision frequency and therefore cannot influence the overall rate, because the presence and release of strain does not have a direct effect on the *collision frequency* of the reaction. If the overall rate is set by the collision frequency, then strain can clearly not lead to a rate acceleration.

If many enzymes are time-dependent catalysts, their rates cannot be determined by the energy involved in transition state formation. As such, the existence and release of strain during transition state formation cannot be directly involved in the catalytic rates achieved by enzymes.

Having said that, we should add a qualifier. Strain *may* have an *indirect* effect on enzymatic rates in that it could conceivably shorten the *residence time* during which the substrate is bound and available for catalysis. That is, by binding in a strained configuration, the substrate may remain bound to the enzyme for a lesser time than would occur in the absence of such strain. This effect could certainly influence the reaction rate. However, it cannot be compared with the rate accelerations achieved in the nonbiological ring-opening of strained cyclic monomers, which are energy- and not time-dependent processes.

7.3.2 Theories Based on Reaction Entropy Are Valid Explanations for Rate Accelerations but Do Not Seem to be Behind the Rate Accelerations of Many Enzymes

Many biochemistry textbooks refer to enzymes as "entropy traps" and rationalize their catalytic properties in these terms [27,39]. Entropy is proposed to contribute to a decrease in the activation energy, E_a, of the transition state, thereby speeding up the rate [27,39]. However, if formation of the transition state is not rate-determining, then an increase in its stabilization can surely not influence the overall catalytic rate. That is, if the rate is dependent on the *collision frequency*, then entropy can only influence the overall rate if it somehow enhances the collision frequency.

In cases where the collision frequency and the energetic efficiency of an enzyme are close to being balanced, it is certainly true that the entropy of reaction may tilt the balance and push the enzyme out of the realm of *energy-dependence* and into the realm of *time-dependence*. That is, the additional stabilization imparted to the transition state by an entropic contribution may result in the collision frequency becoming rate-determining.

However, once time-dependence is established, an entropic contribution of this type can surely not influence the catalytic rate.

Entropy can, of course, influence the rate of *reactant binding* and activation and thereby affect the overall catalytic rate. A binding entropy of this type ($\Delta S_{binding}$) would alter the free energy of reactant binding ($\Delta G_{binding} = \Delta H_{binding} + T\Delta S_{binding}$), thereby increasing the likelihood of concurrent reactant binding and activation, amplifying the collision frequency and increasing the rate. Such a contribution may not, however, constitute an "entropy trap" *per se*.

Studies relating the influence of reaction entropy in model nonbiological homogeneous catalysts to enzymatic rate accelerations are therefore generally flawed by an unsuitable choice of control. Nonbiological catalysts are largely energy-dependent and subject to the controlling influence of E_a. Entropic changes will necessarily have a significant influence on their overall rates. But such changes will not directly affect many enzymes, which are time-dependent and therefore unaffected by thermodynamic enhancements of this type.

7.3.3 Experiments Studying Intramolecular Reaction Rates Were Probably Often Conceptually Contradictory

A remarkable feature of the experiments employed to test the *effective molarity* concept in intramolecular rate studies is that they were, in many cases, probably internally contradictory.

By this is meant that the *intramolecularity* concept sought to demonstrate that the catalytic rates of enzymes were dependent in a fundamental way on reactant concentration. In other words, they sought to demonstrate that reactant concentration was involved in the rate-determining step in catalysis by enzymes.

Reactant concentration does, indeed, affect the *collision frequency*, which is rate-determining in time-dependent catalysis. But it does not formally affect the *activation energy* E_a, which is rate-determining in the energy-dependent, nonbiological chemical systems that were generally used in the studies employed to develop the concept. Reactant concentration does, of course, affect the overall rate of an energy-dependent processes, but this is an incidental "flux-generating" process, not a fundamental determinant as it is in a time-dependent process.

In effect, therefore, intramolecular rate studies can be construed as having tried to prove a philosophical concept that did not exist in most of the examples tested.

This is not to say that all intramolecular reactions were energy-dependent. Many of the acid- or base-catalysts employed in such reactions may well have bound the substrate only very briefly [2–6], thereby making the reactions time-dependent. The resulting mixture of energy-dependent and time-dependent catalysis clearly confused the situation and contributed to the wide range and variability of the observed EM parameter. In a following section, we will discuss the origin of the observed fluctuations of the EM parameter in greater detail.

7.3.4 Theories of "Coupled" Protein Motions and Machine-Like Catalytic Actions Seem to Be Generally Accurate Descriptions of Enzymatic Catalysis

As noted theories explaining enzymatic catalysis in terms of dynamic and coupled protein motions fundamentally describe time-dependent and not energy-dependent catalytic actions. As such, they must be considered broadly accurate descriptors of the catalytic action of many enzymes.

We can now formally explain the involvement of protein motions in enzymatic catalysis. Such motions mediate and mechanically create the collisions between activated reactants during the catalytic process. A successful collision is possible in each cycle of the conformational flexing of the enzyme. Whether it occurs depends only on whether all of the relevant substrate functionalities are bound to the enzyme at the precise instant of collision. Thus, as noted in Chapters 5 and 6, enzyme rates are formally set by the extent to which enzyme flexing and substrate binding are synchronized.

Protein motions are consequently a critically important component of the rate-determining step in enzymatic catalysis. They do not simply *promote* catalysis. Nor do they occur *in tandem* with catalysis. Rather, they are an *integral part* of the catalytic action. They participate in the rate-limiting step.

As such, protein motions also determine, in part, the overall properties of enzymatic catalysis. The more rapid these motions, the faster can be the rate-limiting step and the quicker the catalysis. Exactly how fast it will be depends, of course, on the extent of its synchronization with substrate binding.

This is not to say that the exclusion of solvent or the gating of the active site or events of similar ilk referred to previously do not take place in enzymatic catalysis. They may well do. And in many cases, they may be the most important process in particular, individual enzymes. The same is true for the step of product release, which is rate-determining in several enzymes. However, the *common feature* of the catalytic action of enzymes is an underlying dependence on the collision frequency; this is determined, at least in part, by protein motions. In other words, if a noncommon action is not rate-limiting in a particular enzyme, then the collision frequency generally will be.

7.4 STUDIES VERIFYING PAULING'S THEORY IN MODEL SYSTEMS ARE CORRECT, BUT DESCRIBE ENERGY-DEPENDENT AND NOT TIME-DEPENDENT CATALYSIS

In the last few decades, numerous studies have been published that seem to confirm the general validity of Pauling's theory of transition state complementarity in nonbiological systems [2–6]. Most of these studies have involved nonbiological intramolecular organic reactions.

Experiments involving such reactions produce drastic rate increases as the system is made more structurally complementary to its transition state. As noted, this is largely because of the control that the catalyst exerts on the trajectories and orientations of the reactant during their approach to each other immediately before collision.

Although perfectly accurate, these studies involve manmade, model, catalytic systems. As such, they largely describe energy-dependent rather than time-dependent catalysis. Thus, they indicate the accuracy of Pauling's concept in the realm of energy-dependent catalysis rather than in time-dependent catalysis.

This point is illustrated in Figure 7.2, which schematically plots the overall catalytic rate (the dark curve) and the constituent underlying rates (the light curves) of an intramolecular catalytic system as a function of its structural complementarity for the transition state.

On the left-hand side of Figure 7.2, the system is poorly complementary to its transition state. As noted in Sections 5.6.2–5.6.3 of Chapter 5, under such circumstances, transition state formation is necessarily the slowest step and it therefore determines the overall rate. That is, energy-dependence prevails.

In moving toward the right of Figure 7.2, the system becomes more complementary to its transition state. This movement decreases the *activation energy* E_a. The rate of transition state formation therefore increases, leading to an accompanying increase in the overall rate. In this region, large, exponential rate accelerations will be seen as predicted by the exponential term in Equation (7.1), which determines the reaction rate of an energy-dependent catalyst.

$$k_2 = k_0 \exp\left(\frac{-E_a}{kT}\right) \qquad (7.1)$$

7.4 STUDIES VERIFYING PAULING'S THEORY IN MODEL SYSTEMS

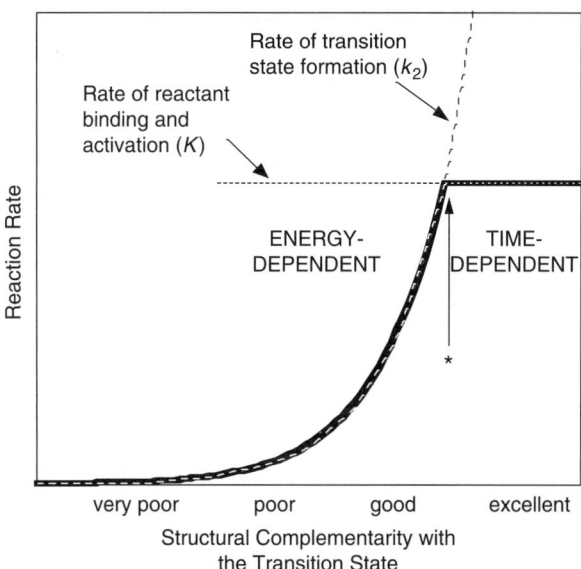

Figure 7.2. Schematic depiction of the overall reaction rate (dark curve) of a catalyst as a function of its structural complementarity with its transition state. The broken curves show the rates of the constituent, underlying steps. The constants K and k_2 refer to the reactions shown in Equations (5.4) and (5.5) of Chapter 5 and Equations (4.1) and (4.2) in Chapter 4. The asterisk marks the change from energy-dependence (left) to time-dependence (right).

However, these increases reflect the influence of energy-dependence and not of time-dependence. As such, they do not describe the conditions under which many enzyme catalysts operate.

At the asterisk in Figure 7.2, the rate of successful transition state formation equals the collision frequency. This point, therefore, marks the changeover from energy-dependence (on the left of Fig. 7.2) to time-dependence (on the right of Fig. 7.2).

To the right of the asterisk, the transition state is sufficiently stabilized that the collision frequency becomes rate limiting. Many enzymes operate within this realm of catalysis.

These studies therefore confirm Pauling's proposal of exceedingly well-stabilized transition states in enzymatic catalysis. But they do not explain just how well stabilized they actually are, nor their role in the catalysis, because they do not take account of dynamism in the catalyst–reactant interactions. In fact, they do not explain what the catalytic process and, more importantly, what the external properties of the catalyst will look like once a proper mimicry is achieved.

An important point must be made here concerning attempts to mimic enzymatic action in manmade catalysis. The stronger and more sustained the individual catalyst–reactant interactions employed in a manmade, biomimetic catalyst, the further to the right will lie the changeover point denoted by the asterisk in Figure 7.2. This is significant because the further to the right that this point lies, the more *difficult* it will be to achieve time-dependence in a biomimetic homogeneous catalyst. That is, the *more perfectly complementary* the catalyst will have to be to its transition state in order to be time-dependent.

Key Point. **Thus, researchers wishing to design authentic biomimetic catalysts must use the weakest individual catalyst–reactant binding interactions that can reasonably be used. These interactions can, obviously, not be so weak that they provide insufficient activation to initiate the catalytic process. In such a case, the catalytic effect would be eliminated entirely.**

In a later chapter, we will discuss in detail the interplay between structural complementarity and catalyst–reactant dynamism in time-dependent catalysts.

7.5 THE ANOMALY DESCRIBED IN THE SPATIOTEMPORAL HYPOTHESIS ORIGINATES, IN PART, FROM THE ONSET OF TIME-DEPENDENCE

As noted, Menger identified an anomaly in the theories seeking to model enzymatic catalysis in nonbiological systems. Rate acceleration data from individual intramolecular organic systems can be readily rationalized according to the extent of transition state stabilization. However, compilations of the *effective molarity* values in separate systems [34] display no clear relationship in this regard. Data are, instead, apparently randomly distributed. The *spatiotemporal hypothesis* was developed in an attempt to resolve this contradiction [2].

The anomaly can now be understood to orignate at least in part, by the onset of time-dependence, whose existence has not been previously considered. The point at which a system changes from energy-dependence to time-dependence (the asterisk in Fig. 7.2) will necessarily vary according to the strength of the individual catalyst–reactant binding interactions employed. Time-dependence will consequently commence at different catalytic rates for each of the acid- or base-catalysts used in the many different intramolecular organic reactions studied.

Thus, it is possible to observe, in each system examined, a large and exponential rate increase as the structural complementarity for the transition state is improved. At the same time, it is possible for the largest rate/s in each case to be seemingly unrelated to other, apparently similar cases, because, although the smaller rates are energy-dependent, the larger rates, and especially the largest rate, are often time-dependent. They are therefore not influenced by the activation energy E_a, and no relationship exists in this regard. Instead, they are determined by the collision frequency, and this is different in each system.

As described by Menger [2], the only possible general explanation for the anomaly in EM values is that the catalytic rate is proportional to the time that the participating functionalities reside in van der Waals contact with each other. Although appearing vague, this statement is, in fact, a brilliantly simple expression of time-dependence. Indeed, it must be said that Menger's *spatiotemporal hypothesis* is a good example of Sherlock Holmes's dictum for detective work: "Once you eliminate the impossible, whatever remains, no matter how improbable, must be the truth" [47].

ACKNOWLEDGMENTS

The author thanks for their insightful comments to: Chuck Dismukes and Damian Carrieri (Princeton University), Steve Benkovic (Pennsylvania State University), and Bob Williams (Oxford University).

REFERENCES

1. (a) Pauling, L. *Chem. Eng. News* **1946**, *24*, 1375; (b) Pauling, L. *Nature* **1948**, *161*, 707.
2. (a) Menger, F. M. *Acc. Chem. Rev.* **1985**, *18*, 128, and references therein; (b) Menger, F. M. *Acc. Chem. Rev.* **1993**, *26*, 206, and references therein; (c) Menger, F. M. *Pure Appl. Chem.* **2005**, *77*, 1873.
3. Page, M. I. Chap. 1 in *Enzyme Mechanisms*, Eds. Page, M. I.; Williams, A. Royal Society of Chemistry, **1989**, p. 1, and references therein.
4. (a) Bruice, T. C. *Acc. Chem. Res.* **2002**, *35*, 139, and references therein. See also: (b) Bruice, T. C.; Lightstone, F. C. *Acc. Chem. Res.* **1999**, *32*, 127; (c) Lau, E.; Bruice, T. C. *J. Am. Chem. Soc.* 2000, *122*, 7165; (d) Bruice, T. C.; Benkovic, S. J. *Biochemistry* **2000**, *39*, 6267.
5. Kirby, A. J. *Angew. Chem. Int. Ed. Engl.* **1996**, *35*, 707, and references therein.
6. (a) Ziang, X.; Houk, K. N. *Acc. Chem. Res.* **2005**, *38*, 379, and references therein; (b) *Chem. Eng. News* **2005**, *83*, 35.
7. (a) Hammes-Schiffer, S.; Benkovic, S. J. *Annu. Rev. Biochem.* **2006**, *75*, 519, and references therein; (b) Benkovic, S. J.; Hammes-Schiffer, S. *Science* **2003**, *301*, 1196, and references therein; (c) Rajagopalan, P. T.; Benkovic, S. J. *Chem. Rec.* **2002**, *2*, 24; (d) Alper, K. O.; Singla, M.; Stone, J. L.; Bagdassarian, C. K. *Protein Sci.* **2001**, *10*, 1319; (e) Tousignant, A.; Pelletier, J. N. *Chem. Biol.* 2004, *11*, 1037; (f) Hammes-Shiffer, S. *Biochem.* **2002**, *41*, 13335.
8. Jencks, W. P. *Catalysis in Chemistry and Enzymology*, McGraw Hill, 1969.
9. Jencks, W. P.; Page, M. I. *Proc. Eighth FEBS Meeting, Amsterdam*, **1972**, *29*, 45; *Biochem. Biophys. Res. Commun.* **1974**, *57*, 887.
10. Bruice, T. C. *The Enzymes, Vol. 2 (3rd ed.)*, Ed. Boyer, P. D. Academic Press, 1970, p. 217.
11. Winstein, S.; Lindgren, C. R.; Marshal, H.; Ingraham, L. L. *J. Am. Chem. Soc.* **1953**, *75*, 147.
12. DeLisi, C.; Crothers, D. M. *Biopolymers* **1973**, *12*, 1689.
13. Milstein, S.; Cohen, L. A. *Proc. Natl. Acad. Sci. USA* **1970**, *67*, 1143.
14. Reuben, J. *Proc. Natl. Acad. Sci. USA* **1971**, *68*, 563.
15. Firestone, R. A.; Christensen, B. G. *Tetrahedron Lett.* **1973**, 389.
16. Cook, D. B.; McKenna, J. *J. Chem. Soc. Perkin Trans. 2* **1974**, 1223.

17. Ferreira, R.; Gomes, M. A. F. *Symp. Theor. Phys. (Proc.) 6th, Vol. 2,* **1980**, p. 281.
18. Low, P. S.; Somero, G. N. *Proc. Natl. Acad. Sci. USA* **1975**, *72*, 3305.
19. Warshel, A. *Proc. Natl. Acad. Sci. USA* **1978**, *75*, 5250.
20. (a) Hol, W. G. J.; van Duijnen, P. T.; Berendsen, H. J. C. *Nature* **1978**, *273*, 443; (b) van Duijnen, P. T.; Thole, B. T.; Hol, W. G. *J. Biophys. Chem.* **1979**, *9*, 273.
21. Henderson, R.; Wang, J. H. *Ann. Rev. Biophys. Bioeng.* **1972**, *1*, 1.
22. Wang, J. H. *Proc. Natl. Acad. Sci. USA* **1970**, *66*, 874.
23. Olavarria, J. M. *J. Theor. Biol.* **1982**, *99*, 21.
24. Dewar, M. J. S.; Storch, D. M. *Proc. Natl. Acad. Sci. USA* **1985**, *82*, 2225.
25. Mock, W. L. *Biorg. Chem.* **1976**, *5*, 403.
26. Nowak, T.; Mildvan, A. S. *Biochemistry* **1972**, *11*, 2813.
27. Stryer, L. *Biochemistry (3rd ed.)*, W. H. Freeman and Company, 1988, p. 191; Kinetic "perfection" in an enzymatic system is described in: Knowles, J. R.; Albery, W. J. *Acc. Chem. Res.* **1977**, *10*, 105.
28. Moss, D. W. *Enzymes*, Oliver and Boyd, 1968, p. 75, and references therein.
29. (a) Haldane, J. B. S. *Enzymes*, Longmans, Green and Co. 1930, p. 182; (b) Fersht, A. *Enzyme Structure and Mechanism (2nd ed.)*, W. H. Freeman and Company, 1977, p. 331, and references therein.
30. Wolfenden, R.; Frick, L. Chap 7 in *Enzyme Mechanisms*, Eds. Page, M. I.; Williams, A. Royal Society of Chemistry, 1989, p. 97, and references therein.
31. Koshland, D. E. *Proc. Natl. Acad. Sci. USA* **1958**, *44*, 98.
32. Fersht, A. *Enzyme Structure and Mechanism (2nd ed.)*, W. H. Freeman and Company, 1977, p. 341, and references therein.
33. Bruice, T. C.; Pandit, U. K. *J. Am. Chem. Soc.* **1960**, *82*, 5858.
34. Kirby, A. J. *Adv. Phys. Org. Chem.* **1980**, *17*, 183.
35. Storm, D. R.; Koshland, E. E. *Proc. Natl. Acad. Sci. USA* **1970**, *66*, 445; Storm, D. R.; Koshland, D. E. *J. Am. Chem. Soc.* 1972, *94*, 5805; Dafforn, A.; Koshland, D. E. *Proc. Natl. Acad. Sci. USA* **1971**, *68*, 2463.
36. Bruice, T. C.; Brown, A.; Harris, D. O. *Proc. Natl. Acad. Sci. USA* **1971**, *68*, 658.
37. Hoare, D. G. *Nature* **1972**, *236*, 437.
38. Westheimer, F. H. *Adv. Enzymol.* **1962**, *24*, 441.
39. Page, M. L.; Jencks, W. P. *Proc. Natl. Acad. Sci. USA* **1971**, *68*, 1678.
40. Bruice, T. C. *Ann. Rev. Biochem.* **1976**, *45*, 331.
41. (a) Eisenmesser, E. Z.; Bosco, D. A.; Akke, M.; Kern, D. *Science* **2002**, *295*, 1520; (b) Eisenmesser, E. Z.; Millet, O.; Labeikovsky, W.; Korzhnev, D. M.; Wolf-Watz, M.; Bosco, D. A.; Skalicky, J. J.; Kay, L. E.; Kern, D. *Nature* **2005**, *438*, 117.
42. Jogl, G.; Rozovsky, S.; McDermott, A. E.; Tong, L. *Proc. Natl. Acad. Sci. USA* **2003**, *100*, 50, and references therein.
43. (a) Bruice, T. C.; Lightstone, F. C. *Acc. Chem. Res.* **1999**, *32*, 127; (b) Lau, E.; Bruice, T. C. *J. Am. Chem. Soc.* **2000**, *122*, 7165; (c) Bruice, T. C.; Benkovic, S. J. *Biochemistry* **2000**, *39*, 6267; (e) Agarwal, P. K. *Microb. Cell Fact.* **2006**, *15*, 2, and references therein; (f) Agarwal, P. K. *J. Am. Chem. Soc.* **2005**, *127*, 15248; (g) Workshop on future directions of catalysis science, *Catal. Lett.* **2001**, *76*, 111.

44. (a) Williams, R. J. P. *Trends Biochem. Sci.* **1993**, *18*, 115; (b) Williams, R. J. *Europ. J. Biochem.* **1989**, *183*, 9; (c) Williams, R. J. P. *Eur. Biophys. J.* **1993**, *21*, 393.
45. For example, see Agarwal, P. K. *Microb. Cell Fact.* **2006**, *15*, 2, and references therein.
46. Moss, D. W. *Enzymes*, Oliver and Boyd, 1968, p. 70, and references therein.
47. Doyle, A. I. C. *The Sign of Four* (First published: *Lippincott's Magazine,* February 1890*)*; First Book Edition: Spencer Blackett, 1890.

8

SYNERGY IN HETEROGENEOUS, HOMOGENEOUS, AND ENZYMATIC CATALYSIS. THE "IDEAL" CATALYST

GERHARD F. SWIEGERS

8.1 INTRODUCTION

In previous chapters, we alluded to the issue of *synergy* in catalysis. When the whole of a system exceeds the simple sum of its constituent parts, the system is said to exhibit synergy [1]. Synergy is a characteristic that is inherent to many mechanical systems, which typically rely on coordinated, synchronized (read: *synergistic*) interactions between their parts to achieve particular, desirable effects. One need only think of the way in which the tiny cogs and wheels within a finely balanced watch work together to provide an accurate measure of time. If even one of those cogs or wheels were absent, or if it was imperfectly machined, the watch would be rendered inoperable. It would then be just a useless collection of bits of intricately machined metal.

We have shown that this is also true of some catalytic systems, with, for example, many enzymes displaying drastically higher orders of synergy relative to comparable nonbiological homogeneous catalysts (see Section 5.5.7 in Chapter 5). These synergies were noted to be highly distinctive. In effect, they constitute a pronouncement on the nature and character of the action employed by the catalyst.

Up to this point, this series has examined the phenomena of catalysis as it manifests itself in so-called *time-dependent* and *energy-dependent* catalysis.

Most nonbiological, multicentered homogeneous catalysts have been shown to be *energy-dependent* in their action. By this is meant that the rate-limiting step in the catalytic process involves the formation of the transition state and the likelihood that

---

Mechanical Catalysis: Methods of Enzymatic, Homogeneous, and Heterogeneous Catalysis,
Edited by Gerhard F. Swiegers
Copyright © 2008 John Wiley & Sons, Inc.

its threshold *activation energy* E_a will be overcome, to thereby yield products. Since E_a is, fundamentally, an energetic quantity, the reaction is *energy-dependent*. Catalysis of this type is controlled by the character and structure of the transition state and the underlying energy landscape of the reaction according to transition state theory.

An alternative form of catalysis involves a *time-dependent* action. The rate-limiting process in a time-dependent catalyst does not involve the formation of, or the stability of the transition state. Instead, it depends on the frequency with which the reactants collide with each other during the course of the reaction. In a catalyzed reaction, the collision of the reactants is mediated by the catalyst, so that one refers to the *catalyst-mediated collision frequency*. Since frequency is a measure of the number of events that occur per unit time, such reactions are termed *time-dependent*. Catalysis of this type is necessarily characterized by a low-energy barrier for reaction since a catalytic effect would otherwise not be observed. Time-dependent catalysis is, consequently, independent of the energy of the collision.

A key distinguishing feature of time-dependent catalysis is the presence of a rapidly equilibrating reactant–catalyst intermediate in the rate expression. In general, the more quickly a reactant associates and dissociates from the catalyst, the lower will be the catalyst-mediated collision frequency. When the association–dissociation process is sufficiently dynamic, the collision frequency will become rate-limiting, making the catalyst time-dependent. A rapidly equilibrating catalyst–reactant intermediate will then appear in the rate expression since this process will then represent the slowest step in the catalytic process.

Many enzymes are time-dependent catalysts, as is demonstrated by the fact that they display Michaelis–Menten kinetics, which involves a rapidly associating–dissociating intermediate in their rate expression.

This is further confirmed by the general structure sensitivity of enzymatic catalysis; that is, many enzymes rapidly lose their catalytic properties if their primary, secondary, or tertiary structure is altered in any way. This is consistent with a catalytic process that is path-dependent and spatiotemporal in character. It also seems to reflect a need for coordinated, synchronized actions within enzymes during catalysis.

Although we have thus far explored the nature of time- and energy-dependence as it pertains to broad classes of catalysis, we will, from this chapter onward, take a more forward-looking perspective. Our intention in doing so is to start thinking about the rational design and development of *new* catalysts that make use of the principles we have described in previous chapters.

The topic of synergy is useful in this respect. It not only provides researchers with a possible means of recognizing time- and energy-dependence in catalysts, but it may also offer a tool that could be employed in the rational design of new, time-dependent homogeneous catalysts. Such catalysts have not been studied to date and must be considered to hold significant promise. To be useful in this respect, however, we must thoroughly and systematically understand the nature of synergy in catalysis.

In this work we will examine this issue in detail. In so doing, we will revisit some of the catalytic systems described in earlier chapters. We will show that synergy offers a powerful and versatile design criterion for developing active and selective new homogeneous catalysts. The origin of catalytic synergy is therefore of considerable fundamental and practical importance.

We will also critically examine what the target of rational design should be. We will ask the question: What should chemists be aiming at when they design new, multicentered homogeneous catalysts? What does the "ideal" catalyst look like? In Chapter 10, we will consider how the features that have been identified in this chapter may be implemented.

8.2 SYNERGY IN HETEROGENEOUS CATALYSTS

Many metal alloy and bimetallic thin film heterogeneous catalysts display properties that are far superior to their unimetallic analogs [2]. For example, the petroleum industry routinely employs highly dispersed, supported, bimetallic clusters of Pt–Rh/Ir to accelerate reforming reactions. Such species exhibit higher catalytic activities and are more durable than pure Pt catalysts [2]. They also produce fuels with higher octane ratings. Their discovery, arguably, made the production of unleaded gasoline feasible.

The origin of such effects has recently been reviewed by Adams [2]. It seems that they derive from the ability of highly dispersed bimetallic and certain binary mixtures to segregate their elements, thereby forming micro-interfaces of the different metals on the surface of the catalyst. Annealed Pt–Ru alloys have been shown, for example, to segregate into Pt and Ru micro-islands. At or near the interface of these islands, *synergistic* catalysis is possible. That is, dissimilar intermediates on or near opposite sides of the interface can interact ("collide") with each other to form novel products that are simply not generated on unimetallic catalysts. The resulting synergies are believed to be achieved in three possible ways [2]:

(1) *Multicentered Activation.* In this mechanism, one reactant is activated on one metal and a second reactant is activated on the other metal [Fig. 8.1(a)]. At the interface of the two metals, the two activated species encounter each other and react. Thus, products may be obtained that cannot be readily generated by each metal separately.

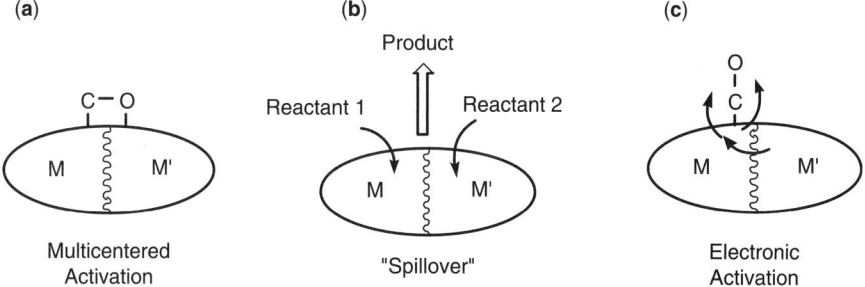

Figure 8.1. Three mechanisms by which synergy is thought to be achieved in bimetallic heterogeneous catalysts whose metals (M and M′) have separated into distinct micro-islands with micro-interfaces (illustrated by the curvilinear lines).

This mechanism has been proposed as an explanation for the superior properties of certain bimetallic supported catalysts in the production of CO relative to their unimetallic analoges.

(2) *Spillover*. Alternatively, each reactant may be activated on a different metal, with the activated reactants then migrating toward one another [Fig. 8.1(b)]. "Spillover" is the phenomenon by which an activated species is transferred from one surface across an interface to a second surface. The final reaction of the two activated reactants may occur near to, but not necessarily at, the interface.

Many hydrides, alkynes, and alkenes are known to migrate from one metal atom to another during catalysis.

(3) *Electronic Activation*. A third mechanism is suggested to involve an electronic effect in which a metal adds or removes electron density from an adjacent metal atom at the interface, thereby enhancing its ability to activate and bring about the reaction of small molecules [Fig. 8.1(c)].

Charge transfer, and orbital rehybridization effects, as well as the modification of valence states are known to play important roles in the adsorption of CO and ethylene on bimetallic thin films.

In all three of these mechanisms, dissimilar metals act *cooperatively* to achieve catalytic outcomes that are, otherwise, not possible. As such, these systems display true synergies. That is, in acting together, they create an effect that is more than the simple sum of their properties.

Because catalysis takes place on a two-dimensional (2-D) surface, the interactions that create this synergy are generally limited by the 2-D nature of the reactive face. A limited three-dimensional (3-D) interaction is, of course, also possible in such instances.

As noted in Section 2.2.1 of Chapter 2, most heterogeneous catalysts are "multisite" in character; that is, catalysis occurs at a variety of dissimilar step, edge, defect, and other sites. Given this diversity of catalytic action and the difficulty of elucidating the mechanism of the catalysis in all of its aspects, one cannot simply and definitively describe or assign the synergies present. Although we can broadly recognize when there are synergies of action in heterogeneous catalysis, we cannot—at this stage at least—characterize them in detail.

A similar problem does not exist for homogeneous catalysis, which is, by contrast, single site in character. That is, all of the catalytic sites in homogeneous solution are, effectively, identical. Their synergies are, therefore, less diverse and complicated. They can, potentially, be more easily considered.

8.3 SINGLE-CENTERED NONBIOLOGICAL HOMOGENEOUS CATALYSTS AND THEIR *'MUTUALLY ENHANCING'* SYNERGIES

8.3.1 Facial Selectivity in Single-Centered Catalysts

As noted in Section 2.3.1 of Chapter 2, some homogeneous catalysts involve a single atom carrying out all of the functions required during catalysis within the active site.

8.3 SINGLE-CENTERED NONBIOLOGICAL HOMOGENEOUS CATALYSTS

Such species are termed *single-centered* catalysts. Species in which two or more atoms within the active site carry out the catalytic functions are termed *multicentered* homogeneous catalysts (Section 2.3.2 of Chapter 2).

In Section 2.3.1 of Chapter 2 we also described the *single-centered* homogeneous catalyst **1**, which converts propylene into so-called *isotactic* polypropylene as shown in Figure 8.2 [3]. Catalyst **1** contains two indenyl ligands that sandwich the Zr ion. A closely related catalyst, **2**, converts propylene into a different form of polypropylene, namely *syndiotactic* polypropylene. Catalyst **2** contains one phenyl and one fulvenyl ligand sandwiching the Zr.

As noted in Chapter 2, catalysis is undertaken in these species by the unsaturated Zr^+ ion, which has a vacant coordination site and is, consequently, exceedingly electrophilic. It readily binds electron-rich species, like propylene monomers, in this vacant coordination site. The propylene monomer binds in a side-on manner. In so doing, it disturbs the stabilization of the other alkyl group already bound to the Zr ion, with the result that the bound propylene becomes "inserted" into this Zr–C bond [3]. That is, the Zr–C bond is broken and reformed with the propylene between the Zr and the C atoms. In doing so, a new, vacant coordination site is created on the opposite side to the original one, thereby allowing the process to repeat itself.

Because these are single-centered catalysts, the transition state for their reactions must form in the plane (the face) immediately in front of the Zr ion. Selectivity can, consequently, be achieved in this particular catalytic process by tailoring the steric shape of this face with sterically bulky groups.

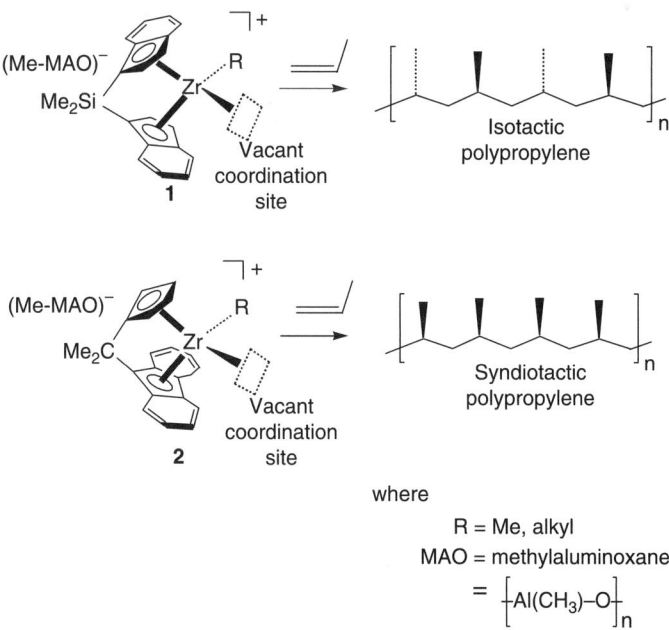

Figure 8.2. Catalytic propylene polymerization using zirconocene catalysts.

Figure 8.3 depicts in schematic form the front face of the Zr ion during catalysis by **1** and **2**. As shown, in **1** the face of the Zr ion is symmetrically hindered with the bulky indenyl ligand (thick black lines) occupying the bottom left and top right quadrants [2]. The bottom right and top left quadrants are, however, unoccupied. The growing polymer chain is therefore favored to flip between these unoccupied quadrants during catalytic turnovers. This steric arrangement has the effect of favoring insertion leading to the *isotactic* form of polypropylene.

By contrast, the front face of the Zr ion in **2** is sterically hindered in the bottom left and bottom right quadrants, with the top left and top rights quadrants unoccupied [3]. The growing polymer chain is, consequently, favored to flip between the top left and top right quadrants during polymerization, thereby leading to the *syndiotactic* product.

The selectivity of these catalysts therefore derives from the steric arrangement about the catalytically active face of the Zr ion.

Other prominent examples of facially selective, single-centered catalysts include species like the Pd ion in its complex with chiral auxiliary **3**, which is, similarly, the sole point of contact during the catalytic transformation depicted in Figure 8.4 [4]. As in **1** and **2**, the steric arrangement about this face governs the selectivity (and activity) of its catalysis.

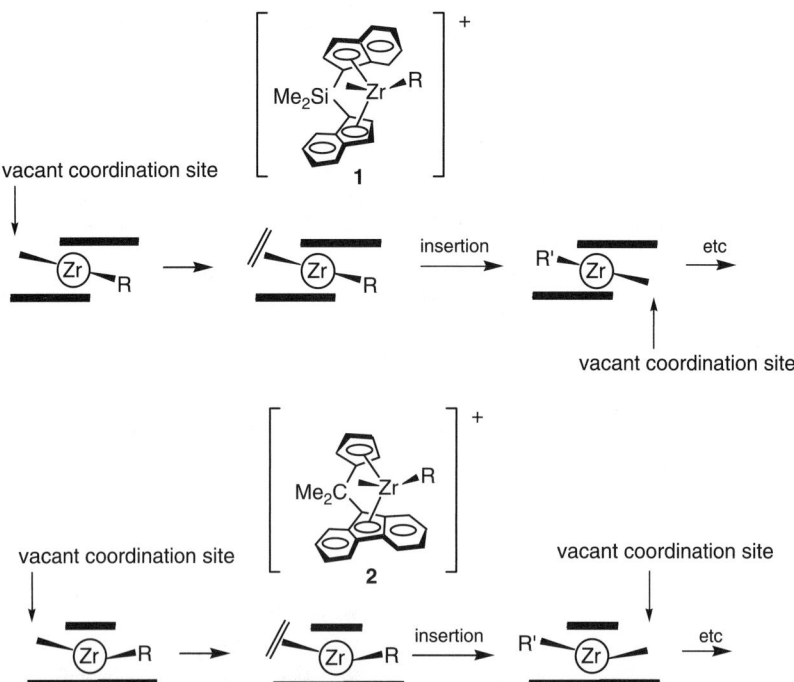

Figure 8.3. Schematic illustration of the catalytically active face of the Zr ion in **1** and **2** showing the positions of the indenyl and fulvenyl ligands as thick black lines.

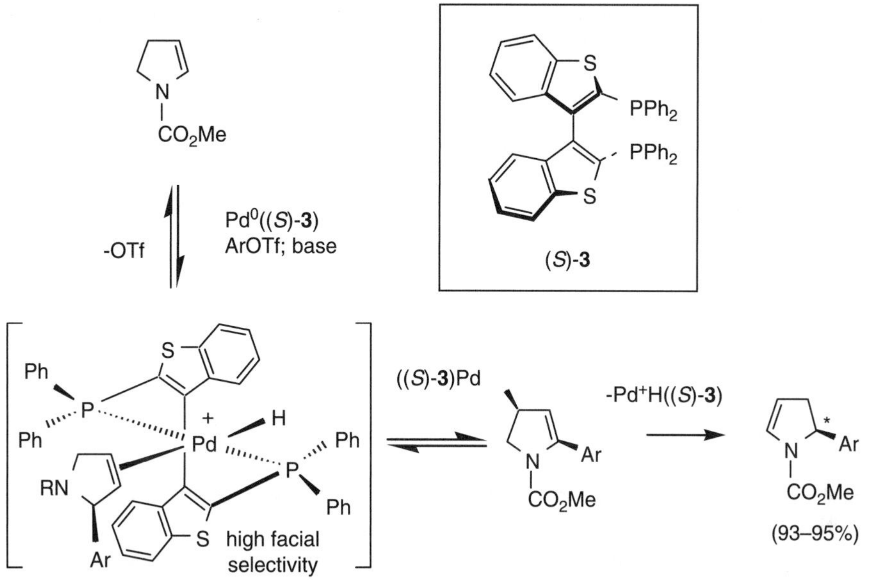

Figure 8.4. *Single-centered homogeneous catalysis.* Stereospecific intermolecular arylation catalysis by the palladium complex of chiral auxiliary **3**.

8.3.2 Energy-Dependent, Single-Centered Homogeneous Catalysts Display "Mutually Enhancing" Synergies

The examples depicted in Figures 8.2–8.4 illustrate an important point about the influence of *synergy* in catalytic processes. We see in these examples that the catalytic action involves an interplay in which structural elements of the catalysts interact with structural constraints of the reactants and/or intermediates to yield a particular stereoisomer of the product.

That is, by blocking particular quadrants in the active face of the catalyst, the bulky steric groups dramatically increase the energy of those transition states that would partially or completely occupy these quadrants. By contrast, those transition states that lie within the sterically unhindered quadrants are unaffected. These particular transition states therefore become relatively more favored, with desirable results in the products obtained.

In effect, by cleverly manipulating the steric arrangement about the reactive face of the catalyst, one may thermodynamically disfavor a particular product by raising the threshold, activation energy of its transition state.

Synergies in which system components interact with each other in an indirect way to block selectively or favor certain outcomes are well known in complex systems science. They are commonly referred to as *mutual enhancement* [1], because by dint of their simple and mutual presence, the system components indirectly favor particular courses of action.

Thus, the indenyl and fluorenyl groups in **1** and **2**, respectively, do not participate directly in the catalytic action, which is undertaken entirely by the

Zr-ion in **1** and **2** [3]. Instead, they are simple bystanders to the processes that occur about them. However, they are significant bystanders whose presence and steric bulk dramatically influences, and even controls, the outcome of the process.

Mutually enhancing synergies of this type are also observed in numerous other systems outside of chemistry and catalysis. For example, the phenomenon of *positive cooperativity* in biology, in which the binding of one substrate leads to the binding of other substrates, is arguably also an example of mutually enhancing synergies [5].

A good way to illustrate mutually enhancing synergy is to cite the example of a man building a cabinet by doing both the sawing and the nailing, while his partner looks on passively. If the partner does nothing whatsoever to help in the construction of the cabinet, no synergy is achieved at all. However, if the partner assists passively by, for example, standing in a position that makes it difficult, or even impossible for the cabinet to be put together incorrectly, then a form of mutually enhancing synergy is achieved. That is, the cabinet will be built better and more quickly than if the partner had not been there. Of course, if the partner goes on to provide more active assistance, then higher level synergies will be achieved. In the following sections, we will describe such higher level synergies.

In somewhat the same way, one can see how the rational design of a single-centered homogeneous catalyst to improve its stereoselectivity largely involves manipulating the available *mutually enhancing* synergies to maximum advantage.

A substantial number of rationally designed, stereoselective, single-centered molecular catalysts have been prepared using such insights [6–9]. Several catalysts are employed industrially to prepare high-value added products, such as enantiomeric pharmaceutical species and designer polymers.

8.3.3 The Synergies in Time-Dependent, Single-Centered Homogeneous Catalysts

As noted, most nonbiological, single-centered catalysts seem to be energy-dependent, not time-dependent, in their catalytic action. That is, they select the *thermodynamically most stable transition state* rather than the *transition state that is mechanically formed first*. However, a handful of manmade homogeneous catalysts are known to display Michaelis–Menten kinetics and must therefore necessarily employ a time-dependent catalytic action [10–17]. What type of synergy do such catalysts display?

To answer that question we must consider the examples that are available. One example of what must be a time-dependent, single-centered homogeneous catalyst is that shown in Figure 8.5, which involves a Diels–Alder reaction that displays Michaelis–Menten kinetics [11]. The reaction is somewhat selective for the *endo*-form **6**, rather than the *exo*-form. Whereas the dienophile **4** probably binds the Cu^{2+} metal ion catalyst relatively strongly, the diene **5** will likely bind it only weakly. This weak binding prospectively explains the time-dependent nature of the catalysis. However, the mechanism has not been elucidated in detail.

Other examples of what are, likely, time-dependent, single-centered catalysts include 1) the hydrolysis of 4-nitrophenyl diphenyl phosphate by copper-loaded polymers [12], and 2) a crown-ether flavin mimic [13]. Both catalysts also display

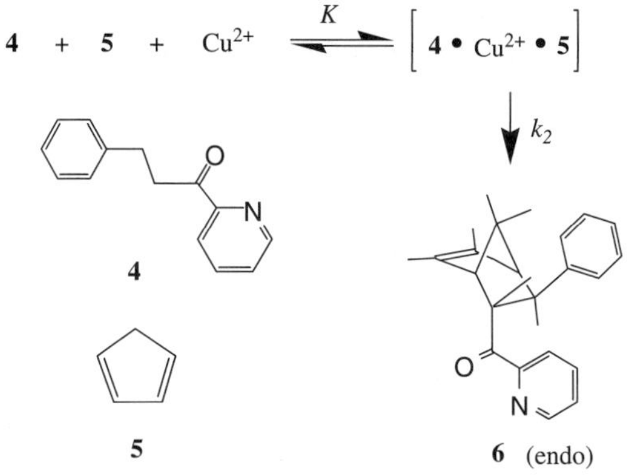

8.3 SINGLE-CENTERED NONBIOLOGICAL HOMOGENEOUS CATALYSTS

Figure 8.5. *Time-dependent, single-centered homogeneous catalysis.* A stereospecific Diels–Alder reaction that displays Michaelis–Menten kinetics.

saturation kinetics, but neither have been examined intensively in terms of their catalytic mechanism and action.

Several homogeneous catalysts that have been reported to exhibit Michaelis–Menten kinetics are, arguably, multicentered in their action; that is, multiple atoms within the active site undertake the catalysis. Examples include Corey's bis-cinchona alkaloid catalysts of the asymmetric dihydroxylation of olefins [10] and Pirrung's dirhodium(II) carboxylate catalysts of various carbenoid reactions [14]. Multicentered, time-dependent homogeneous catalysis will be discussed in a later section.

Other than the above examples, none of which have been thoroughly studied, no authoritative examination has been conducted on the mechanistic commonalities in manmade homogeneous catalysts that display saturation kinetics.

Because of this scarcity of definitive information, it is not possible, at this stage, to describe and define unequivocally the type of synergies that exist in single-centered, time-dependent catalysts. However, given the need for a single-center to which the reactants must be bound, and the consequent limitation to a facial active site, such catalysts are also likely to display mutually enhancing synergies of the type that typify energy-dependent, single-centered catalysis.

8.3.4 The Selectivity of Single-Centered Catalysts

A feature of single-centered homogeneous catalysis is its generally imperfect selectivity. Thus, although **1**–**2** and Pd-**3** are impressively selective in their product generation, they are not *perfectly* selective. By contrast, the catalysis displayed by, for example, many enzymes is routinely *perfectly* stereoselective.

This issue is significant because, despite many thousands of man-years spent on developing mutually enhancing single-centered homogeneous catalysts, truly perfect selectivity has never been achieved. Even the syndiotactic and isotactic polypropylene depicted in Figure 8.3 do not display 100% stereoregularity [3].

Imperfect selectivity is also observed in time-dependent, single-centered catalysts like the Cu(II) ion in Figure 8.5. Thus, selectivity in single-centered catalysts seems to be connected with the single-centered nature of the catalysis rather than with the catalytic regime. That is, it seems to be a function of the fact that the catalysis must occur in a single face of the active site, rather than the fact that the catalyst is energy- or time-dependent.

The presence of multiple, not single, catalytically active atoms within the active site of a homogeneous catalyst potentially eliminates the constraint of having to facilitate a reaction within a 2-D face of the catalyst. Instead, catalysis may then occur in the 3-D space between the catalytic groups present in the active site. This action necessarily drastically alters the synergies that are possible.

In a later section, we will examine the synergies of multicentered homogeneous catalysts.

8.4 MULTICENTERED, ENERGY-DEPENDENT HOMOGENEOUS CATALYSTS AND THEIR *FUNCTIONALLY COMPLEMENTARY* SYNERGIES

As noted in previous chapters, some homogeneous catalysts involve two or more atoms acting together during catalysis. Such species are termed *multicentered* catalysts.

Figure 8.6 presents an example that has been cited several times previously in this series. The two Cr ions in catalysts **7a–c** and **8** facilitate the ring-opening reaction of an epoxide with an azide [18]. The active site lies between the two metal centers. A detailed account of catalysis by **7a–c** and **8** is provided in Chapter 6 (Section 6.3).

A critical feature of the catalytic action of **7a–c** and **8** is that the tasks carried out by the two Cr ions are *complementary* to each other and not *mutually enhancing* (see Chapter 5) [1]. Thus, one Cr ion binds and activates the azide, whereas the other Cr ion binds and activates the epoxide. A catalytic effect is achieved when these two tasks are combined.

In other words, the system components do not *indirectly* influence each other to achieve a beneficial outcome, as is the case in *mutual enhancement*. Rather, they actively carry out tasks that physically *complement* each other. It is this complementarity that generates the net catalytic effect.

As very briefly noted in Section 5.5.7 of Chapter 5, synergies of this type are known in complex systems science as *functional complementarity* [1]. Figure 8.7(a) schematically depicts a functionally complementary catalyst.

Functional complementarity is a defining characteristic of energy-dependence in multicentered catalysis. Virtually all nonbiological, multicentered homogeneous catalysts are functionally complementary in their mode of action and therefore energy-dependent. How do we know that?

We know it because in being complementary in their action, such catalysts can be harnessed in any physical structure that provides for the required complementarity. This includes as monomeric species in open solution (e.g., **8**) or as linked species involving tethers of various lengths (e.g., **7a–c**). Catalysis is, essentially,

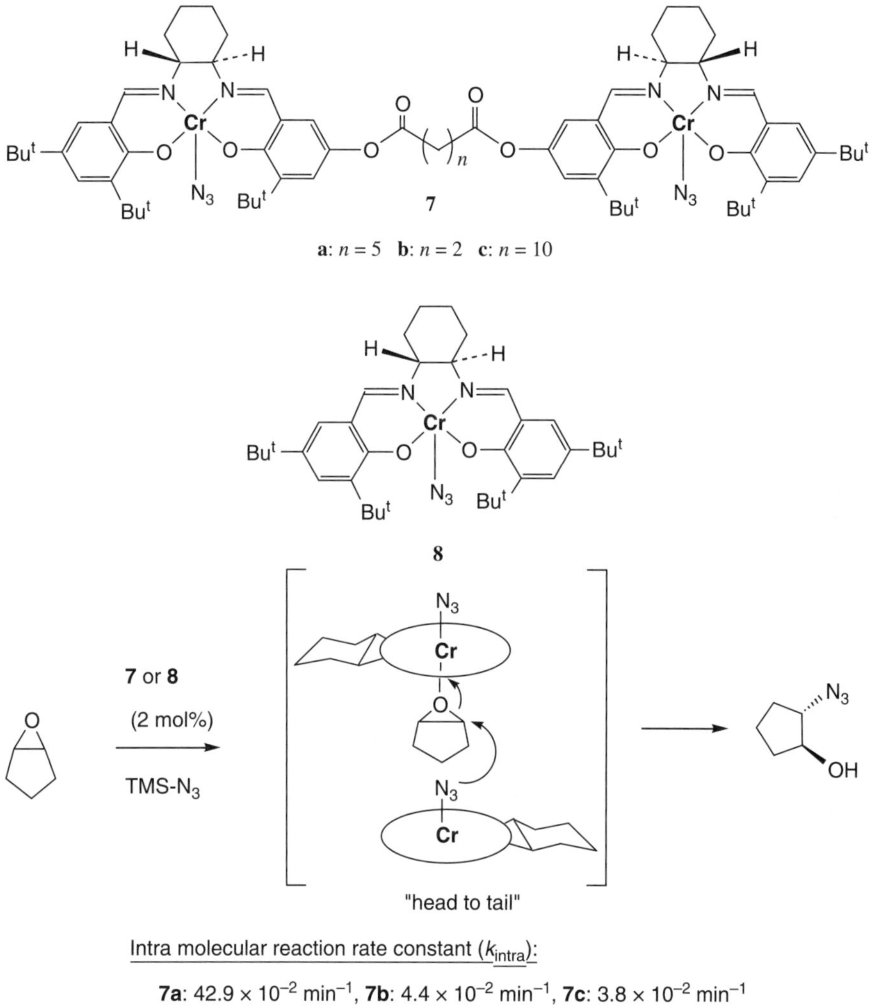

8.4 MULTICENTERED, ENERGY-DEPENDENT HOMOGENEOUS CATALYSTS

Figure 8.6. *Multicentered homogeneous catalysis.* Functional complementary synergies during an asymmetric catalytic ring-opening of an epoxide.

a: $n = 5$ b: $n = 2$ c: $n = 10$

Intra molecular reaction rate constant (k_{intra}):

7a: 42.9×10^{-2} min^{-1}, **7b**: 4.4×10^{-2} min^{-1}, **7c**: 3.8×10^{-2} min^{-1}

independent of the spatial arrangement of the catalytic groups, provided that the catalytic groups are free to approach each other regularly and therefore to mediate collisions between any bound reactants.

In effect, there is almost no need for the individual catalytic groups to coordinate or synchronize their actions, because there is, essentially, no limitation to the time available to mediate collisions between bound reactants. In such catalysts, the reactants bind and stay bound to their respective catalytic groups, undergoing many collisions with each other, until they finally collide in a reactive manner and form products. Such "reactive" collisions may be few and far between, being only a tiny fraction of the average numbers of collisions that occur before reaction.

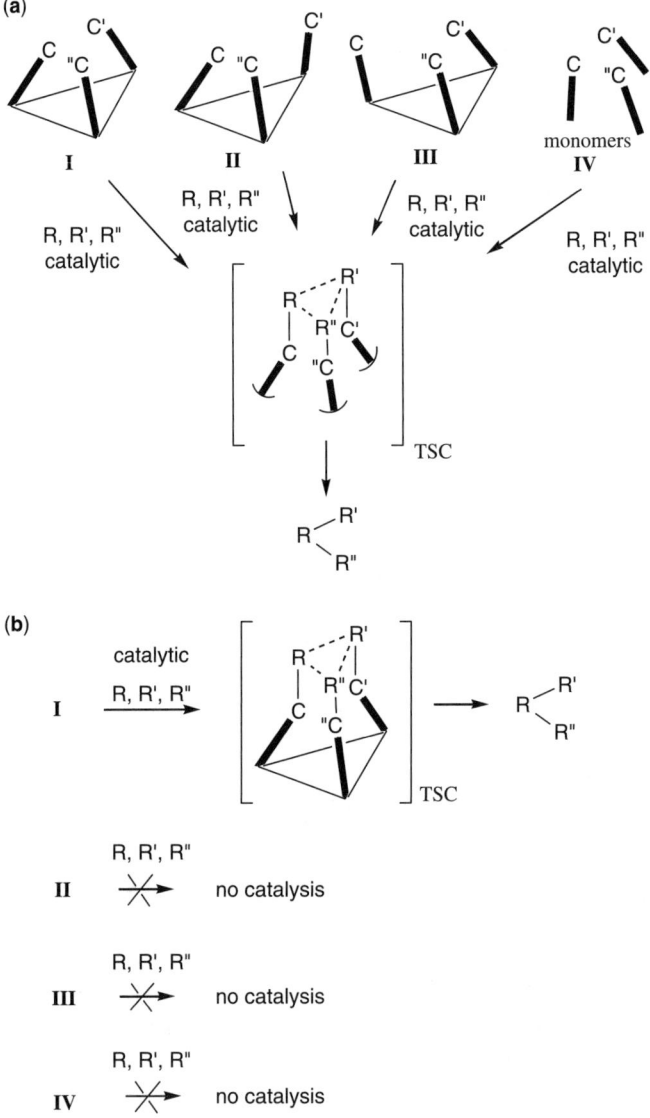

Figure 8.7. *Synergy in multicentered catalysis* [1]. Schematics depicting (a) functionally complementary catalysis and (b) functionally convergent catalysis. In (a), the catalytic groups carry out complementary tasks and therefore do not need to spend most of their time in particular spatial arrangements. Provided sufficient conformational flexibility exists, all of I–III are catalytically active, as are the monomeric catalytic groups IV, dissolved in open solution. In (b), each catalytic group (C, C′, and C″) depends on the other catalytic groups to carry out its catalytic function. To display an effect, the catalytic groups must spend most of their time in a structure that complements the transition state (as in I). Variations in this respect, even within highly flexible molecules, result in a complete loss of catalytic activity (as in II, III, and IV).

Catalysts of this type consequently require such limited synchronization that the rarity of statistical probability is sufficient to generate an overall catalytic effect.

An interesting and important feature of this type of catalysis is the dramatic influence that even very minor *coordination* between the catalytic groups has on the overall activity.

Thus, in dilute open solution, **8** is a slow catalyst of the reaction because at any one instant, collisions having an optimum spatial arrangement for reactions are not highly populated. However, when the two catalytic groups are tethered, as in **7a–c**, this optimally arranged collision becomes distinctly more favored, with an accompanying dramatic increase in the catalytic rate.

In effect, the tether increases the likelihood that all of the elements needed for catalysis will be simultaneously present. That is, it accentuates the probability that the bound reactants will be ideally oriented for reaction when they collide with each other. In other words, it increases the likelihood of *coordinated action* between the two Cr ion catalytic groups. Coordinated actions are more correctly termed *convergent* since the catalytic groups are more disposed to act in a synchronized way; that is, their movements *converge* to create a coordinated action.

The relatively minor increase in convergent behavior in **7a–c** relative to **8** results in a significantly higher catalytic rate in the order k_{intra} 42.9×10^{-2} min^{-1} (**7a**), 4.4×10^{-2} min^{-1} (**7b**), and 3.8×10^{-2} min^{-1} (**7c**) (Fig. 8.6).

The very strong relationship among convergence, structure, and rate is illustrated by the fact that the tether is too short in **7b** and too long in **7c**, which leads to lower catalytic rates than is observed in **7a**. This is because there is a lower likelihood of simultaneous action in **7b** and **7c** than there is in **7a**.

Even in species where the catalytic groups are overwhelmingly *complementary* in their mode of action, the extent of *convergence* therefore still plays a significant and important role in the efficiency of the catalysis.

It is critically important to recognize that this convergence does not create the catalytic effect in **7a–c** and **8**. The effect is there regardless of the convergence present. Thus, the convergence in **7a–c** is not a true form of synergy since it is not essential to the achievement of a catalytic effect. That is, the *catalytic effect* is not created by, nor does it rely on, the *convergence* in the same way that the actions of the cogs in a mechanical watch *must* converge in order for it to operate properly.

Instead, the convergence in **7a–c** is *coincidental*. It improves the catalysis, but it is not essential to it. All it does is to bring about a higher likelihood of simultaneous, complementary action.

It is ironic that the entire field of comparative *intramolecular* organic reactions (described in Section 7.2.6 of Chapter 7) [19] has, effectively, been devoted to the study of this minor and incidental form of convergence. It is truly remarkable that this field nevertheless succeeded in accurately identifying several very important relationships in multicentered homogeneous catalysis [19].

Key Point. **In summary, although functional complementarity is an important form of synergy in homogeneous catalysis, it has a profound inherent inefficiency. The quickest and most effective way of facilitating reactions in a multicentered** catalyst is for each of the catalytic groups to perform their tasks simultaneously, in a coordinated, synchronized (convergent) manner.

But, such behavior is not a requirement, per se, of functionally complementary catalytic systems. The catalytic groups in such systems can and will perform their tasks when it is most convenient and efficient for them to do so, not when it is most efficient overall.

Thus, convergence in a functionally complementary system can only ever be a coincidental feature of the system. Convergence in such cases is a "nice-to-have" feature, but it can never be a "must-have" feature.

Functionally complementary systems can potentially facilitate many types of reactions since they do not depend on a particular, convergent action to create a catalytic effect. This capability makes them vastly less selective in their catalytic function than a system that did rely on convergence to generate a catalytic effect. A functionally complementary catalyst placed in a mixed reactant stream will, consequently, generate a wide variety of different products. This will *not* be the case for a catalyst that depends on a particular, convergent action to create a catalytic effect. Such a catalyst will, selectively, catalyze only a reaction or reactions that it is capable of convergently facilitating. It will do so even in an extremely diverse, mixed reactant stream.

Moreover, the best that could ever be hoped for in a functional complementary system is that the components can be structured to have a high likelihood of acting simultaneously. But such a high probability of synchronized action *does not guarantee* synchronized actions. Some small fraction of catalytic actions will always be available that are not synchronized. This fraction may form unwanted products.

8.5 ENZYMES AND THEIR *FUNCTIONALLY CONVERGENT* SYNERGIES

As noted in Section 5.5.7 of Chapter 5, a distinctive feature of many enzymes is that they display *functionally convergent* synergies in their catalytic action [1]. Each catalytic group depends on the other catalytic groups to carry out its task. That is, the different tasks carried out by the various catalytic groups must all be performed correctly, at the same time, in unison. If any one group fails to carry out its task at the point of reaction, all of the others also fail. Thus, a high degree of interdependence exists in which all catalytic groups either succeed together or fail together.

Synergies of this type are of an exceedingly high order. They involve not only the participation of multiple catalytic groups but also an *absolute dependence* on spatial organization and pathway during catalysis.

As noted in Chapters 5 and 6, enzymes achieve such synergies by:

1. Maintaining a high level of structural complementarity for their reaction transition state during conformational flexing.
2. Requiring synchronization between enzyme flexing and binding to generate a catalytic effect. This feature comes about because of the dynamism of enzyme–substrate interactions and of conformational flexing in the enzyme.

8.5 ENZYMES AND THEIR *FUNCTIONALLY CONVERGENT* SYNERGIES

These constraints ensure simultaneous, coordinated, and inter-related action by all catalytic groups.

Key Point. The key feature of functionally convergent synergies is that convergent behavior is an absolute requirement in order for catalysis to take place at all [1]. Thus, the catalytic process does not, and cannot proceed unless all of the groups simultaneously bear their respective substrate functionalities at the instant of collision. The enzyme must, moreover, also be complementary to the transition state at this moment.

As depicted in Figure 8.7(b), it is for this reason that enzymes are highly "structure-sensitive" catalysts; that is, they display nonlinearity in the relationship between their structure and their catalysis (see Chapter 5). Even minor changes in the spatial arrangement of their catalytic groups destroy the necessary convergence and eliminate the catalytic effect entirely.

Functional convergence is a characteristic of *time-dependent*, multicentered homogeneous catalysts (see Chapter 5). This is because such systems are controlled by the catalyst-mediated collision frequency. To achieve a successful collision between bound substrate functionalities, the reactants must be optimally disposed and arranged during the very brief occasions available for transition state formation.

In Chapter 5 of this series, we described the general, observable properties of a functionally convergent, multicentered catalyst. They are as follows:

1. Weak and dynamic individual catalyst–reactant binding interactions
2. Structural complementarity with the transition state during conformational flexing
3. Highly structure-sensitive catalysis
4. The transformation of unconventional catalytic groups into potent catalysts
5. High selectivity and activity
6. Bidirectional catalysis
7. Michaelis–Menten kinetics

These features are illustrated somewhat in a report by Corey on bis-cinchona alkaloid catalysts of the type depicted in Figure 8.8 [10]. These species display Michaelis–Menten kinetics. They facilitate the dihydroxylation of olefins inside a chiral, bowl-shaped active site within which an OsO_4 is complexed.

Figure 8.8 illustrates a typical reaction sequence. The allyl-4-methoxybenzoate substrate **10** is dihydroxylated in 98% ee [10]. Perfect selectivity was therefore very nearly achieved. Substantial changes to the structure of the catalyst destroy its activity and selectivity.

In this example, the substrate "binds" to the catalyst active site by the formation of extensive π-contacts between both faces of the allyl and 4-methoxybenzoate moieties and the two methoxyquinoline units. Edge contacts are also made with the pyridazine ring and with the Os(VIII) moiety. Interactions of this type are weak and dynamic, as is common in time-dependent catalysts.

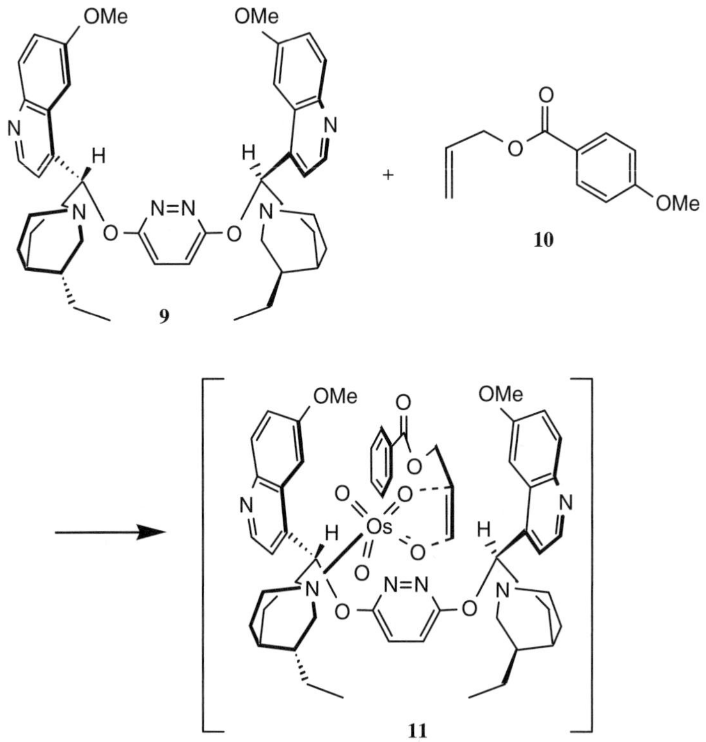

Figure 8.8. *Time-dependent, multicentered homogeneous catalysis.* Bis-cinchona alkaloid catalyzed dihydroxylation of olefins.

The shape of the **9**-OsO$_4$ active site is, moreover, complementary to the transition state of the reaction. Annulated aromatic and heteroaromatic groups are, effectively, also turned into catalytic groups that bind and orient the substrate.

Thus, we have in this example a highly active and selective, structure-sensitive form of catalysis that is dependent on the bowl-shaped arrangement of the active site.

Some interesting and potentially important findings came out of this work. One was the observation of a positive correlation between the extent of overall catalyst–substrate binding ($1/K_M$) and the extent of enantioselectivity. Thus, the more binding contacts made between the catalyst and the substrate, the more selective was the reaction. In other words, molecular recognition plays an important role in the selectivity of the catalytic process.

A correlation also seemed to exist between the catalytic rate at saturation (V_{max}) and the extent of enantioselectivity. The faster the reaction rate, the greater was the enantioselectivity displayed.

In effect, the system seemes to work like a very efficient molecular machine. The faster it turns over, the more efficiently it functions. This is, arguably, also true of many enzymes.

A final point that must be addressed is the question as to why Corey's catalysts do not display *perfect selectivity* of the type displayed by many enzymes? After all, if a catalytic effect is only possible in a time-dependent catalyst if all of the catalytic groups participate, then why do Corey's catalysts generate mixtures of products?

The answer is, of course, that multiple time-dependent catalytic pathways are available in these catalysts. One pathway may involve a certain selection of π-contacts, whereas others involve different selections of π-contacts. Thus, the presence of a time-dependent action does not *guarantee* perfect selectivity. To achieve high selectivity, one needs a *single* time-dependent pathway available, or one needs selective binding between the catalyst and the reactant that strongly favors only one of the available time-dependent pathways. Good molecular recognition between the catalyst and the reactant, as occurs in many enzymes, is therefore a critically important aspect of efficient time-dependent catalysis (as discussed in Section 2.5 of Chapter 2). The better the molecular recognition, the faster will be the catalysis because there will be a lower likelihood of competing reaction pathways. It is in the intersection of good molecular recognition and a time-dependent catalytic action that enzymes achieve their remarkable selectivities and activities.

8.6 BIOMIMETIC CHEMISTRY AND ITS *PSEUDO-CONVERGENT* SYNERGIES

Accurate emulation of enzymatic catalysis in manmade systems is of inestimable importance. It offers the only realistic prospect of facilitating a wide range of important reactions that cannot currently be catalyzed on an industrial scale. Mimicry of enzymatic catalysis is the central aim of *biomimetic chemistry*, a field first defined and developed by Breslow [20].

To imitate enzymes properly, it is, of course, necessary to understand how they operate. Thus far, biomimetic chemistry has taken as its key goal the development of catalysts that are structurally complementary to their transition states. This goal stems from Pauling's proposal that the power of enzymes derives from their structural complementarity for the transition states [21] (see Chapter 5).

Although correct in part, this view omits the crucially important role of dynamic catalyst–reactant interactions in enzymatic catalysis. Weak and dynamic individual binding interactions are equally essential to achieving an authentic mimicry of enzymatic action. They are, for example, critical to achieving high selectivity in mixed reactant streams.

Consistent with Pauling's viewpoint, however, extensive efforts have been made in biomimetic chemistry to develop catalysts whose binding sites closely complement reaction transition states [22–24].

Some techniques, effectively, try to do this as a form of hybrid heterogeneous–homogeneous catalysis [25–27]. For example, the process known as *molecular imprinting* seeks to synthesize polymers containing cavities whose shapes complement reaction transition states. This synthesis is achieved by polymerizing monomers about molecules having shapes similar to those of the transition state. These molecules are later removed from the polymer.

Although certainly an interesting concept, molecular imprinting has not led to dramatic improvements in catalysis. This is likely because no cognizance has been taken of the need for conformational dynamism in the catalyst, nor of the need for dynamic catalyst–reactant binding.

Considerably greater success has been achieved, however, with carefully designed homogeneous catalysts in open solution. We will now discuss an example in this respect.

8.6.1 Cyclodextrin-Appended Epoxidation Catalysts: Pseudo-Convergence in a Nonbiological, Multicentered Catalyst

Perhaps the crowning achievement of biomimetic chemistry to date has been the development of homogeneous catalysts that display significant, albeit imperfect, convergence.

One example already mentioned in Section 2.5 of Chapter 2 is the Mn porphyrin **11**, which is laterally (*trans*-) appended with two β-cyclodextrins as shown in Figure 8.9 [28].

When treated with a linear olefin substrate **12** of appropriate length, the long-chain tails of the olefin are usually included in the two opposite cyclodextrin cavities, with the C–C double bond then held atop the Mn ion. Oxidative epoxidation consequently occurs mainly at this bond. However, it does not occur exclusively at this bond.

By contrast, the comparable *cis*-appended porphyrin **13** catalyzes the same reaction, but it does so entirely unselectively. This must be because a single olefin cannot simultaneously bind to both cyclodextrins in **13**.

The selectivity of **11** in the reaction therefore depends on the olefin interacting with the *trans*-cyclodextrin groups and thereafter being transformed by the Mn ion. Component functioning is certainly convergent on those occasions that these actions take place. But the actions are not essential to the catalysis. In other words, there is *no absolute requirement* for such actions in order for the Mn ion to catalyze the reaction. The Mn ion can and does act on its own on some occasions. In **13**, the Mn ion acts more or less exclusively on its own; the cyclodextrins play no obvious role at all.

The actions of the cyclodextrin and Mn catalytic groups in **11** can consequently be considered to converge in most, but not all, turnovers. It is not a genuine convergence since the catalytic groups do not depend on each other to carry out their respective tasks in order to achieve a catalytic effect. Although coordinated and simultaneous action occurs for much of the time, it is not essential to catalysis and no requirement for it exists. Product selectivity is therefore far from perfect.

Because it displays a form of convergence without capturing its full essence, **11** may be termed a *pseudo-convergent* catalyst. It is, almost certainly energy-dependent rather than time-dependent in its action. This is made clear by the fact that **11** can operate *without* the involvement of the cyclodextrin catalytic groups. As noted (Section 6.3 in Chapter 6), such an outcome is formally impossible in a time-dependent catalyst where each catalytic group relies on the others (see Chapter 5).

8.6 BIOMIMETIC CHEMISTRY AND ITS *PSEUDO-CONVERGENT* SYNERGIES

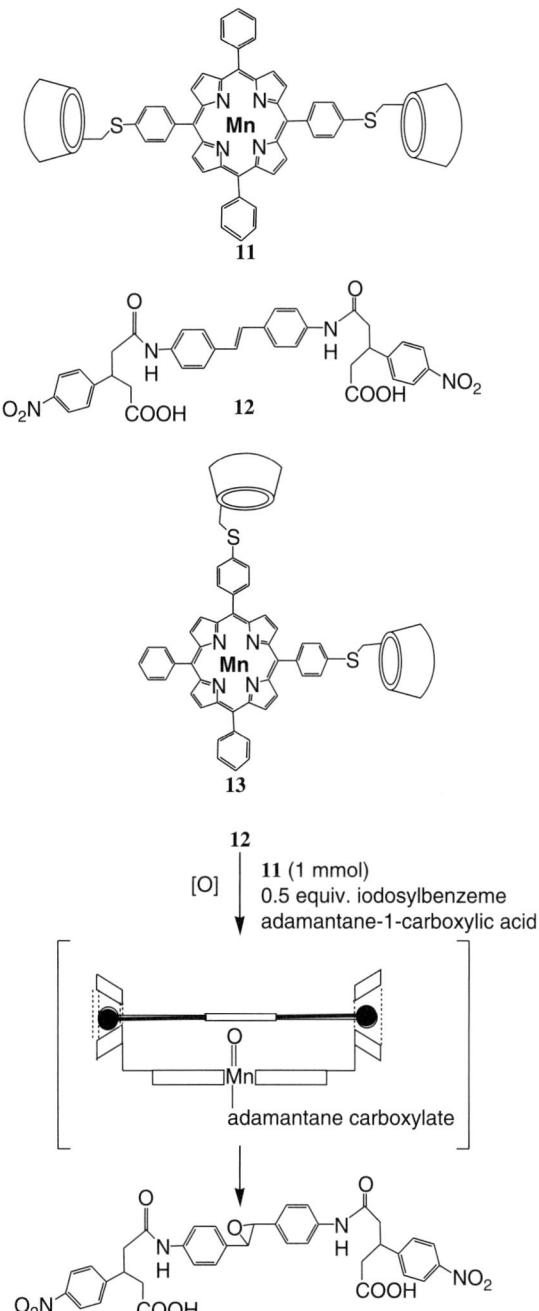

Figure 8.9. *Pseudo-convergence in a biomimetic catalyst.* Mn porphyrin **11** oxidizes substrate **10** with high, but not perfect, selectivity. Porphyrin **12** shows little selectivity under comparable conditions. Catalyst **11** therefore displays functional convergence on most, but not all, turnovers.

Time-dependent catalysts generate no catalytic effect whatsoever if all of the catalytic groups do not act simultaneously and in a coordinated, synchronized fashion.

Moreover, it is clear that **11** realizes its selectivity by *energetically* (thermodynamically) favoring a particular transition state. The cyclodextrin groups offer an *additional* stabilization to the desired transition state, which is what leads to it being favored. However, it does not guarantee the formation of this particular, desired transition state.

The strategy employed by **11** in this respect is, in fact, somewhat similar to that employed in **1** and **2**. That is, the catalyst is structured so as to favor thermodynamically some transition states and disfavor others. The only difference is that whereas the ligands in **1–2** act to destabilize competing transitions states selectively, the appended cyclodextrins in **11** act to increase selectively the stabilization of the desired transition state (thereby also disfavoring the competing transition states).

Thus, although **11** undoubtedly has a structure that is close to complementary to its transition state, it does not achieve optimum selectivity, which is due, in no small measure, to the fact that it is not time-dependent in its action. In other words, it does not depend on simultaneous, coordinated, synchronized actions by the Mn and cyclodextrin catalytic groups for its catalysis. The Mn=O oxo functionality that participates in transition state formation is simply too strongly bound and too strongly activated by the Mn ion to require the involvement of the cyclodextrin groups.

Biomimetic catalysis as it has been carried out to date therefore achieves a result that is *superficially* similar to that of many enzymes. The catalytic action is, however, quite different in character.

8.7 THE SPECTRUM OF SYNERGISTIC ACTION IN HOMOGENEOUS CATALYSIS

What becomes clear from the preceding discussions is that several forms of synergy exist in homogeneous catalysis. Indeed, a continuum of synergistic action can be observed. Figure 8.10 shows this spectrum. The stick figures at the top of Figure 8.10 are intended to provide a practical illustration of the various types of synergy. They depict the synergies as they would be observed in the comparable situation of the building of a cabinet by one or two men.

On the left, in Figure 8.10(a), is illustrated the case where only one man builds the cabinet. This man must do all of the sawing and nailing needed to build the cabinet. There is, consequently, no possibility of synergy in this case, since only one person is involved.

In catalysis, this equates to the situation where there is a single catalytic center and the entire reaction is facilitated only by this center. No other group plays a role in the catalytic action, either passively or actively.

Figure 8.10(b) depicts the situation where one man builds the cabinet while another man looks on. The observer does not actively participate in the building of the cabinet but assists in some passive way. He may, for example, hand the builder his tools or make suggestions that improve the work. The synergy involved here is

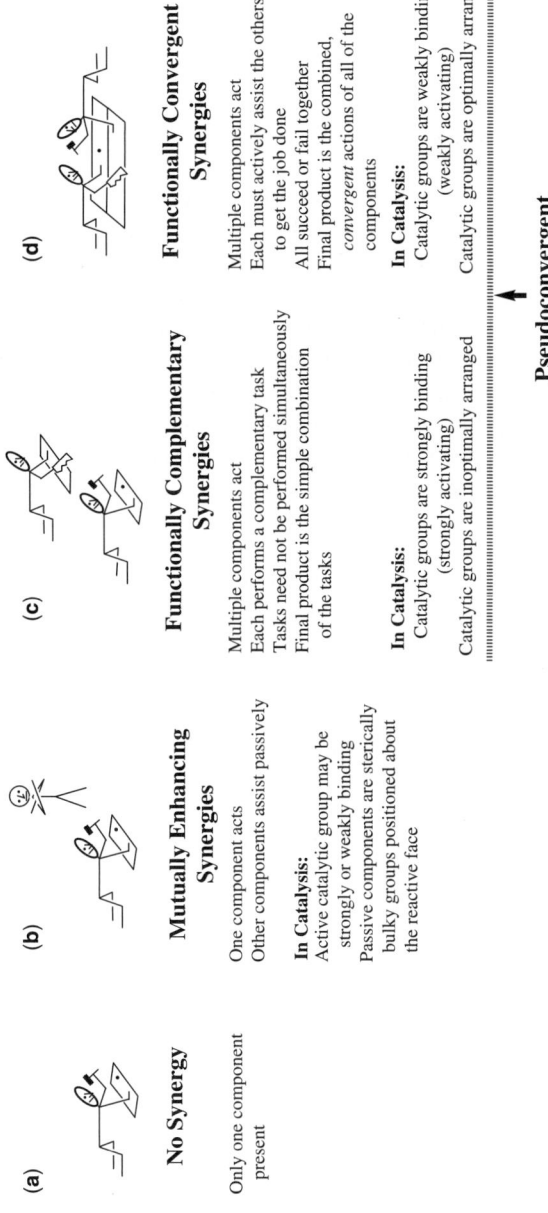

Figure 8.10. The hierarchy of synergy in catalysis. [Synergy increases in the order (a) to (d)].

mutual enhancement [1]. The effect of the passive observer is that a better cabinet is produced.

In catalysis, this equates to a single-centered catalyst in which bulky steric groups positioned about the active site improve the selectivity of the process. The presence of these groups disfavors possible competing transition states, thereby resulting in a more desirable outcome.

Figure 8.10(c) depicts the situation where both men are building the cabinet, with one man doing all of the nailing and the other doing all of the sawing. Each therefore performs complementary tasks and the cabinet is created by a simple combination of these tasks. As noted, synergies of this type are entitled *functional complementarity* [1].

In catalysis, functional complementarity is observed in multicentered species that combine strongly binding catalytic groups with a poor average spatial arrangement for transition state formation. The catalytic groups must be strongly binding because they need to hold their reactants for extended periods of time before conformational flexing brings the catalytic groups into a suitable collision for transition state formation. They also need to activate their reactants robustly (that is, polarize them in order to make them conducive for reaction) because such ideally oriented collisions are rare and they are seldom perfectly optimum. To form a successful transition state leading to products, the activation must be extreme.

Figure 8.10(d) depicts the case where both men are building the cabinet with the one man doing the sawing and the other the nailing. The difference here is that each cannot do his job without the assistance of the other. Thus, the man who is sawing needs the man who is nailing to hold steady his piece of timber, or help him in some other way. Similarly, the man who is nailing cannot do his task properly without the active assistance of his partner. In effect, both either succeed or fail together in their tasks.

Such synergies are titled *functional convergence* [1]. The term *convergence* means that the activities undertaken by the different men converge in a mutually reinforcing and amplifying way.

In catalysis, functional convergence occurs in a multicentered species when weakly binding and activating catalytic groups are combined with structural complementarity for the transition state. It is this complementarity that allows the use of weakly binding and activating groups.

8.7.1 The Relationship Between Complementary and Convergent Synergies

Near the bottom of Figure 8.10 is a broken line that spans both *functional complementarity* and *functional convergence*. This line is intended to indicate that a continuum of action exists between these two synergistic extremes. On the left, *Functional complementarity* combines strong individual binding interactions with poor structural optimization. On the right, the situation is exactly reversed: *Functional convergence* combines weak individual binding interactions with a near-optimum structural arrangement.

This continuum makes an important point. Because of the poor structural optimization of most nonbiological, multicentered catalysts, chemists have, thus far, been generally limited to using strongly binding and activating catalytic groups such as metal ions. However, the more optimally the catalyst can be structured, the more weakly binding and dynamically activating can be their catalytic groups.

Thus, a wide variety of new catalytic species that have hitherto been considered unsuitable for catalysis can, in fact, be gainfully employed in this respect. Chemists need to identify and study such groups because, in general, the more weakly binding the catalytic groups are, the greater will be their dynamism and, therefore, the more active and specific will be the catalysis.

Of course, strongly binding catalytic groups may also be combined with near-optimum structural arrangement. Such systems form the basis of the *pseudoconvergent* synergies referred to in the previous section. Synergies of this type are, in very broad terms, not as powerful as convergent synergies. The problem is that the binding interactions in pseudo convergent catalysts are *too strong*, which gives them the opportunity to participate in processes that are not desirable, thereby diminishing their catalytic selectivity. However, their strong binding and near-ideal structures will, certainly, generate high activities, which may be accompanied by significant selectivities; such catalysts cannot be ignored. We will discuss more fully their significance in the next section.

Finally, we must note that weakly binding and activating catalytic groups may also be combined with poor structural complementarity for the transition state. Such a combination will, of course, generate no catalytic effect whatsoever. This is the reason that functionally convergent systems rely on simultaneous, coordinated, synchronized actions to create a catalytic effect.

8.7.2 The Ideal Catalyst

One of the objectives of this series was to understand, in principle, the constitution of the perfect homogeneous catalyst and how it could be designed.

It is now clear that an excellent guide for the rational design of efficient homogeneous catalysts is to aim to achieve the greatest possible catalytic synergies.

In very broad terms, multicentered catalysts will generally be more selective and active than single-centered catalysts because they are capable of displaying higher synergies of action. This develops because of the 3-D nature of the active site, compared with the facial, 2-D active sites in single-centered catalysts.

Moreover, *time-dependent*, multicentered catalysts will, in broad terms, be generally more active and selective than *energy-dependent*, multicentered catalysts because they require convergent synergies merely to exhibit a catalytic effect. If the aim is to seek the most efficient possible catalyst, then a good target would be to design a multicentered, time-dependent catalyst.

All of the above are, of course, extremely broad generalizations. They are mere guidelines. The most efficient single-centered catalyst will, certainly, be far superior to a poorly performing multicentered, time-dependent catalyst.

Nevertheless, the realm of time-dependent, multicentered catalysis is the place *most likely* to yield an excellent catalyst.

What, then, would the very best possible homogeneous catalyst look like? To answer this question, we need to go back to Sabatier's principle and the volcano plots that derive from it, as discussed in Chapter 4.

If we could draw a volcano plot for a reaction in homogeneous catalysis, it would probably look something like Figure 8.11.

From the point of view of catalytic activity, the "best" catalyst will, of course, lie at the peak of the volcano plot, where the dynamism of catalyst–reactant interactions are perfectly balanced with the thermodynamic efficiency of transition state formation.

Very few catalysts will achieve such a balance. Most will lie to one side or the other. That is, the catalyst will be dominated either by the dynamism of the catalyst–reactant interactions (time-dependent) or by the efficiency of transition state formation (energy-dependent).

What, then, is the next best option?

A catalyst that lays near the peak of the volcano plot on the right-hand side of the dividing line would display *pseudo convergent* synergies. As mentioned, such a catalyst would certainly display impressive activity, but this would probably be accompanied by imperfect selectivity.

On the other hand, a catalyst that lays near the peak of the volcano but on the left-hand side of the dividing line would display what could be called *highly efficient functional convergence*. Such a species would, arguably, be the "ultimate" catalyst. It would differ from the comparable pseudoconvergent catalyst in that the strength of its individual binding interactions would be slightly less than optimum.

But the outcome of this difference would be to create the best possible catalytic selectivity. Its selectivity would be better than the comparable pseudoconvergent catalyst because the weakness of its binding interactions would leave the reactants with insufficient time to participate in undesirable side reactions.

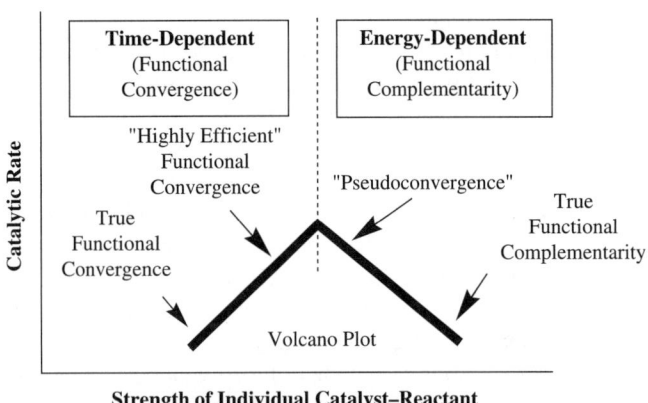

Figure 8.11. Hypothetical volcano plot for multicentered homogeneous catalysis.

At this stage we cannot definitively predict how selectivity is likely to vary down the left-hand side of the volcano plot in Figure 8.11. That is, we do not know if "true functionally convergent" catalysts (bottom left of Fig. 8.11) are typically more or less selective than "highly efficient functionally convergent" catalysts.

However, the very limited evidence available, which consists really only of Corey's bis-cinchona alkaloid catalysts [10], suggests that the more active a time-dependent catalyst is, the more selective will be the catalytic process.

This is certainly true on the right-hand side of Figure 8.11. Pseudo-convergent catalysts are generally much more selective than the "true functionally complementary" catalysts on the bottom right of Figure 8.11.

Thus, it seems that high activity and high selectivity go hand in hand. The greater the activity of the catalyst, the larger will be its catalytic selectivity. This finding seems to be true within both the time-dependent and the energy-dependent realms.

Another noteworthy feature is thereby illustrated. Although, pseudoconvergent biomimetic catalysts do not achieve their stated aim of imitating enzymatic action, they do significantly advance the drive to highly efficient homogeneous catalysis. They only do it from the "wrong" side of the volcano curve. Whereas enzymes achieve high efficiency by lying toward the upper left-hand side of Figure 8.11, pseudoconvergent catalysts seek to approach the same peak performance but from the upper right-hand side of Figure 8.10.

The target is the same in both cases: to achieve a balance between catalyst–reactant binding dynamism and thermodynamic efficiency, as prescribed by Sabatier's principle. Enzymes and pseudo convergent catalysts simply approach this target from two different directions.

8.8 SYNERGY IN CATALYSIS IS CONCEPTUALLY RELATED TO OTHER SYNERGISTIC PROCESSES IN HUMAN EXPERIENCE

As a final remark to this chapter, we must note that the concept of synergy cuts across numerous fields of nature and of human experience. Catalysis is only one of those fields. The synergies achieved in catalysis, therefore, have counterparts in other aspects of human endeavor. We have, for example, already mentioned their similarities to the synergies that may be observed when two men build a cabinet.

But the connections go much further than that. Synergy provides a conceptual connection with all manner of action–reaction sequences outside of chemistry and catalysis, including extremely powerful and important ones. For example, *evolution* and *laissez-faire economics*, among others, also seem to often display convergent synergies [1]. That is, these systems seem often to operate by harnessing coordinated, synchronized activities to bring about truly extraordinary outcomes that can simply not be achieved in any other way. Our understanding of these synergies and the way that they shape events is extremely rudimentary at present. Yet, they may govern large and important issues in nature and in human society. Indeed, they may cut to the core of numerous natural and human phenomena.

We can cite two examples to illustrate just how far-reaching the issue of synergy may be. We cannot vouch for the validity of these examples, but their potential significance is undeniable.

The first example is political and economic: It is argued that the apparent efficiency of *capitalism* (laissez-faire economics) over *communism* (centrally planned economics) may have to do with the higher synergies achieved in the former system. Communism, being essentially bureaucratic and "top-down" in character, could never generate the sort of synchronized, mutually beneficial, convergent behavior that is characteristic of capitalism. In effect, although communism explicitly aspires to such convergent behavior, it does not recognize, nor set up, the conditions to bring it about.

The second example is religious: It is argued that the concept of *forgiveness*, which is advocated by several religious orders, derives its power from the change in synergy that results when individuals break, rather than propagate, negative, behavioral, action–reaction chains. In effect, by forgiving an enemy, one may create a new, positive synergy in society that encourages and better allows for synchronized, mutually beneficial, convergent behavior.

The topic of synergy is, therefore, a very broad subject indeed. It is also a deeply fundamental one. The synergies present in catalysis describe only a very limited aspect of this field.

Key Point. **The extraordinary catalytic power that develops as a result of the high orders of synergy in many enzymes, consequently, has a profound and far-reaching origin that goes far beyond mere chemistry and catalysis.** In effect, such enzymes seem to have harnessed a very basic and powerful driver for efficiency. This driver is seen also in other, unrelated, but significant systems involving sequential action–reaction chains.

REFERENCES

1. Corning, P. A. *Systems Research* **1995**, *12*, 89 (published on the web at http://www.complexsystems.org/publications/pdf/synselforg.pdf).
2. Adams, R. D. *J. Organometal. Chem.* **2000**, *600*, 1, and references therein.
3. (a) Brintzinger, H. H.; Fischer, D.; Mulhaupt, R.; Rieger, B.; Waymouth, R. M. *Angew. Chem. Int. Ed. Engl.* **1995**, *34*, 1143; (b) Bochmann, M. *J. Chem. Soc., Dalton Trans.* **1996**, 255; (c) Ewen, J. A.; Elder, M. J.; Jones, R. L.; Curtis, S.; Cheng, H. N. *Int. Symp. Catalytic Olefin Polymerisat*, Tokyo, Japan, 1989.
4. (a) Grubbs, R. H.; Coates, G. W. *Acc. Chem. Res.* **1996**, *29*, 85; (b) Blaser, H. *Chem. Rev.* **1992**, *92*, 935.
5. Pfeil, A.; Lehn, J.-M. *J. Chem. Soc. Chem. Commun.* **1992**, 838.
6. Brintzinger, H. H.; Fischer, D.; Mulhaupt, R.; Rieger, B.; Waymouth, R. M. *Angew. Chem. Int. Ed. Engl.* **1995**, *34*, 1143.
7. Ewen, J. A.; Elder, M. J.; Jones, R. L.; Curtis, S.; Cheng, H. N. *Int. Symp. of Catalytic Olefin Polymerisation*, Tokyo, Japan, 1989.

REFERENCES

8. Bochmann, M. *J. Chem. Soc., Dalton Trans.* **1996**, 255.
9. Grubbs, R. H.; Coates, G. W. *Acc. Chem. Res.* **1996**, *29*, 85.
10. Corey, E. J.; Noe, M. C. *J. Am. Chem. Soc.* **1996**, *118*, 319.
11. Otto, S.; Bertoncin, F.; Engberts, J. B. F. N. *J. Am. Chem. Soc.* **1996**, *118*, 7702.
12. Menger, F. M.; Tsuno, T. *J. Am. Chem. Soc.* **1989**, *111*, 4903.
13. Shinkai, S.; Ishikawa, Y.; Shinkai, H.; Tsuno, T.; Makishima, H.; Ueda, K.; Manabe, O. *J. Am. Chem. Soc.* **1984**, *106*, 1801.
14. Pirrung, M. C.; Liu, H.; Morehead, A. T. *J. Am. Chem. Soc.* **2002**, *124*, 1014.
15. (a) Hui, B. C.; James, B. R. *Can J. Chem.* **1974**, *52*, 3760; (b) James, B. R.; Wang, D. K. W. *Can. J. Chem.* **1980**, *58*, 245; (c) Joshi, A. M.; MacFarlane, K. S.; James, B. R. *J. Organometal. Chem.* **1995**, *488*, 161.
16. Hanabusa, K.; Ye, X.; Koyama, T.; Kurose, A.; Shirai, H.; Hojo, N. *J. Mol. Catal.* **1990**, *60*, 127.
17. (a) Su, C.-C.; Reed, J. W.; Gould, E. S. *Inorg. Chem.* **1973**, *12*, 337; (b) Arakawa, H.; Moro-oka, Y.; Ozaki, A. *Bull. Chem. Soc. Jpn.* **1974**, *47*, 2958.
18. Konsler, R. G.; Karl, J.; Jacobsen, E. N. *J. Am. Chem. Soc.* **1998**, *120*, 10780.
19. (a) Menger, F. M. *Acc. Chem. Rev.* **1985**, *18*, 128, and references therein; (b) Menger, F. M. *Acc. Chem. Rev.* **1993**, *26*, 206, and references therein.
20. Breslow, R. *Acc. Chem. Res.* **1995**, *28*, 146.
21. Pauling, L. *Chem. Eng. News* **1946**, *24*, 1375; Pauling, L. *Nature* **1948**, *161*, 707.
22. Breslow, R.; Dong, S. D. *Chem. Rev.* **1998**, *98*, 1997.
23. Stoddart, J. F. In *Enzyme Mechanisms*, Eds. Page, M. I.; Williams, A., The Royal Society of Chemistry, 1987, p. 35–55; Breslow, R. *Chem. Rec.* **2001**, *1*, 3; Behr, J. P.; Lehn, J. M. In *Structural and Functional Aspects of Enzyme Catalysis*, Eds. Eggerer, H.; Huber, R., Springer-Verlag, 1981, p. 24–32; Klotz, I. In *Enzyme Mechanisms*, Eds. Page, M. I.; Williams, A., The Royal Society of Chemistry, 1987, p. 14–34.
24. Saenger, W. In *Structural and Functional Aspects of Enzyme Catalysis*, Eds. Eggerer, H.; Huber, R., Springer-Verlag, 1981, p. 33–42; Bender, M. In *Enzyme Mechanisms*, Eds. Page, M. I.; Williams, A., The Royal Society of Chemistry, 1987, p. 56–66.
25. Brady, P. A.; Sanders, J. K. M. *Chem. Soc. Rev.* **1997**, *26*, 327.
26. Wolfenden, R.; Frick, L. In *Enzyme Mechanisms*, Eds. Page, M. I.; Williams, A., The Royal Society of Chemistry, 1987, p. 97–122.
27. Rowan, S.; Sanders, J. K. M. *Curr. Opin. Chem. Biol.* **1997**, *1*, 483.
28. Breslow, R.; Zhang, X.; Xu, R.; Maletic, M.; Merger, R. *J. Am. Chem. Soc.* **1996**, *118*, 11678.

9

A CONCEPTUAL UNIFICATION OF HETEROGENEOUS, HOMOGENEOUS, AND ENZYMATIC CATALYSIS

GERHARD F. SWIEGERS

9.1 INTRODUCTION

In Chapter 2 we noted that the field of catalysis is currently divided into three subdisciplines: heterogeneous, homogeneous, and enzymatic catalysis. Each subdiscipline has an apparently distinct set of principles, which, often, do not readily translate into the other disciplines. This lack of consistency seems to be amplified by the fact that each subdiscipline is largely the preserve of a different group of specialist researchers, with heterogeneous catalysis being dominated by chemical engineers, homogeneous catalysis by chemists, and enzymatic catalysis by biochemists.

Catalysis science has long sought a means to unify coherently the different subdisciplines [1,2]. However, the common strand that relates these fields has proved difficult to identify.

A key problem has been the physical diversity of the catalytic actions both within the different subdisciplines and between them. Whereas heterogeneous species employ two-dimensional (2-D) surfaces comprising large collections of atoms to act as catalysts, homogeneous catalysts involve three-dimensional (3-D) active sites distributed within homogeneous media. This distinction has made it all the more problematic to unearth the underlying commonality that connects the different subdisciplines.

What, then, is the connection? What is missing from our conceptualization of catalysis that will allow us to unify these subdisciplines into a single, coherent whole?

Mechanical Catalysis: Methods of Enzymatic, Homogeneous, and Heterogeneous Catalysis,
Edited by Gerhard F. Swiegers
Copyright © 2008 John Wiley & Sons, Inc.

In this chapter, we will examine this question. We will show that the missing element in our understanding of catalysis is a recognition of time- and energy-dependence. Once these realms are taken into account, the fundamental unity of catalysis, across the entire spectrum of catalytic action, becomes clear.

9.2 DIFFUSION-CONTROLLED AND REACTION-CONTROLLED CATALYSIS

Before discussing the key topics of this chapter in detail, we need to revise briefly the concept of *diffusion-controlled* or *diffusion-limited* catalysis that we touched on in Section 1.5.3 in Chapter 1 and Section 3.3.4 of Chapter 3. These terms originate in the field of heterogeneous catalysis, where they are used to describe industrial catalytic processes whose overall rate is determined by the rate at which reactant feedstock is fed to the catalyst.

As mentioned, the rate at which some catalytic systems facilitate a reaction is limited by how quickly the reactants diffuse to the catalyst. This will typically happen when relatively little reactant is present about each catalyst. In such a case, the average time required for the reactants to diffuse to the catalyst will normally be substantially longer than the average time taken by the catalyst to generate products once the reactants have bound or adsorbed to it.

Virtually any catalytic process will ultimately become diffusion-limited if the concentration of reactants about it is systematically decreased. Of course, the actual reactant concentration at which diffusion control sets in will vary from catalyst to catalyst. It will depend on the efficiency of the catalyst and the rapidity with which that catalyst can generate products after binding/adsorbing to the reactants. The faster this occurs, the higher will be the reactant concentration below which the process becomes diffusion controlled.

In the extreme case, where the catalyst turns over bound reactants into products very quickly indeed, catalysis may be diffusion controlled at, essentially, any reactant concentration. That is, the speed of the catalyst is then so great that the overall rate, at any reactant concentration, will depend on how fast the reactants can get to the catalyst.

Catalysis of this type is rare, but it is observed, in particular, in the field of enzymatic catalysis [3]. Several enzymes, including triosephosphate isomerase and carbonic anhydrase, are diffusion controlled at all available substrate concentrations. Catalysis of this type is referred to as being *kinetically perfect*. It typically involves k_{cat}/K_m factors in the order of 10^8 to $10^9 \, M^{-1} \, s^{-1}$ [3].

The opposite side of the coin is catalysis in which the process of reaction on the catalyst is slower than the rate at which the reactants diffuse to the catalyst. Catalytic systems of this type are described in the field of heterogeneous catalysis as being *reaction controlled*. That is, the overall rate of the catalytic process is dependent on how quickly the catalyst facilitates the reaction (i.e., how quickly it turns over any attached reactants into products). It is not affected by the rate at which reactant feedstock is fed to the catalyst.

Except for the *kinetically perfect* enzymes and related examples mentioned above, virtually all catalysts may be either *diffusion-* or *reaction-controlled* depending only on the concentration of reactants present about the catalyst.

The term *diffusion-controlled*, therefore, does not provide any information in respect of the nature of the catalytic action and whether it is fundamentally time- or energy-dependent. We have, consequently, not considered the distinction between diffusion-controlled and reaction-controlled catalysis in this series. Instead, we have examined only the actual catalytic action during the reaction.

9.3 THE DIVERSITY OF CATALYTIC ACTION IN HETEROGENEOUS CATALYSTS

In Chapter 4, we examined the incidence and properties of time- and energy-dependent catalysts in the field of heterogeneous catalysis. We showed that both *time-* and *energy-dependence* are observed in heterogeneous catalysis. These realms are created by the *residence time* τ, during which the reactants remain fully bound to the catalyst on each occasion of binding.

When the residence time is long, the catalytic process tends to be *energy-dependent* because the reactants will remain attached to the catalyst during many different collisions with other reactants. The overall rate of catalysis, therefore, depends only on how often a successful collision is undertaken. In other words, it depends on the average threshold energy that must be overcome during the collisions. In the absence of diffusion control, the reaction rate is, consequently, determined by the activation energy E_a of the catalyzed reaction, and this is reflected in the kinetics of the process.

However, when the residence time of the reactants on the catalyst is very short on each occasion of binding, then the number of collisions between bound reactants falls to the point that it becomes rate-limiting. This result occurs because the reactants spend most of their time binding and releasing the catalyst, not colliding with each other on the catalyst surface. In such a *time-dependent* case, a substantial catalytic effect is observed only if the collisions that occur have a high likelihood of generating products. In the absence of diffusion control, the reaction rate is consequently determined by the rate at which the reactants bind the catalyst and by the residence time during which they remain bound. This effect is reflected in the kinetics of the catalysis, which is dominated by the presence of a rapidly equilibrating intermediate.

On a heterogeneous catalyst, all of the collisions between attached reactants must occur on a 2-D surface. Thus, reactant collisions will occur when, for example:

(1) An unbound reactant collides with a reactant that is already adsorbed to the catalyst surface
(2) A reactant adsorbs to a catalyst site immediately adjacent to an already occupied site

(3) An adsorbed reactant reorientates itself within a catalytic site to thereby come into contact with an adjacent reactant
(4) A reactant migrates from atom to atom across a catalyst surface until it contacts another attached reactant

In all collisions, the catalyst brings about the reaction by activating the attached reactant, that is, by polarizing their electronic structure to thereby make it conducive to reaction. In the case of, for example, a Pt catalyst, this activation is believed to derive from the presence just above the Pt surface, of a layer of unoccupied Pt d-orbitals. Adsorbed reactants interact with, and are polarized by, these orbitals. In so doing, they are predisposed to react with other activated reactants into which they bump on the catalyst surface.

9.4 THE DIVERSITY OF CATALYTIC ACTION IN NONBIOLOGICAL HOMOGENEOUS CATALYSTS

In nonbiological homogeneous catalysts, many of the same processes observed in heterogeneous catalysis take place. Thus, reactants bind to the catalytic group/s of the catalyst. In so doing, they are activated and their electronic arrangements are polarized. The catalyst then brings them into contact with other activated reactants, thereby creating the possibility of generating a product molecule. Alternatively, it pulls them apart into separate chemical entities or rearranges them into a new chemical entity.

A key difference between a heterogeneous catalyst and a nonbiological one is the manner in which binding and activation occurs. In heterogeneous catalysis, reactants adsorb to the surface of the catalyst and interact with a collective electronic structure that is created by the presence of many catalyst atoms all packed together. In homogeneous catalysis, they undergo chemical binding to a specific catalytic group, which then alters their electronic structure.

The way in which collisions are brought about is also different. In a homogeneous catalyst, collisions are mediated when, for example:

(1) A reactant physically contacts another reactant during binding to a catalytic site that already contains an attached reactant (e.g., in single-centered catalysis).
(2) An attached reactant reorientates itself within a catalytic site, thereby contacting another attached reactant in the same (single-centered) or an adjacent (multicentered) catalytic site.
(3) The catalyst undergoes a conformational change to thereby bring two or more reactants into contact with each other (in the case of multicentered catalysts).

9.4 THE DIVERSITY OF CATALYTIC ACTION IN NONBIOLOGICAL CATALYSTS

Thus, the physical mechanism and the intricacies of binding, activation, and collision are different in homogeneous catalysts compared with heterogeneous catalysts. But the overall process that takes place is not. It also involves reactant binding, activation, and collision.

A significant difference, however, in nonbiological homogeneous catalysis relative to its heterogeneous counterpart is the kinetics of the process. Whereas heterogeneous catalysts display kinetics that is characteristic of both time- and energy-dependence, nonbiological homogeneous catalysts display, overwhelmingly, only energy-dependent kinetics. That is, exceedingly few nonbiological homogeneous catalysts are observed to contain a rapidly equilibrating intermediate in their rate expressions.

This result occurs because, as we stated in Chapters 5 and 6, the only way to achieve a catalytic effect in the face of rapid binding and dissociation of a reactant is for the catalyst to structurally complement the optimum transition state of the reaction. If the catalyst does not have this structure, then no reaction can occur because there is simply insufficient time on each occasion of binding for the catalyst to bring the reactants into successful contact with each other.

One may ask at this stage, why is there such a sharp and critical difference in this respect between *nonbiological* heterogeneous and homogeneous catalysts?

Several reasons are likely, but the key ones probably have to do with (1) the multisite nature of heterogeneous catalysis (versus the single-site nature of homogeneous catalysis) and (2) the 2-D character of catalysis in heterogeneous catalysis (versus the 3-D character in most homogeneous catalysis).

Because heterogeneous catalysts simultaneously employ numerous, different face, edge, defect, and other catalytic sites (see Section 2.2.1, Chapter 2), they offer a variety of geometries to facilitate catalysis. Only one of these geometries needs to be suitable for transition state formation when rapidly associating–dissociating reactants are involved. Thus, an inherent tendency in heterogeneous catalysis exists to facilitate reactions by reactants that enjoy both strong and weak individual interactions with the catalyst.

Homogeneous catalysts are, by contrast, single site in character (see Section 2.2.1, Chapter 2). Therefore, only a single geometry of active site is available. If this geometry is not complementary to the desired transition state, the catalyst cannot facilitate a reaction involving weak, individual catalyst–reactant interactions. Time-dependent catalysis is then impossible.

This difficulty is made all the more demanding by the fact that the reactants bind and must be brought into collision in 3-D space in many homogeneous catalysts. Thus, the reactants are not necessarily constrained to bind to the catalyst in a 2-D plane, or near thereto, as they are in a heterogeneous catalyst. The removal of this constraint imparts weakly interacting reactants with more orders of freedom, thereby diminishing the likelihood that they will coincidentally bind in an arrangement that allows for catalysis.

Thus, nonbiological homogeneous catalysts are generally energy-dependent and not time-dependent.

9.5 THE DIVERSITY OF CATALYTIC ACTION IN ENZYMES

The catalysts of biology (enzymes) form a subgroup of homogeneous catalysts that mediate the key processes of binding, activation, and collision in the same way that it occurs in nonbiological homogeneous catalysts. However, they are distinguished from nonbiological homogeneous catalysts in several significant ways, one of which is that many enzymes display kinetics consistent with a time-dependent catalytic action. That is, they exhibit a rapidly equilibrating intermediate in their rate expression.

As noted in Section 5.6.4 of Chapter 5, this property of time-dependence derives from brevity in the residence time τ, during which substrates are fully bound to such enzymes. That is, whereas substrates are often strongly associated with such an enzyme (as testified by the long periods during which they are *partially bound* and their high $1/K_m$ values [4]), they are usually *fully bound and properly activated* for only very brief periods on each occasion of binding. The resulting brevity in the residence time creates a time-dependent action in precisely the same way that it does in time-dependent heterogeneous catalysts (see Section 4.2.3.1 in Chapter 4).

Because of the short residence time, a catalytic effect can only be generated in a protein if the active site flexes about a structure that is complementary to its transition state. For this reason, many enzyme active sites are structurally complementary to their transition states [5].

The weakness of the individual interactions of biology (namely, hydrogen bonding, hydrophobic–hydrophilic, ion-pairing, and van der Waals interactions) and the resulting dynamism of substrate interactions with proteins has meant that time-dependent catalysis is favored in biological systems. This weakness places severe restrictions on the types of species that may undertake catalysis; they must have very particular 3-D geometries in order to display a catalytic effect. These geometries must complement the structure of the reaction transition state during conformational flexing.

Fortunately, the many different proteins in nature demonstrate an almost infinite variety of 3-D geometries in their many folds, crevices, and other physical structures. Despite the stringency of the above demand, numerous proteins have therefore proved able to act as powerful and efficient catalysts.

9.6 HETEROGENEOUS CATALYSIS AND ENZYMATIC CATALYSIS HAS, EFFECTIVELY, INVOLVED COMBINATORIAL EXPERIMENTS THAT HAVE PRODUCED TIME-DEPENDENT CATALYSTS. NONBIOLOGICAL HOMOGENEOUS CATALYSIS HAS NOT

Key Point. In effect, then, biology, with its innumerable protein folds and creases, has acted as a giant combinatorial experiment in which proteins having structures suitable for time-dependent catalysis have found themselves capable of carrying out specific catalytic transformation. This combinatorial experiment is similar to the combinatorial experiment that occurs on the many step, defect, and other sites on the surface of heterogeneous catalysts. The biological version of time-dependent catalysis has differed from that in heterogeneous catalysis in that it has been limited to a narrower band of individual catalyst–reactant interactions because of the predominance of amino acid groupings in proteins (although proteins also contain a diversity of metal ions). The fact that these two classes of catalyst employ a combinatorial approach seems to have made time-dependence common in them, although it seems to be more prevalent in enzymes than in heterogeneous systems.

Similar combinatorial experiments have not been developed or attempted in manmade, nonbiological homogeneous catalysts, which have instead been limited to single species, examined in isolation. It is arguably for this reason that time-dependence is rare in abiological homogeneous catalysis.

A question that we will try to answer in the next chapter is as follows: How can one design a combinatorial experiment that yields a practical, manmade, time-dependent homogeneous catalyst?

9.7 HOMOGENEOUS AND ENZYMATIC CATALYSTS ARE THE 3-D EQUIVALENT OF 2-D HETEROGENEOUS CATALYSTS

As described in Chapter 4, we first recognized the practical presence of time-dependent catalysis in the field of heterogeneous catalysis. The key impediment to such catalysis involved the limited time available for the bound reactants to come into physical contact (collision) with each other. Thus, time-dependent heterogeneous catalysis is controlled by the catalyst-mediated collision frequency.

In the comparable energy-dependent heterogeneous systems, reactant binding is not dynamic. As such, the key barrier to catalysis involves the energetic efficiency of the catalytic process. In other words, energy-dependent heterogeneous catalysis is controlled by the activation energy E_a of the system.

Identical processes are clearly at play in multicentered homogeneous catalysts. Thus, the properties of time-dependent homogeneous catalysts, like many enzymes, are also governed by the collision frequency of the bound reactants. Similarly, energy-dependent homogeneous catalysts, like most nonbiological multicentered catalysts, display properties that derive from the need to overcome the activation energy of the reaction.

Key Point. **In effect, the most fundamental catalytic processes that take place in heterogeneous, homogeneous, and enzymatic catalysis are the same. They differ only insofar as they play out in different settings. In heterogeneous catalysts, they generally take place on what are, effectively, 2-D or near-2-D surfaces of large agglomerations of atoms. In many single-centered homogeneous catalysts, they also involve 2-D space but around single atoms that must carry out all of the catalytic functions. In multicentered homogeneous catalysts and in enzymes, however, they usually take place in 3-D space.**

The availability of 3-D space explains the many advantages that homogeneous and enzymatic catalysts have over heterogeneous catalysts (see Sections 2.2.2 and 2.2.3 in

Chapter 2). Obviously, many more options and variables are available in 3-D space than in 2-D space. In particular, 3-D space provides opportunities for synergies in catalytic actions that are not generally available to heterogeneous catalysts, which are constrained to a 2-D or a near-2-D environment. This is particularly true of the synergies employed by enzymes. Although these synergies are substantially more difficult to achieve in 3-D space, they are that much more powerful. Moreover, nature has conveniently acted as a combinatorial experiment with which to realize such synergies. Proteins, with their large and varied 3-D structures, have been able to create the necessary synergies in the form of many enzymes and have therefore been able to bring about time-dependent catalysis.

Similar, but presumably lower order, synergies are achieved on the diverse surface features of heterogeneous catalysts. These features have also offered a combinatorial experiment that makes possible time-dependent catalysis.

Thus, the most fundamental processes, properties, and features of heterogeneous, homogeneous, and enzymatic catalysis are conceptually indistinguishable. They differ only in their setting.

Key Point. The common conceptual thread that links the subdisciplines of heterogeneous, homogeneous, and enzymatic catalysis is, therefore, the existence of time- and energy-dependent catalytic actions. Heterogeneous catalysis seems to employ routinely both time- and energy-dependent processes. Nonbiological homogeneous catalysis is mainly energy-dependent, whereas biological catalysis is largely time-dependent.

Homogeneous and enzymatic catalysis are, therefore, in both principle and practice, the 3-D analogs of the 2-D surface catalysis present in heterogeneous systems. Each subdiscipline of catalysis consequently does not have its own set of unique catalytic principles. Instead, universal catalytic principles are merely applied differently and selectively in the various classes.

9.8 A CONCEPTUAL UNIFICATION OF HETEROGENEOUS, HOMOGENEOUS, AND ENZYMATIC CATALYSIS

By explicitly recognizing time- and energy-dependence, it is possible to demonstrate a conceptual unification of heterogeneous, homogeneous, and enzymatic catalysis. The hierarchy of this unified view of catalysis is illustrated in Figure 9.1.

As shown, the highest level of the hierarchy divides the field of catalysis into biological ("Catalysis in Biology") and nonbiological catalysis ("Catalysis in Chemistry"). The latter separates into two subdisciplines: homogeneous and heterogeneous catalysis.

Heterogeneous catalysis comprises catalysts that contain multiple different active sites (*multisite catalysis*). These sites can be divided into time- and energy-dependent catalysts. The former catalysts display a catalytic effect because of the diversity of active site structures present on the surfaces of many heterogeneous catalysts.

9.8 A CONCEPTUAL UNIFICATION OF CATALYSIS

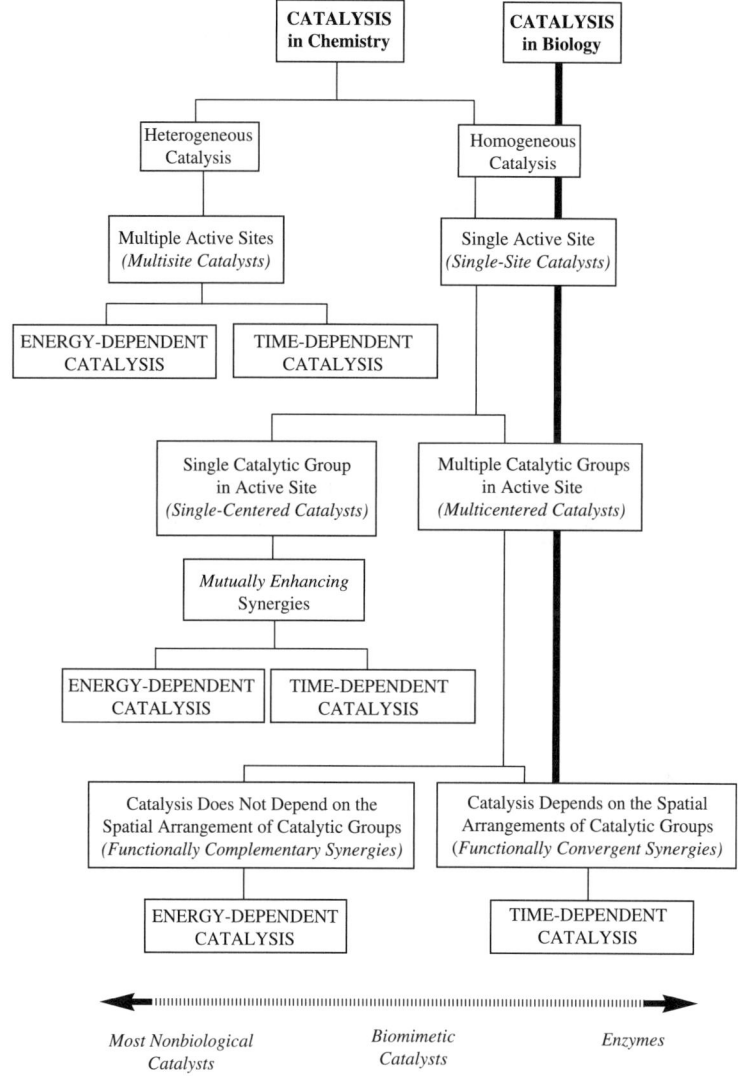

Figure 9.1. *The conceptual overlap of catalysis in chemistry and biology.* This depiction considers only reaction-controlled catalytic systems.

Homogeneous catalysis comprises catalysts with a single active site (*single-site catalysts*). The catalysts of biology are also single-site homogeneous catalysts.

Such catalysts can be divided into *single-centered* species, which contain a single atom that performs all catalytic functions, and *multicentered* species, which involve more than one atom acting catalytically in the active site. Most biological enzymes are *multicentered* in character.

Single-centered homogeneous catalysts generally display so-called *mutually enhancing* synergies, as described in Section 8.3 of Chapter 8. They can be divided into energy- and time-dependent catalytic actions, although the latter seem to be rarer than the former. Very little is, indeed, known about time-dependent, single-centered homogeneous catalysts. Only a few examples of such catalysts are known, and they have not been thoroughly studied.

Multicentered homogeneous catalysts can also be divided into energy- and time-dependent catalysts. The former catalysts display so-called *functionally complementary* synergies. Time-dependent, multicentered catalysts are, by contrast, *functionally convergent* in their action. Many enzymes, but certainly not all, seem to be time-dependent catalysts with functionally convergent synergies.

REFERENCES

1. (a) Author not stated: *Catal. Lett.* **2001**, *76*, 111, and references therein; (b) Bercaw, J., *Opportunities for Catalysis in the 21st Century, A Report from the Basic Energy Sciences Advisory Committee* (Workshop Chair: White, J. M.), May 14–16, 2002, U.S. Department of Energy.
2. Thomas, J. M.; Williams, R. J. *Phil. Trans. R. Soc.* **2005**, *363*, 765.
3. Kinetic "perfection" in an enzymatic system is described in: Knowles, J. R.; Albery, W. J. *Acc. Chem. Res.* **1977**, *10*, 105.
4. Stryer, L. *Biochemistry (3rd ed.)*, W. H. Freeman and Company, 1988, p. 182–183, p. 191.
5. Pauling, L. *Nature* **1948**, *161*, 707.

10

THE RATIONAL DESIGN OF TIME-DEPENDENT ("MECHANICAL") HOMOGENEOUS CATALYSTS. A LITERATURE SURVEY OF MULTICENTERED HOMOGENEOUS CATALYSIS

Junhua Huang and Gerhard F. Swiegers

10.1 INTRODUCTION

In a series of seminal articles and reviews, Breslow coined the term and described the principles of "biomimetic chemistry." This field seeks to design rationally nonbiological catalysts that mimic the action of enzymes [1–3]. The practical intention is to develop new catalysts whose efficiency and utility are comparable with enzymes. The successful achievement of this aim is widely considered to be one of the "Holy Grails" of chemistry [4].

But what are the fundamental processes involved in enzymatic catalysis? Debate has raged about this most fundamental of questions for decades.

In Chapters 5 and 6, we established a fundamental attribute that seems to distinguish enzymes from comparable nonbiological homogeneous catalysts. Many enzymes seem to be *time-dependent* ("mechanical") catalysts, whereas most manmade homogeneous catalysts are *energy-dependent* ("thermodynamic").

Time-dependent catalysts are limited by their ability to bring about mechanical collisions between bound substrate functionalities and the frequency with which

---

Mechanical Catalysis: Methods of Enzymatic, Homogeneous, and Heterogeneous Catalysis, Edited by Gerhard F. Swiegers
Copyright © 2008 John Wiley & Sons, Inc.

they do so (the "collision frequency"). (In reactions where substrates are pulled apart, "collision frequency" refers to the rate per unit time at which this occurs.) The collision frequency is, in turn, dependent on the dynamism of catalyst binding and flexing. Collisions occur when the catalyst brings dynamically bound reactants into physical contact with each other during conformational flexing. The term "time-dependent" derives from the fact that frequency (as in the *collision frequency*) involves the quantification of time.

The overwhelming majority of nonbiological homogeneous catalysts are, by contrast, controlled by the character and energy of the transition states that are formed during collision by catalyst-bound reactants. Catalysis of this type is, consequently, directed by the thermodynamics of the collision and by the underlying energy landscape of the reaction. Such catalysts are energy-dependent in character.

In Chapter 8, we also showed that time- and energy-dependent catalysis could be distinguished by the *synergy* of their catalytic processes. *Synergy* is achieved when the whole of a system exceeds the sum of its parts.

Time-dependent catalysts, like many enzymes, display so-called *functionally convergent* synergies, whereas their energy-dependent counterparts display *functionally complementary* synergies.

Functional complementarity comes about when each catalytic group in a multicentered catalyst independently performs a complementary task. The simple sum of these tasks generates the catalytic effect.

Functional convergence is, however, produced when the catalytic groups act in a coordinated, interdependent, concerted manner. The catalytic effect then derives from the resulting *convergence* of action.

The above differences between biological and nonbiological catalysts have not previously been recognized. Enzymatic catalysis has, instead, been incorrectly thought to generally comprise an unusual form of energy-dependent catalysis. The key feature was believed to involve an extreme stabilization of the transition state originating from structural complementarity in the catalyst for its transition state. The imitation of this feature has been the central aim of biomimetic chemistry. Numerous reviews have described studies dedicated to this goal [5–11].

Authentic biomimetic chemistry involving true nonbiological, time-dependent catalysis has, consequently, not been reviewed. Moreover, the continuum of catalytic action, from energy-dependent to time-dependent homogeneous catalysis, has also not been explored. Thus, catalysts that are unintentionally or nonobviously biomimetic (time-dependent) have not been recognized. Nor have catalysts whose properties are partially biomimetic. Such moieties include species that are *probably* time-dependent in character (many catalytic mechanisms are not known in full detail). They also incorporate energy-dependent catalysts that lie close to the time-dependent realm, thereby displaying significant biomimetic character.

This chapter is intended to address these deficiencies by identifying and examining selected examples of biomimetic catalysts. In so doing, it also aims to clarify, as far as is possible at present, the properties of time-dependent and energy-dependent, multicentered, homogeneous catalysts.

10.2 RATIONAL DESIGN OF TIME-DEPENDENT HOMOGENEOUS CATALYSTS

Because biomimetic chemistry and enzymatic catalysis are formally subsets of homogeneous catalysis, the terminology employed and the perspective of this chapter is largely that of chemical catalysis. The term "artificial enzyme" [12,13], which has been widely used (with different meanings) in both biochemistry and biomimetic chemistry, is not employed here. Instead, we refer to "nonbiological, time-dependent, homogeneous, catalysts."

This chapter also: 1) reviews the general topic of nonbiological, multicentered, time-dependent homogeneous catalysts, 2) describes pertinent features of their rational design, and 3) highlights the importance and growth of this field.

It also addresses how the principles of *authentic* biomimetic chemistry inform and enhance the development of new catalysts. Such species should provide access to chemical transformations that cannot be currently achieved in nonbiological systems. Selected examples in this respect are discussed in detail, so as to illustrate their inherently biomimetic or bioinspired character and the shades of biomimicry that are possible.

A particular emphasis is placed here on catalysts of reactions involving kinetically inert, small gaseous molecules, such as H_2 and O_2, which are not readily catalyzed in homogeneous media. A range of hydrogenase, oxygenase, and nitrogenase enzymes facilitate transformations of such molecules highly efficiently. However, only a very limited set of such reactions can be catalyzed in manmade systems.

A critical challenge addressed by this chapter is to examine systematically how the fundamental requirements for catalysts of such reactions may be practically achieved.

10.2 THE RATIONAL DESIGN OF TIME-DEPENDENT HOMOGENEOUS CATALYSTS

10.2.1 Design Criteria for a Time-Dependent Homogeneous Catalyst

In Chapters 5 and 6, it was noted that time-dependent homogeneous catalysts display two key properties:

1. Catalytic groups that are arrayed so as to rapidly flex about a structure that is highly complementary to the transition state
2. Weak and highly dynamic individual catalyst–reactant interactions that are, nevertheless, sufficient to activate the reactants and make them conducive to reaction

Attribute 1 diminishes the activation barrier of the reaction by constraining to near-optimal the approach pathways of the catalyst-bound reactants during collision. It also ensures that the system spends a significant proportion of time at or near the point of reactive collision during conformational flexing.

Attribute 2 limits the duration of reactant binding and thereby restricts the time available for a reactive collision between the reactant functionalities. It also ensures that, when the bound reactants are brought into optimum collision with each other, their electronic structure is sufficiently polarized that they will react to form products.

The combined effect of these measures is to create a system whose thermodynamic efficiency allows for a faster reaction than its kinetic dynamism. The system is therefore time-dependent.

If one wishes to achieve time-dependence in a homogeneous catalyst, one must simultaneously reproduce these attributes in the catalyst.

There are only a few qualifiers to the above requirements. As noted in Section 7.4 of Chapter 7, the *first qualifier* is that the individual catalyst–reactant binding interactions should be as weak as reasonably possible. The catalytic groups must, of course, still be capable of activating the reactant functionalities sufficiently strongly that they will undergo a chemical reaction when brought into successful collision. The process of activation involves the catalyst polarizing or "stretching" bonds in the bound reactants. This result is typically achieved by electron withdrawal from reactant bonding orbitals or by electron donation into reactant anti-bonding orbitals [14]. Figure 10.1 schematically illustrates the activation of a representative reactant, dioxygen O_2. As shown, binding to the catalyst has the effect of lengthening the O–O bond.

Currently, no means exist to determine what level of activation is needed in a time-dependent catalyst, because of the *second qualifier*, which states that the extent of activation required depends entirely on how complementary the catalyst becomes to its transition state during flexing (see Sections 6.3 and 6.4.2 in Chapter 6). The more ideal its structure, the smaller need be the activation created in the reactant by binding to the catalyst (that is, the smaller need be the initiating impulse for the catalytic process).

A key problem in the rational design of nonbiological, time-dependent catalysts, therefore, involves simultaneously optimizing both of the above attributes. This problem is not trivial.

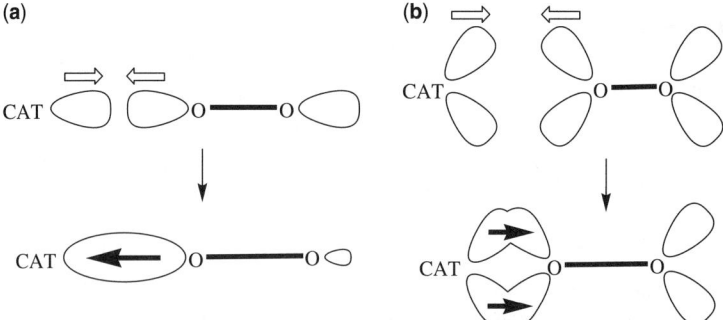

Figure 10.1. *Reactant activation by a catalyst.* Schematic illustrating a catalyst (CAT) activating a representative reactant (O_2) by (a) withdrawing electron density from its σ-bonding molecular orbital and (b) increasing electron density in its π^*-anti-bonding orbital. In each case, the O–O bond is lengthened, facilitating cleavage.

10.2 RATIONAL DESIGN OF TIME-DEPENDENT HOMOGENEOUS CATALYSTS

The challenge is that the catalytic rate in a time-dependent species is dependent on the extent to which catalyst flexing (attribute 1) is synchronized with catalyst–reactant binding (attribute 2). If even a slight mis-synchronization occurs, then the average opportunity for catalysis rapidly declines to zero. Thus, if either attribute 1 or 2 are even moderately imperfect, no catalytic effect will be observed whatsoever.

Key Point. In other words, authentic mimicry of enzymatic catalysis and the realization of time-dependent catalysis requires the concurrent optimization of two inter-related variables:

1. *The catalytic groups to be employed. These must be suitably weakly and dynamically binding and activating.*
2. *Their structural arrangement with respect to each other. They should be driven by conformational flexing about a structure that is complementary to the transition state of the reaction.*

Efficient, time-dependent catalysis will only be observed when each of the above variables approaches its optimum in tandem with the other. If only one variable is at its optimum, then no catalytic effect will occur. Both variables must be simultaneously approaching their optimum. The only way to know whether they are simultaneously approaching their respective optimum is to see whether they generate a catalytic effect.

To design a time-dependent homogeneous catalyst, one must therefore identify suitable catalytic groups and their active spatial arrangement. However, a suitable catalytic group will not yield a catalytic effect unless it is disposed to flex about an optimum spatial arrangement. The optimum spatial arrangement can, similarly, not be identified, except by the fact that it generates a catalytic effect when suitable catalytic groups are used.

10.2.2 The Problem of Simultaneously Identifying Suitable Catalytic Groups and Their Active Spatial Arrangement

The first problem that must be faced in seeking to design a time-dependent homogeneous catalyst is, therefore, the identification of suitable catalytic groups. Weakly binding and activating groups will not be widely recognized in the chemical literature as being catalytically effective, because they are only active when they are optimally arrayed and flexed. Therefore, no record or precedent is available on which to draw in this respect. How, then, does one identify such groups?

As importantly, how does one *simultaneously, and in parallel*, identify the structural arrangements within which they will display significant catalytic activity?

To illustrate the severity of this problem, consider Figure 10.2(a), which schematically illustrates the formation of a transition state, R $\cdots$ R, by two catalytic

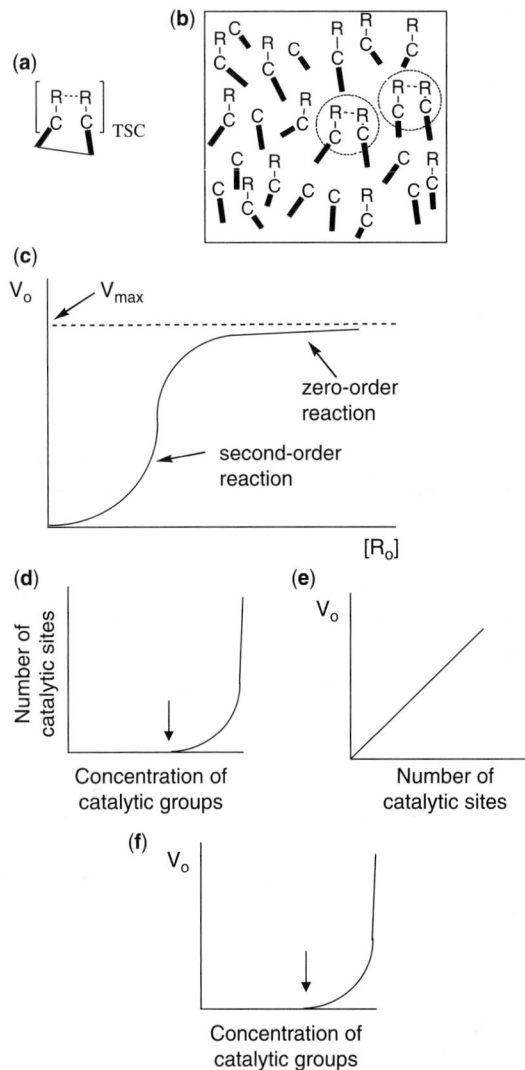

Figure 10.2. *Schematic depiction of a combinatorial time-dependent (statistical proximity) catalyst involving dynamically binding catalytic groups C and reactants R.* (a) Transition state formation during flexing of a conventional, time-dependent dicentered intramolecular catalyst. (b) Instances of transition state formation (broken circles) in an equivalent time-dependent, *combinatorial* ("statistical proximity") catalyst consisting of a highly concentrated mixture of monomeric groups. At each of these catalytic sites, two monomeric catalytic groups adventitiously bring their bound reactants into collision along pathways and trajectories suitable for reaction. (c) The expected kinetic properties of such a catalyst. At low initial reactant concentrations $[R]_o$, the combinatorial catalysts will display a second-order increase in the initial catalytic rate V_o. However, at higher concentrations, saturation kinetics will set in (zero-order kinetics). A key feature of such a time-dependent, homogeneous, combinatorial

(*Continued overleaf*)

10.2 RATIONAL DESIGN OF TIME-DEPENDENT HOMOGENEOUS CATALYSTS

groups C during flexing of a dicentered molecular catalyst. If this catalyst is to be time-dependent, it will have to

1. Regularly and rapidly become structurally complementary to the R $\cdots$ R transition state during flexing
2. Employ a dynamically binding catalytic group C that, nevertheless, activates the reactant R sufficiently strongly to bring about reaction when it is carried along the optimum pathways and trajectories into collision.

Requirement 1 can potentially be met by performing an iterative, trial-and-error study of many closely related molecular structures so as to identify the optimum arrangement of the catalytic groups. But this approach will have to be repeated for *every* prospective catalytic group, C, because the only way to know that both variables are optimum is by the physical observation of a catalytic effect.

Out of all this effort, it is possible that *one* catalytic group that acts in *one* structural arrangement *may* be discovered. This outcome would be very significant. But it is unlikely to be practically useful, because the preparation of this particular catalyst could then easily prove to be unworkable or uneconomic.

Despite the arduous nature of the above process, we will show in Section 10.4.1 (*vide infra*) that several manmade, time-dependent homogeneous catalysts have been discovered and, in fact, optimized. As will be seen, such discoveries have typically involved extensive and, sometimes, truly extraordinary research efforts.

10.2.3 Time-Dependent Homogeneous Catalysis May Conceivably Be Achieved by Mimicry of a Natural Time-Dependent Catalyst

Perhaps the real problem to be overcome in rationally designing a time-dependent homogeneous catalyst is the question of where to start? If one could start with a structure and a set of catalytic groups that generated some sort of a catalytic effect—even a very weak one—then one would have a basis on which to optimize systematically each variable in isolation. However, if one does not have any sort of a catalytic effect at all, then it is impossible to implement an optimization strategy. Every experiment would be a complete, blind, shot in the dark.

One possible starting point is to examine the shape, structure, and actions of an enzyme that catalyzes the desired reaction. As noted in Chapters 5 and 6, many enzymes seem to employ a time-dependent catalytic action. As such, their active

---

Figure 10.2. (*Continued*) catalyst is that the number of catalytic sites will be dependent on the concentration of the monomers within the catalyst. (d) At low monomer concentrations, there will be zero catalytic sites. Above a certain high monomer concentration (marked with an arrow), the number of catalytic sites will increase rapidly. The number of catalytic sites in such a combinatorial time-dependent catalyst is therefore dynamic, not fixed. Moreover, (e) the initial catalytic rate V_o will increase in direct proportion to the number of catalytic sites present in the combinatorial catalyst. As shown in (f), the catalytic rate will therefore be zero below a certain high threshold concentration of the monomeric catalytic groups (marked as an arrow). Above the threshold, the catalytic rate will increase in a nonlinear way.

site will typically be complementary in structure to the transition state of the reaction. Thus, their active site prospectively offers an indication of the optimum structure required in the catalyst.

Moreover, in facilitating turnover, the catalytic groups of the enzyme are clearly capable of providing the necessary reactant activation. Use of the same or similar catalytic groups, should, therefore, conceivably, solve the problem or at least provide an insight into the requirement for a dynamically binding, but suitably activating, catalytic group.

Key Point. In effect, this strategy involves initially setting the key variables according to those present in a known, time-dependent catalyst (albeit a natural one). This approach is taken in the hope that, in the initial manmade catalyst, the variables will, concurrently, be sufficiently close to optimum that some sort of a catalytic effect will be achieved. From there, it should then be possible to alter systematically one variable and then the other, in order to move slowly to a truly optimized homogeneous catalyst of the reaction. In other words, this approach offers a possible starting point for a trial-and-error study of the type described in the previous section.

Although such a strategy may sound somewhat unlikely—and it certainly will be in many, if not most cases—there is no fundamental conceptual reason that it should not work. One would think that it should be especially worth considering for reactions in which the comparable enzyme displays an action that is 1) simple, 2) thoroughly studied, and 3) well characterized.

In Chapter 11, we will describe a practical example that seeks to employ this approach. We will show that a model Mn-oxo complex whose shape and composition is similar to the active site in the water-oxidizing center in Photosystem II, and which dynamically self-assembles in solution, also acts as a photocatalyst of water-oxidation.

10.2.4 Time-Dependent Homogeneous Catalysis May Conceivably Be Achieved in the form of a Combinatorial Experiment Involving a "Statistical Proximity" Effect

For the discovery of simple two- or perhaps three-centered, time-dependent catalysts, a potential alternative to a trial-and-error approach is to screen discrete monomers containing individual catalytic groups using a combinatorial approach. As noted in Section 9.6 of Chapter 9, time-dependence seems to have been achieved in both biology and heterogeneous catalysis by what is, effectively, a combinatorial methodology.

Key Point. How would this work? If suitable, weakly binding and activating monomeric catalytic groups can be sufficiently concentrated in a limited volume, some small but significant proportion of them may be adventitiously correctly arrayed to form the transition state during thermally induced motion, as shown in Figure 10.2(b). Each such event will generate a single catalytic turnover. That is, it will generate a single product molecule or set of molecules. However, many

10.2 RATIONAL DESIGN OF TIME-DEPENDENT HOMOGENEOUS CATALYSTS

such single turnovers, generated over an extended period of time, will produce a noticeable catalytic effect. Given the efficiency of such catalysis, this effect may happen even if a minuscule proportion of the groups happen to be optimum at any one instant.

In such an experiment, the necessary proximity and disposition in the catalytic groups would be achieved by the *statistical* influence of high concentration, rather than by covalently connecting them together in a structure that complements the transition state during flexing. The term *statistical* is used here in the same fundamental way that it is used in mechanics (see Sections 1.1 and 1.3 in Chapter 1). That is, it describes a certain statistical likelihood that some proportion of the reactants bound to such monomeric catalytic groups will collide along trajectories and pathways that are optimum for transition state formation leading to products.

Any catalytic effect that is obtained in such an experiment could therefore be termed a *statistical proximity* effect. That is, it would involve a *proximity effect* of the same type that is observed in many enzymes, but the effect would be created *statistically* and not by covalent interconnection as is the case in these enzymes.

The key advantage of this approach is that it is not necessary to know what the optimum structural arrangement for catalysis is since *a library of all possible spatial arrangements* should be present. The system will therefore be a combinatorial experiment in which potential catalytic groups and *all of their possible spatial arrangements* are simultaneously tested for catalytic activity.

Indeed, if catalysis of significant activity can be achieved in this way, the combinatorial experiment may itself be a practically useful catalyst. Such a catalyst would then be a *statistical proximity catalyst*.

The selectivity and activity of such a catalyst would derive from the same path-dependence and synchronization in the interactions of bound reactants that are observed in many enzymes. Unlike such enzymes, however, these features would be realized by statistical chance, not by the flexing of a covalent structure. A *statistical proximity* catalyst would therefore employ the *same underlying, fundamental principles* as conventional time-dependent homogeneous catalysts. However, these features would be achieved in an unconventional and more practical manner.

Indeed, such a system will, effectively, be the three-dimensional and molecular counterpart of two-dimensional heterogeneous catalysts that employ high-surface concentrations of reactants and irregular successions of single turnovers at individual surface atoms of the catalyst. In providing a multiplicity of different spatial arrangements of the catalytic groups, it will, to some extent, mimic the many different geometries of active site present on the surface of a typical heterogeneous catalyst.

In Chapters 12 and 13, we will describe two practical examples that seek to employ such a combinatorial approach to developing new, *time-dependent* homogeneous catalysts, albeit as hybrid homogeneous–heterogeneous catalysts.

It should be noted that there are several possible elaborations to the above combinatorial approach. For example, instead of concentrating monomeric catalytic groups in a limited volume, one could design supramolecular associations between appropriate monomers. Such associations may potentially be used to induce the monomeric catalytic groups into close, but dynamic, association with each other. In effect, one would be creating many small and discrete active sites (regions of high local concentration) dispersed in the solution. A proportion of catalytic groups may thereby also be brought into suitable proximity and disposition at any one instant.

The observation of even a small catalytic effect in a combinatorial experiment of this type will signal that the species being tested are potentially suitable as weakly binding and activating catalytic groups for the reaction. An iterative trial-and-error study could then be carried out using the equivalent covalently bound species to identify the ideal structural and conformationally flexible arrangement of these groups during catalysis.

It should also be noted that, since catalytic groups that are not optimally disposed *cannot* generate a catalytic effect in a time-dependent system, the groups that are incorrectly arranged (most groups) will be passive observers. They will not interfere in the catalysis or generate unwanted by-products. In other words, the selectivity of the catalysis will be preserved.

10.2.4.1 A Time-Dependent Combinatorial Catalyst May Display Unique Kinetics.
If it is possible to create a time-dependent catalyst using the combinatorial, "statistical proximity" approach described above, then such a system will display unique kinetics according to Equation (10.1):

$$2C + 2R \xleftrightarrow{K'} 2[CR] \xleftrightarrow{K''} [CRRC] \xleftrightarrow{K_2} 2C + P \qquad (10.1)$$

where C = catalyst, R = reactant, and P = product.

The initial step described by the equilibrium K' will be identical to the initial step in a Michaelis–Menten kinetic scheme [see Equation (6.2) in Section 6.4.5 in Chapter 6]. However, the following step described by the *mechanical equilibrium* K'' will be second order, not first order, as it is in the Michaelis–Menten scheme. That is, it will involve two $[CR]$ intermediates combining to form a single $[CRRC]$ transition state.

A "statistical proximity" combinatorial, time-dependent catalyst will therefore display second-order kinetics at low reactant concentrations, but become saturated at high reactant concentrations [as shown in Fig. 10.2(c)].

A particularly distinctive feature of such a catalyst is that the catalytic sites are created by statistical fiat, rather than through a mutually interconnected, covalent structure. Thus, the number of sites within such a combinatorial catalyst is not fixed. Rather, it will vary according to the physical conditions employed. As the monomeric catalytic groups depicted in Figure 10.2(b) are highly concentrated within a limited volume, there will be a nonlinear increase in the number of active sites beyond a certain threshold [depicted as an arrow in Fig. 10.2(d)], because the likelihood of synchronization between reactant binding and collision will increase disproportionately in accordance with the average proximity of the catalytic groups to each other. As the groups are packed increasingly closely, there will be more opportunities, as a proportion of the total, for time-dependent collisions to take

10.2 RATIONAL DESIGN OF TIME-DEPENDENT HOMOGENEOUS CATALYSTS

place. Since the overall catalytic rate will depend on the number of active sites present [Fig. 10.2(e)], the overall rate of catalysis will increase nonlinearly as the concentration of monomeric catalytic groups is increased above a certain threshold concentration [marked as an arrow in Fig. 10.2(f)]. Below this threshold no catalytic effect will be observable.

In summary, a "statistical proximity" combinatorial catalyst will employ the same principles of time-dependence as any other time-dependent homogeneous catalyst. However, it will differ in that it will not be necessary to build into each catalyst a covalent structure that is restricted to flex conformationally about a particular spatial arrangement. Instead, the catalytic groups will be monomeric in character. As a result, the number of catalytic sites in such a combinatorial catalyst will not be fixed, but it will, rather, vary according to the physical conditions present. At very high concentrations of monomeric catalytic groups, many catalytic sites will be created adventitiously and, therefore, a high catalytic rate. It will be because, at high concentrations, reactant binding and collision have a higher likelihood of being synchronized since the average time interval between collisions will be small. As the concentration of the catalytic groups is decreased, however, this time interval will increase, causing the necessary synchronization to decline precipitously, thereby resulting in a rapid decline of the catalytic rate to zero below a certain threshold concentration [shown as an arrow in Fig. 10.2(f)].

10.2.4.2 Previous Attempts at Concentration-Based Biomimetic Catalysis Involved Energy-Dependent Systems.

The use of high concentrations of monomeric catalytic groups has previously been widely studied as an unconventional means of imitating enzymatic catalysis. But this was done without a full understanding of the need for, and role of, dynamic binding and activation.

For example, interfacial catalysis, involving the inclusion and reaction of monomeric organic species within aqueous dispersions of hydrophobic micelles, has been shown to lead to dramatic increases in the reaction rate [1].

To the best of our knowledge, however, all previously studied systems have been *functionally complementary*, and therefore energy-dependent, in character. This is unequivocally indicated by the fact that catalysis was also observed in such cases, in open solution (albeit at a substantially slower rate) [11]. As noted in Section 6.3 of Chapter 6 and Section 5.6.2 of Chapter 5, it is impossible for monomeric catalytic groups to display a time-dependent catalytic effect when they are dispersed in dilute open solution because there can be no synchronization between reactant binding and collision. If monomeric species act catalytically in dilute open solution, they must therefore necessarily employ an energy-dependent action.

Interfacial catalysis of a similar type but using weakly and dynamically binding catalytic groups (suitable for a time-dependent action) may, nevertheless, offer important possibilities as a combinatorial screening and catalyst discovery technique, provided that sufficiently high concentrations can be achieved (see Section 10.2.4 above).

10.2.5 Time-Dependent Catalysis May Be Useful in Transformations of Small Gaseous Molecules

An alternative to identifying weakly and dynamically binding catalytic groups is to turn the problem around and employ reactants that are invariably weakly and dynamically bound. This approach ensures the presence of the weak individual catalyst–reactant interactions required for a time-dependent catalytic action.

The small gaseous molecules H_2, O_2, and N_2 are neutral, nonpolar species that have few "handles" with which to be bound. For this reason, Collman has described them as "kinetically inert" [15]. Although this property has, thus far, been considered extremely problematic in attempts to transform them catalytically, reactions of such molecules are inherently more likely to be achieved in time-dependent, *functionally convergent* systems than in energy-dependent, *functionally complementary* ones.

10.2.6 Why Do We Need New Time-Dependent Catalysts?

Although unsurpassed in their basic catalytic functions, enzymes have significant practical disadvantages when used in industrial-scale manufacturing. These disadvantages include high cost, complicated and demanding handling and maintenance procedures as well as sensitivity to the physical conditions employed.

Biotechnological applications of enzymes are nonetheless expanding rapidly, although generally only for the preparation of high-value materials [16,17].

For these reasons, it is highly desirable to develop robust and practical nonbiological catalysts that mimic at least some of the useful features and properties of enzymes.

In the following sections, we will review the main features of the rational design of multicentered, biomimetic catalysts as reported to date in the chemical and catalytic literature.

10.3 ELEMENTS OF RATIONAL DESIGN IN MULTICENTERED CATALYSIS

10.3.1 Modes of Binding in Multicentered Catalysts

In broad terms, reactant molecules and catalytic groups in multicentered catalysts may associate in two basic modes.

Figure 10.3(a) schematically illustrates Feringa's so-called "Class A" type associations [18], in which a single reactant molecule R attaches to, or becomes associated with, two catalytic groups C and C' simultaneously. After activation, the molecule R* may fragment into two or more product molecules R' and R", or it may react with another suitably activated reactant that is associated with another catalytic group. The resulting species can undergo more reactions or be released as products.

Figure 10.3(b) illustrates Feringa's "Class B" type associations [18], in which a reactant molecule R' associates with a single catalytic group C. After activation, the molecule R'* reacts with another activated reactant molecule (R"*) associated

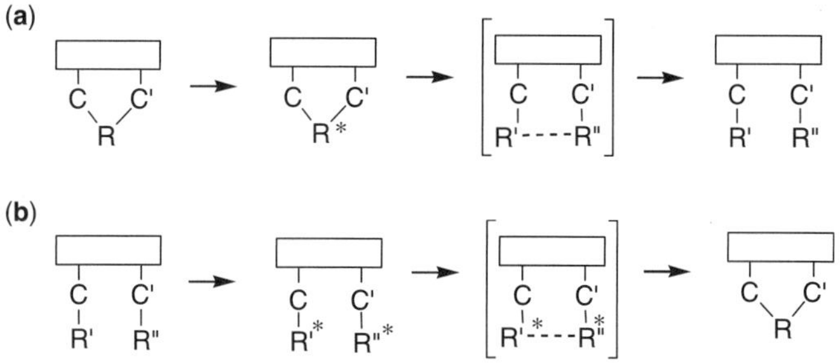

10.3 ELEMENTS OF RATIONAL DESIGN IN MULTICENTERED CATALYSIS

Figure 10.3. *Modes of binding in multicentered catalysts:* Schematic depicting (a) "Class A" type and (b) "Class B" type multicentered catalysts [18]. The symbols C and C' represent catalytic groups; R, R', and R" represent reactants.

with a different catalytic group (C') to form a new species (R) that is simultaneously attached to both catalytic groups. The group R may be released as products or undergo additional reaction.

The association between the reactant and the catalytic group may take a variety of forms, all of which involve some type of donation or withdrawal of electron density (Fig. 10.1) [18].

Several variations on these general themes are known. For example, Sanders has described binding modes entitled "transfer" and "transformation" [19]. In the former of these modes an activated functional group is transferred from one catalytic group to another. In the latter mode, an activated functional group attached to a catalytic group is internally transformed without the involvement of other catalytic groups.

Feringa has also described a "Class C" binding that is a special example of one of the earlier cases [18].

One may also classify multicentered catalysts according to the types of association present. *Multifunctional catalysts* are, for example, multicentered catalysts in which one center performs a different task to another [20,21].

10.3.2 Optimizing the Spatial Arrangement of Catalytic Groups

10.3.2.1 Intramolecular Catalysts. To achieve maximum efficiencies in multicentered catalysts, the catalytic groups must be arranged so that they facilitate reactive collision as often as possible during conformational flexing. This process provides for the greatest possible number of putative collisions mediated by the catalyst. The most common approach employed in this regard has been to secure covalently the catalytic groups to each other within a single molecule. Such molecules are known as *intramolecular catalysts* [Fig. 10.4(a)] [22].

This approach offers several significant benefits:

1. The arrangement of the catalytic groups at each point in time can be precisely and uniformly controlled.

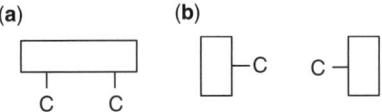

Figure 10.4. *Intra- and intermolecular catalysts in multicentred catalysis:* Schematic depicting (a) an intramolecular catalyst and (b) an intermolecular catalyst, containing two catalytic groups each. The symbols C represent the catalytic groups.

2. Their organization within the molecule during flexing can be systematically varied in order to find the ideal molecular arrangement.
3. The catalytic groups can, most likely, be located more closely to each other than they can in the absence of a covalent or coordinate framework.

Indeed, the use of a covalent framework arguably allows extremely high "concentrations" of catalytic groups (and therefore also of their attached reactants) to be effectively be achieved within the tiny volume of an intramolecular catalyst. As noted in Chapter 7, studies of intramolecular reaction mechanisms indicate that "effective molarity" offers an indirect but often a relatively effective measure of the likely reaction rate [23–25].

The use of intramolecular catalysts also has important practical and theoretical drawbacks. Chief among these drawbacks is their synthetic complexity and the difficulty of constructing, in small molecules, an oscillating three-dimensional arrangement in which the catalytic groups are ideally arrayed relative to each other. The required configuration may, for example, be structurally inaccessible, lying between two different, available arrangements. Structural *quantization* of this type limits the utility of small-molecule structures and arguably explains why enzymes are generally macromolecules [26].

Increasing the flexibility of an intramolecular catalyst may overcome this difficulty. However, it can also reduce the proportion of time that the catalyst spends in the conformer associated with reactive collision, thereby becoming self-defeating [27,28]. That is, it may reduce the number of collisions that the catalyst may mediate per unit time.

Alternatively, the structure of an intramolecular catalyst may be enlarged in order to provide for a better fit of the desired spatial arrangement. Such enlargement typically results in a rapid escalation in synthetic complexity, however, making such catalysts impractical in every sense.

Unpredictable cooperative interactions between closely spaced catalytic groups within an intramolecular catalyst may also cause difficulties. For example, mutual deactivation (*negative cooperativity*) in which catalytic groups within the same molecule deactivate each other has been reported to inhibit certain intramolecular catalysts [27,28]. In other cases, unexpected electronic perturbations of closely located catalytic groups have led to *positive cooperativity*, including dramatically changed physical properties [29]. or new catalytic mechanisms [30].

All such effects—both positive and negative—significantly complicate rational design because they obscure patterns of reactivity and provide no useful information regarding the fundamental utility of the spatial arrangement employed [31].

10.3 ELEMENTS OF RATIONAL DESIGN IN MULTICENTERED CATALYSIS

Finally, intramolecular catalysts have another very significant disadvantage: They do not allow for the ready discovery of new, functionally convergent, multicentered catalysts, because only one very specific conformational interchange of the catalytic groups is typically active in such catalysts. Without the intervention of providence, discovery of this particular interchange generally requires a nontrivial trial-and-error examination of many different molecules. The current rarity of enzyme-like nonbiological catalysts is, undoubtedly, caused by this constraint.

10.3.2.2 Intermolecular Catalysts.
Simpler methods for optimizing the spatiotemporal arrangement of catalytic groups in time-dependent catalysts are clearly desirable.

An alternative approach is that of *intermolecular* catalysis in which the catalytic groups on separate molecules are brought into the optimum arrangement for reactive collision either adventitiously or through the use of concentration or noncovalent, supramolecular interactions between the molecules [19]. Figure 10.4(b) schematically illustrates catalysts of this type. This type of thinking provides the rationale behind the combinatorial, "statistical proximity" catalysts described earlier (in Sections 10.2.3 and 10.2.4).

Intermolecular catalysts are undoubtedly more practical than intramolecular catalysts because a complicated covalent or coordinate framework is not required to limit the movements of the catalytic groups to optimum. However, they present a new challenge: that of finding a convenient way to organize the catalytic groups for maximum efficiency. In effect, chemical bonds are used in intramolecular catalysts to organize the spatiotemporal arrangement of the catalytic groups. For intermolecular catalysts, however, unconventional, nonbonding means must be found.

10.3.2.3 Unconventional Approaches to Optimizing the Spatial Organization of Catalytic Groups.
Several unconventional approaches to the problem of correctly disposing the catalytic groups in multicentered molecular catalysts have been proposed. These approaches will be discussed in greater detail in the examples that follow, but we can, for the moment, summarize them to include:

(1) The use of transient, common bridging atoms, such as μ-*oxo*- or μ-*chloro*- species to correctly array two or more metallic catalytic groups relative to each other (see "Ruthenium-based Water Oxidation Catalysts," Section 10.4.1.5, and "Intermolecular Catalysts," Section 10.4.2.2) [32].

(2) The incorporation of prospective catalytic groups within monomers that are then polymerized into regularly repeating structures in which a certain proportion of the catalytic groups are favored to be optimally organized (see "Macromolecular Catalysts," Section 10.4.1.8).

(3) The use of concentration effects or supramolecular and other interactions to assemble discrete catalytic groups into correct arrangements (see "Intermolecular Catalysts," Sections 10.4.2.2 and 10.4.2.3).

10.3.3 Creating Functionally Convergent Catalysts

As noted, the most challenging aspect of multicentered catalysis will, likely, involve combining the effects of reactant activation, proximity, and disposition to achieve true functionally convergent catalysis. This process involves selecting catalytic groups that dynamically bind and suitably activate reactants and then engineering them into a spatial arrangement that rapidly flexes about an optimum structure. Only in optimizing this spatial arrangement will a catalytic effect be observed.

Little is known in respect to the binding and activating properties of, even, potentially useful metallic catalytic groups, much less organic ones. The challenge of discovering and selecting suitable catalytic groups will therefore require an extended program of study.

The second problem is to identify the ideal arrangement of these catalytic groups. This, again, is likely to be a nontrivial matter. Advanced computational methods that are capable of predicting useful catalytic groups and optimum spatial arrangements may greatly speed up this development.

Several practical approaches can also be prospectively used.

10.3.3.1 Practical Approaches to Achieving Functionally Convergent Catalysis

Intramolecular Catalysts. In intramolecular systems, a potential strategy for achieving functionally convergent catalysis is to start with a functionally complementary catalytic system and then iteratively evolve it into a convergent system by progressively weakening the catalyst–reactant interactions. In this process, the ideal structure for maximum convergence should gradually be identified. Building this arrangement into the original system should then produce highly efficient, functionally convergent catalysis.

This tactic assumes, of course, that a functionally complementary catalyst exists for the particular reaction of interest. Many reactions, such as those involving weakly binding species like small gaseous molecules, are not catalyzed in functionally complementary systems.

This approach makes use of the important relationship that exists between the spatial arrangement of the catalytic groups within a multicentered catalyst and the strength of their interactions with the reactant molecules. Thus, for example, when a time-dependent catalyst has a high likelihood of being structurally complementary to its transition state, very weak binding and activating interactions are sufficient to create a catalytic effect. However, when the catalyst is structurally poorly disposed to transition state formation, then much stronger binding and activation is required to bring about catalysis. Another explanation of this effect can be found in Sections 5.6.2 and 5.6.3 in Chapter 5 and Sections 6.3–6.4.2 in Chapter 6.

As coordination bonding is stronger than hydrogen bonding, catalytic groups that coordinate their reactants will generally need to be arrayed for reactive collision for a lesser proportion of the time than catalytic groups that hydrogen bond their reactants. A wide range of interplays can consequently be engineered in which the extent of catalyst pre-organization is matched to the nature of the catalyst–reactant interactions employed.

Intermolecular Catalysts. A prospectively simpler approach is to employ the combinatorial "statistical proximity" effect described earlier. Such catalysts offer the possibility of simple, practical, and diverse systems capable of facilitating transformations that can currently not be catalyzed, using catalytic groups that are currently not recognized as useful.

By employing concentration or supramolecular effects or densely packed, polymer-bound catalytic groups, it may be feasible to devise catalysts that blend the synthetic simplicity and other advantages of intermolecular catalysis with the efficiency of functionally convergent catalysis.

Since most intramolecular catalysts are too synthetically complex to be practically or commercially useful, finding ways to create efficient, intermolecular, functionally convergent catalysts remains a key issue.

10.4 A REVIEW OF NONBIOLOGICAL, MULTICENTERED MOLECULAR CATALYSTS DESCRIBED IN THE CHEMICAL LITERATURE

To illustrate the concepts discussed above, selected examples of multicentered molecular catalysts are now presented. As the mechanisms of many of these examples remain subject to additional investigation, these discussions are definitive only in terms of present knowledge. Because most known nonbiological, multicentered catalysts capable of sustained catalysis involve metal-ion catalytic groups, the examples below are largely organometallic in character. In principle, however, correctly devised nonmetallic species should also be highly effective multicentered catalysts.

10.4.1 Intramolecular Catalysts

10.4.1.1 Functionally Convergent Catalysis (Class A Type): Cofacial and Capped Metalloporphyrins as Oxygen Reduction Catalysts. Among the best known and most thoroughly studied multicentered organometallic catalysts are the so-called cofacial diporphyrins [15].

As first recognized by Collman in 1977, arranging two metalloporphyrins in a face-to-face manner prospectively allows the two metal atoms to act cooperatively (convergently) in the catalytic transformation of small molecules, like dioxygen O_2, in the gap between the rings [33]. This process was elegantly demonstrated by the dicobalt porphyrin **1** (Scheme 10.1), which catalyzes the four-electron reduction of O_2 to H_2O at pH < 3.5 and at potentials negative of 0.71 V (vs. NHE), when adsorbed on a graphite electrode [34,35]. Under identical conditions, the corresponding monomer of **1** generates the two-electron product H_2O_2, which is the usual product of the reaction using metalloporphyrin catalysts. Compound **1** does not catalyze the transformation of H_2O_2 into H_2O and therefore does not perform successive reductions. The different behaviors of **1** and its monomer indicate that the catalysis in question is molecular.

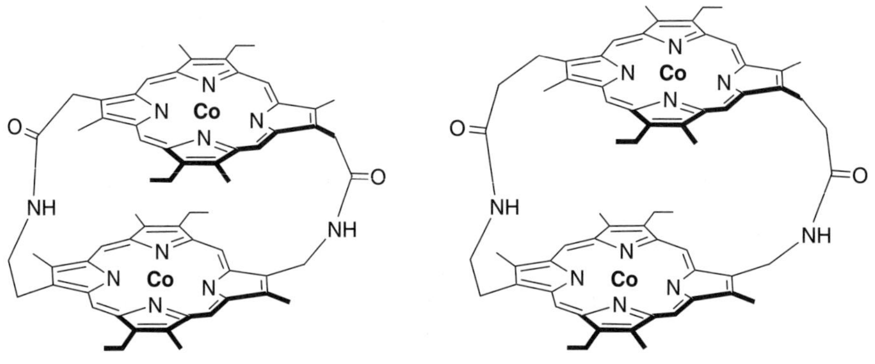

Scheme 10.1.

$$O_2 \xrightarrow{1} H_2O \quad (4\ e^- \text{ process})$$

$$O_2 \xrightarrow{2} H_2O_2 \quad (2\ e^- \text{ process})$$

Extensive comparative studies were performed to clarify the influence of the various components in this system.

This work indicated that the nature of the linker, the metal ions present, and the metal–metal distance were all crucial to the catalysis. For example, species **2** (Scheme 10.1), which differs from **1** only in an extra carbon atom in the linkers, is exclusively a catalyst of 2-electron H_2O_2 formation [35,36]. By contrast, species **3** and **4** (Scheme 10.2) are both catalysts of the four-electron process despite a difference of almost 1 Å in their metal–metal separations [15]. Figure 10.5 illustrates, for representative purposes only, one possible mechanism for the catalysis that is consistent with the data available thus far.

These results illustrate some of the features of rationally designing multicentered, convergent, intramolecular catalysts.

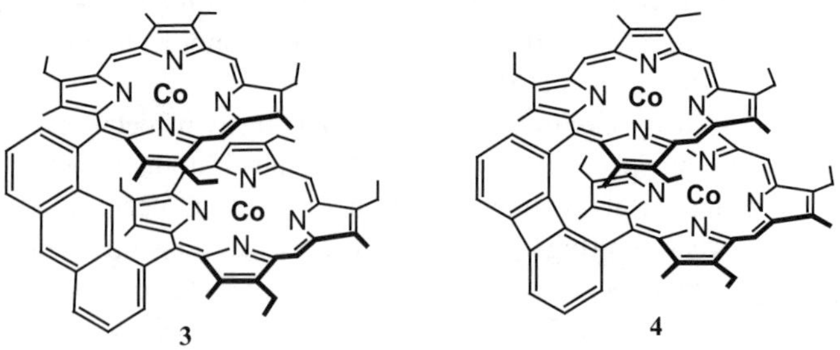

$$O_2 \xrightarrow{3} H_2O \quad (4\ e^- \text{ process})$$

$$O_2 \xrightarrow{4} H_2O \quad (4\ e^- \text{ process})$$

Scheme 10.2.

10.4 NONBIOLOGICAL, MULTICENTERED MOLECULAR CATALYSTS

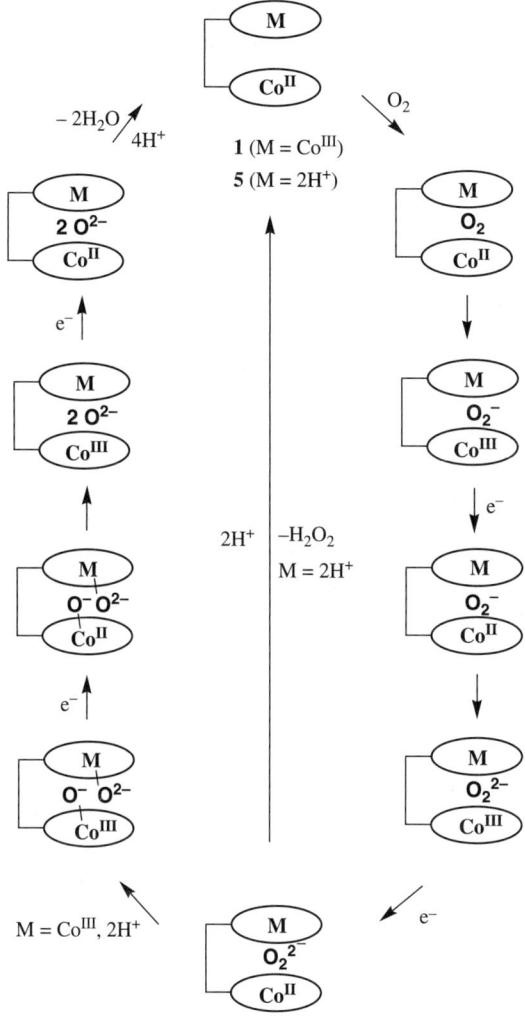

(Overall complex charges not displayed for clarity)

Figure 10.5. *Dioxygen reduction by cofacial metalloporphyrins:* A reaction scheme consistent with the available data for dioxygen reduction by cofacial diporphyrins.

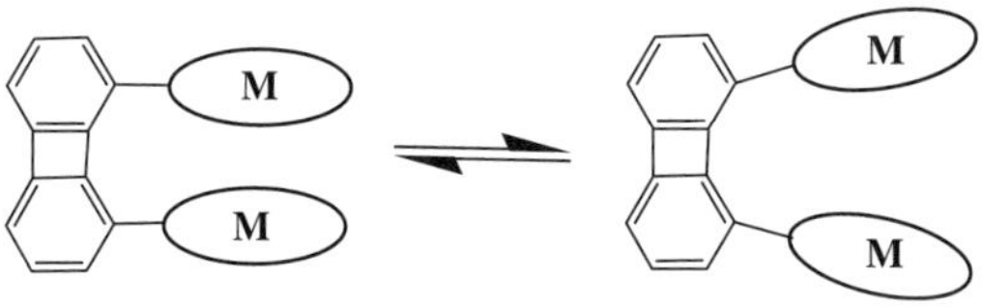

Scheme 10.3.

To achieve maximum selectivity for the four-electron process, it seems that the metal ions in cofacial diporphyrins should eclipse, or almost eclipse, each other during conformational flexing. The aryl systems of the linkers in **3** and **4** are ideal for this task, although some lateral shift does occur, depending on the metals present. Although rigid in the aryl plane, the linkers in **3** and **4** nevertheless allow some flexibility in the metal–metal distance within the binding pocket. The diporphyrins therefore seem to be able to open their "bite" as shown in Scheme 10.3 during flexing [15]. This capability allows binding and activation of O_2 as well as the formation and stabilization of the various intermediates and transition states.

By contrast, **1** depends on the shortness of its linkers to achieve satisfactory face-to-face orientation. Lengthening the linkers, as in **2**, increases the flexibility, thereby diminishing the contribution of the face-to-face conformer and destroying the selectivity of the catalysis.

The extreme sensitivity of the catalysis to this relatively minor structural effect indicates that the key peroxide intermediate in **1**, shown within the diporphyrin pocket at the bottom of Figure 10.5, is weakly bound. It also indicates that this intermediate is correctly activated for O–O bond cleavage only when it is bound to both of the porphyrins and they are in an eclipsed arrangement relative to each other; that is, it is a system of class A in Figure 10.3. The movement to open the "bite" clearly pulls apart the O–O bond.

In **2**, the peroxide is attached too briefly to be present when both porphyrins are also correctly arrayed. At any one instant, therefore, effectively no molecules of **2** simultaneously have the peroxide intermediate attached and the porphyrins face-to-face and opening their "bite." The effect of lengthening the linkers in **2** is, consequently, to weaken simultaneously the overall binding of the peroxide intermediate and decrease the proportion of time that the catalytic groups spend in the conformational interchange leading to reactive collision. This effect greatly reduces the opportunity for catalysis when each porphyrin correctly activates the peroxide for only very brief periods at a time. The fact that the catalytic effect for the reaction leading to O–O cleavage is lost entirely in **2** and not merely diminished by a substantial amount is totally consistent with what would be expected to occur in a functionally convergent molecular catalyst (like many enzymes).

A more complete explanation of these effects was provided by the discovery that only one redox metal is, in fact, required to achieve selectivity for H_2O [37]. Thus, the monocobalt analog, **5** (Scheme 10.4), along with a cobalt–aluminium diporphyrin, also catalyzes four-electron reduction under suitable circumstances, although this occurs in parallel with the competing two-electron process [38]. The second metal in **1** is thereby revealed to act as a simple Lewis acid during the catalysis [15,37].

Studies have shown that when O_2 binds in the pocket between the metals, it is initially activated as a μ-superoxide species [39]. The actual catalyst is the $Co^{II}Co^{III}$ form of **1** depicted in Figure 10.5 [15]. The task of the second metal, that is the Co^{III}, must be to contribute to the binding, activation, and subsequent O–O cleavage of the unprotonated or partially protonated peroxide intermediate shown

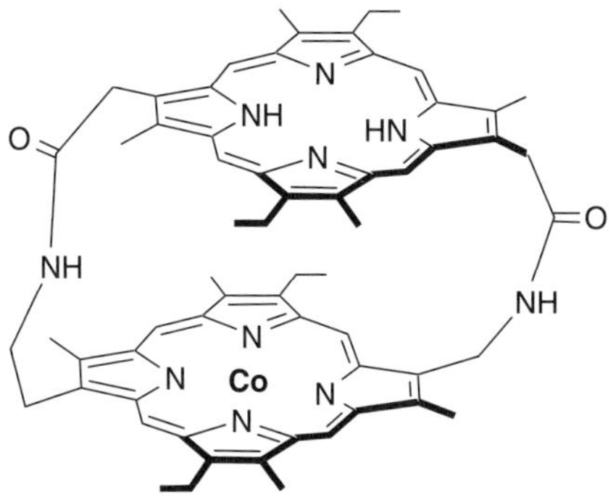

10.4 NONBIOLOGICAL, MULTICENTERED MOLECULAR CATALYSTS

5

$$O_2 \xrightarrow{5} H_2O \quad (4\ e^-\ \text{process})$$

Scheme 10.4.

in the diagram at the bottom of Figure 10.5. It presumably does so by reducing electron density in the O_2 bonding molecular orbitals.

The Co^{II} in the actual catalyst of **1** therefore effectively acts as an electron shuttle, mediating the movement of electrons from the underlying electrode to the bound oxygen. The rate of this process may determine the selectivity of the catalysis. The catalysts only undertake four-electron reduction if they are adsorbed onto edge-plane graphite electrodes. The use of other electrodes or incorporation within thin-layer porous polymers deposited on electrode surfaces leads to the exclusive production of H_2O_2 [37]. As graphite is known to adsorb porphyrins strongly, electron transfer to the porphyrin is suggested to be faster using a graphite electrode than with other electrode materials [40]. If this hypothesis is correct, it supports the contention that the activated state of the key peroxide intermediate in **1** must be extremely short-lived—so short-lived that it can undergo O–O cleavage only when the complex is very rapidly supplied with an electron.

A range of other cofacial diporphyrins catalyze the formation of water from oxygen, although usually only as a coproduct with H_2O_2.

Perhaps the most surprising of these porphyrins involve those that are well separated, such as **6** (Scheme 10.5) [41]. In this molecule, a dibenzofuran linker causes the two rings to be angled at 56.5° with respect to each other. The Co–Co distance is 8.624 Å, which is substantially longer than the 3.73 Å in **4**. Compound **6** nevertheless catalyzes the conversion of 80% of the O_2 reactants to water. The study's authors ascribed this result to a longitudinal flexibility, which allows the molecule to "bite" down on the O_2 substrate. The eclipsed conformation of the porphyrins is clearly still highly populated in molecules like **3**, **4**, and **6**, which contain rigid, orthogonal aryl linkers.

Key Point. **In 4-electron dioxygen reduction by 1, 3, 4, and 6 we see the distinctive features of a mechanical action. This occurs in the form of the reactant being**

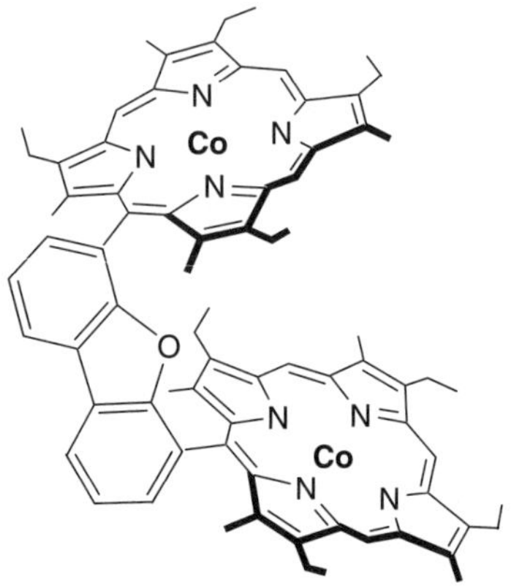

6

$$O_2 \xrightarrow{\text{6}} H_2O + H_2O_2$$
$$(80\%) \quad (20\%)$$

Scheme 10.5.

pulled apart, rather than being put together into a new moiety. Thus, we have two dynamic processes which must be synchronized: (i) transient O_2 binding (at both Co ions simultaneously), and (ii) catalyst conformational flexing. These processes can only be synchronized by constraining the catalyst to rapidly flex about a face-to-face structure that complements the transition state. When this is achieved, as in 1, the catalytic groups act in a concerted, coordinated, and convergent manner. When it is not, as in 2, the collision frequency declines non-linearly to zero. The whole process is driven by the mechanical impulse of conformational flexing. The system is machine-like in that it dynamically takes up reactant molecules, and then mechanically pulls them apart within a structure that complements the desired outcome. It does so in a highly repeatable and very specific way.

A similar effect must presumably exist in the dimanganese porphyrin **7** (Scheme 10.6), which has been reported to catalyze the reverse reaction; that is, it converts H_2O to O_2 in a Class B-type catalytic process [42(a)]. The mechanism of this reaction was suggested to involve coordination of the Mn^{III} centers by hydroxide, followed by their conversion to $Mn^{V}(=O)$ species. Two of the latter moieties then presumably form a $Mn^{IV}-O-O-Mn^{IV}$ complex during flexing of the Mn centers. Replacement of the peroxy bridge by OH^- with simultaneous decomposition was proposed to give O_2 and the starting Mn^{III} complex. This mechanism is supported by comparative studies, and the crucial importance of the spatial organization of the Mn^{III} centers to the catalytic effect [42(b)].

In recent work, a range of copper-imidazole-capped metalloporphyrins have been prepared and shown to be catalysts of the four-electron reduction of oxygen even at physiological pH [43–48]. Cofacial diporphyrins have also been widely used as

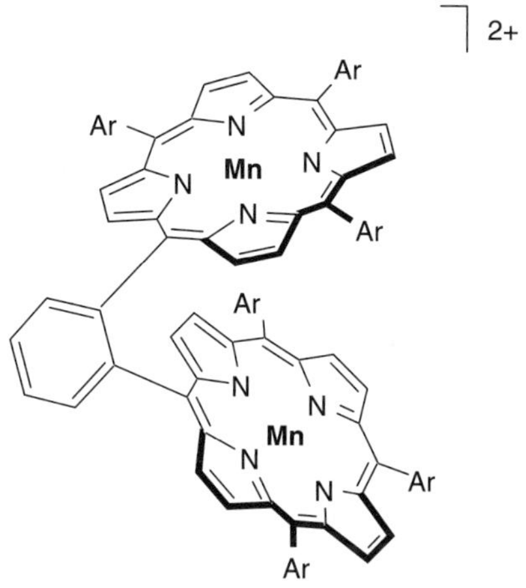

10.4 NONBIOLOGICAL, MULTICENTERED MOLECULAR CATALYSTS

Scheme 10.6.

proton reduction and hydrogen oxidation catalysts, as well as in studies examining the catalytic interconversion of dinitrogen and its nitrogen hydrides [15,49,50].

Other examples of Class A-type, multicentered, time-dependent catalysts include Corey's bis(cinchona) alkaloid catalysts of asymmetric dihydroxylation of olefins [51]. These species were discussed in Section 8.5 of Chapter 8.

10.4.1.2 Functionally Convergent Catalysis (Class B Type): [1.1]Ferrocenophanes and Related Compounds as Hydrogen Generation Catalysts.
In 1973, Bitterwolf and Ling reported that the dimethyl-bridged [1.1]ferrocenophane **8a** reacts quantitatively with the strong nonoxidizing acid HBF_3OH to form dihydrogen H_2 and the corresponding ferrocenophane dication **10** [Fig. 10.6(a)] [52]. The dication could be quantitatively reduced, in situ, back to **8a** by the addition of $SnCl_2$ reductant.

This result was curious because free ferrocene **13** [Fig. 10.6(b)] does not react to produce hydrogen under identical conditions. Nor do 1,1,12,12-tetramethyl[1.1]ferrocenophane **11**, [0.0]ferrocenophane **12**, diferrocenylmethane **14** or diferrocenylethane **15** [52,53]. However, the mono- and dimethylated [1.1]ferrocenophanes **8b–c** undergo the same reaction.

NMR evidence indicated that the individual ferrocenes in **8a** and **13** are protonated in an identical fashion at the Fe centers, with **8a** doubly protonated and **13** singly protonated. It was therefore proposed that H_2 elimination by **8a** involves a homolytic combination of two reduced protons that are present on the ferrocene iron atoms

(a)

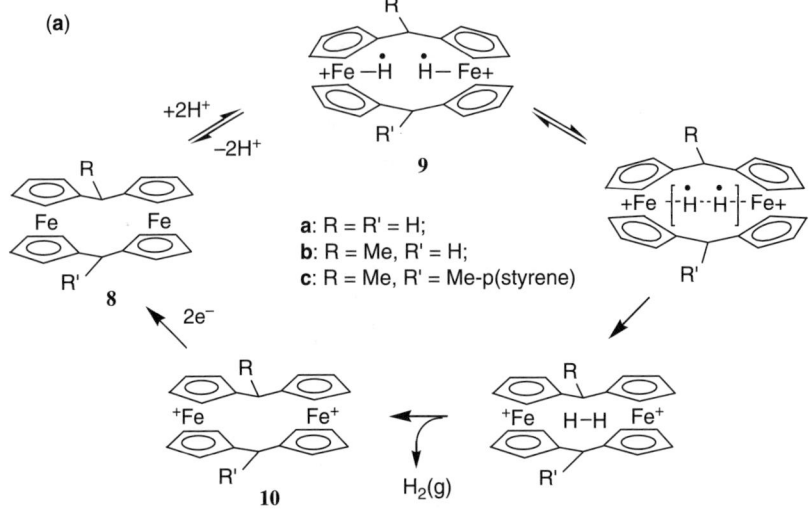

a: R = R' = H;
b: R = Me, R' = H;
c: R = Me, R' = Me-p(styrene)

(b) Comparable Reactivities:

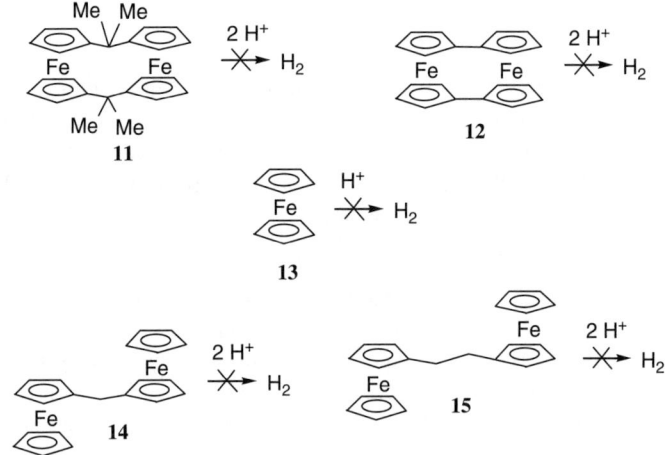

(c) Protonation Equilibrium: at −122 °C

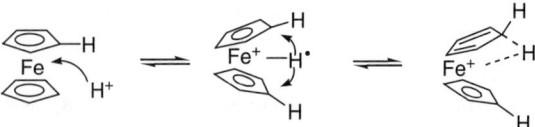

Figure 10.6. *Functionally convergent multicentered catalysis (Class B)*: Hydrogen generation catalysis by [1.1]ferrocenophane **8a** and derivatives.

after protonation [as illustrated in Fig. 10.6(a)] [52]. The extreme selectivity of **8a** was suggested to originate as a result of the proximity and disposition of the ferrocene groups in these molecules, which places the activated iron-bound hydrogens within bonding distance during conformational flexing [52].

Kinetic studies by Hillman and coworkers [54,55] were consistent with this mechanism and not with the competing heterolytic process. A theoretical study involving INDO-SCF calculations confirmed that the ferrocenes in **8a** became correctly proximate for a collision between attached hydrogens leading to a homolytic cleavage of the Fe–H bonds during flexing. It also indicated that the H^+ ions present on the ferrocene iron atoms in protonated **8a** were effectively activated atomic hydrogen ($H^\bullet$), with most of their positive charge residing on the metal [56]. This was verified in subsequent density functional theory (DFT) calculations [57,58]. Cyclic voltammetry studies indicated that **8a** exhibits two, one-electron oxidations centered on the respective metal ions [59]. A single crystal X-ray structure determination, along with molecular modeling, established the Fe $\cdots$ Fe distance in **8a** to be 3.4–4.8 Å during flexing [60].

An important feature of the reaction that became the subject of some debate was the nature of the protonation process. It was recognized early on that ferrocene is also protonated at the ring carbons. Deuteration and mass spectrometry studies on **8a** and **13** indicated that a dynamic equilibrium is established during protonation in which the proton rapidly moves between Fe–H and agostic Cp-ring C–H bonding as illustrated in Figure 10.6(c) [57,58].

The intermediacy of the Fe–H species allows the proton to attach to ring carbons on either side of the metal ion [57,58]. NMR experiments confirmed the presence of two Fe–H species in protonated **8a** and indicated that the Fe—H ↔ Cp C–H exchange processes are too rapid to be resolved on the NMR time scale even at $-122\ ^\circ$C [53,61]. The lifetimes of the higher energy Fe–H species are, consequently, extremely short. NMR and other data are consistent with a tilted, non-coplanar arrangement of the rings in protonated ferrocene, suggesting that H_2 formation in diprotonated **8a** may be favored by the release of steric strain [53,61].

The short lifetimes of the activated Fe–H intermediates explains the unusual reactivity of **8a**. Hydrogen formation occurs only because the two ferrocenes have a high probability of mediating a collision between the attached $H^\bullet$ reactants during the short periods of time that both simultaneously bear an activated H–Fe proton. If this probability is reduced, as in the conformationally less flexible and more hindered **11**, no reaction is observed. Such an effect on catalytic behavior is fully consistent with true time-dependence and functional convergence, as seen in many enzymes.

In the mid-1980s, an improved synthetic route led Mueller-Westerhoff and co-workers to recognize that **8a** could be employed to catalytically generate hydrogen [62]. Thus, the addition of a sacrificial reductant such as lead or tin to a solution of **8a** in HBF_3OH led to the *in situ* reduction of **10a**, thereby establishing the catalytic cycle depicted in Figure 10.6(a) [63]. This system generates H_2 continuously until all of the lead is exhausted (that is, the number of turnovers during the lifetime of an average catalyst = infinity). Metals such as lead are more electropositive than hydrogen and should consequently react with acids. However, their overpotential renders them entirely inert (e.g., the lead-acid battery). In effect, therefore, **8a** eliminates the overpotential of lead in acid during catalysis, allowing complete reaction [63].

The [1.1]ferrocenophanes **8a–b** were subsequently employed in the photoassisted catalytic production of hydrogen in 1 M acids [64]. Tethering of **8b** to a methylpolystyrene coating deposited upon a *p*-type silicon substrate produced the photoelectrode **8c**/Si that catalytically produced hydrogen upon irradiation with sunlight. Hydrogen generation commenced at a potential more than 250 mV anodic of the most positive potential for hydrogen generation on platinum. The uncoated silicon was inactive under identical conditions.

Additional examples of functionally convergent catalysts can be found in the closely related cyclic tetrameric $[1]^4$ferrocenophanes **16** and **17** and in their corresponding [1.1]metallocenophanes **18** and **19** (Scheme 10.7) [53,58]. In the former pair, ferrocene units are tethered in an arrangement that allows opposing ferrocenes still to mediate collisions between attached hydrogen reactants, albeit at a lesser, but still significant rate.

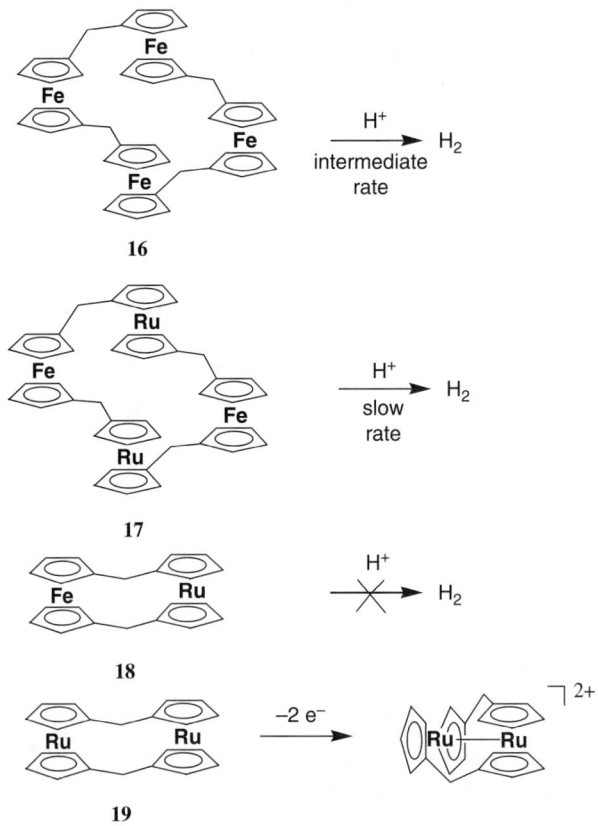

Scheme 10.7.

Protonation of **16** leads to the evolution of hydrogen gas with accompanying formation of tetra-ferrocenium ions but at a noticeably slower rate than observed for **8a**. This decrease in rate could be directly measured by deuteration experiments in DBF$_3$OD, which indicated that proportionally more of the Cp C–H protons in **16**$^{4+}$ undergo H–D exchange than in **8a**$^{2+}$. This can only occur if hydrogen elimination is slower in **16** than in **8a**, as expected from the diminished proportion of time that opposing ferrocenes spend in a structure that complements the transition state. Mass spectrometry indicated that a secondary reaction path involving electron exchange between the ferrocenes and the ferrocenium ions in the intermediate diferrocenium cation, also competed weakly with hydrogen elimination.

The mixed tetramer **17** undergoes a much slower but still clear dihydrogen elimination when protonated [53]. Ruthenocene is substantially more basic than ferrocene and therefore more readily protonated. However, it favors two-electron oxidation and is consequently unable to stabilize the catalytically required +1 oxidation state [59]. In acidic solution, the ruthenocene units become protonated but do not eliminate H$_2$. The resulting tri- or tetra-protonated species is sterically and conformationally unfavorable for formation of H$_2$ by the two ferrocene units on opposite sides of the molecule. It does nevertheless occur but only at the slower rate observed. By contrast, the corresponding dimer **18** is catalytically inert in acid [65].

Interestingly, the diruthenocene dimer **19** undergoes a two-electron oxidation leading to the Ru–Ru bonded dication shown in Scheme 10.7 and not the expected four-electron oxidation [29]. This was shown to be caused by a cooperative effect in which the enforced proximity of the two ruthenium ions within the molecule perturbed the usual electronic structure of the individual ruthenocenes [29].

10.4.1.3 Pseudoconvergent Catalysis: Supramolecular, Bifunctional Catalysts of Organic Reactions.

Bifunctional catalysts incorporate two or more catalytic groups that carry out widely differing functions during catalysis; one typically binds a reactant molecule in a selective disposition, and the other then activates and catalyzes its reaction. The binding interactions employed are usually supramolecular in character, such as hydrophobic–hydrophilic interactions. Several reviews of such catalysts are available [20,21,66–68].

The most common receptor sites in bifunctional catalysts are inclusion-promoting or supramolecular species, like macrocycles, crown ethers, cyclodextrins, and the like. Their task is to bind selectively a substrate in the vicinity of another catalytic group. The structural or chiral selectivity of the ensuing catalysis depends on the spatial shape and size of the binding site as well as on the nature of the binding interaction.

The rationale behind these catalysts is generally not to create new reactivities *per se* but to exploit existing ones in a highly selective manner. Thus, reactions involving at least one long-lived activated reactant are typically accelerated by bringing them into close proximity to each other. It is achieved in a highly discriminating way; one reaction is specifically accelerated over all other competing ones by carefully arraying the catalytic groups within the catalyst.

The π-allylpalladium(II) complex of the chiral ferrocenylphosphine–crown ether ligand **20** is a highly efficient catalyst of the asymmetric allylation of β-diketones, like the one shown in Scheme 10.8 [69]. Not only is the reaction significantly accelerated relative to an unsubstituted control, but high enantioselectivity is also achieved. It seems to be caused by the initial formation of a favored ternary complex **21**, involving the crown ether, a potassium cation, and the enolate anion. Only after prearranging in this way, does the enolate anion attack the π-allylpalladium(II) moeity, thereby generating the corresponding allyl product. Because of the weakness

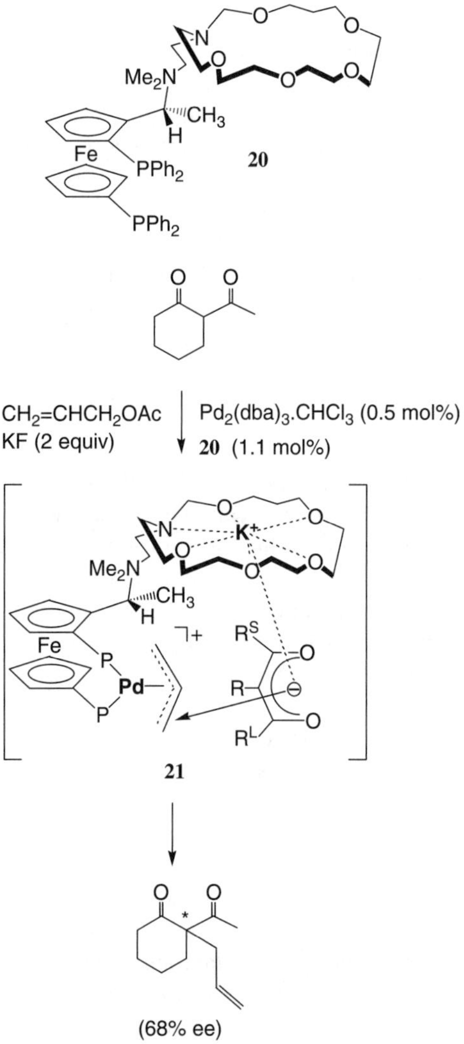

Scheme 10.8.

of the K^+-enolate ion-pairing interaction, changes in the crown ether tether destroy the selectivity of the catalysis but not necessarily the catalysis itself.

Cyclodextrins have also been widely used in biomimetic catalysis. Cyclodextrins are cylindrical, cyclic oligomers of glucose in which the central cavity is strongly hydrophobic. Hydrophobic substrates with suitable spatial geometry are therefore readily included in the cavity [78]. Figure 10.7 illustrates a system reported by Breslow and co-workers that has been discussed in several previous chapters [70].

The manganese porphyrin **22**, which is derivatized with two β-cyclodextrins arranged linearly across the Mn ion, converts the linear olefin substrate **23** to the epoxide product shown in Figure 10.7, with high selectivity. This selectivity derives from the inclusion of the long-chain tails of the olefin in the two opposite cyclodextrin cavities, thereby holding the C−C double bond above the Mn ion in the porphyrin ring. Oxidation consequently results in epoxidation at this bond. The catalyst shows good selectivity and activity, undergoing 40 turnovers on average before degradation. By contrast, the corresponding catalyst **24**, in which two β-cyclodextrins are attached to the porphyrin in a nonlinear fashion, exhibits only poor selectivity [5].

In catalysts **21** and **22**, the system components clearly perform complementary roles. However, they also demonstrate a significant level of convergence: that is, the product obtained is strongly, albeit not exclusively, dependent on the spatial arrangement of the catalytic groups within the molecule. This outcome occurs is because catalysis of the reaction does not necessarily require a multicentered transition state. Single-centered catalysis may also take place, although it is clearly less favored. For this reason, **21** and **22** can be considered to display functional convergence in most turnovers, but on some occasions, they act in a single-centered manner. As such, the synergies in catalysts of this type differ from those of many enzymes in that modest structural changes to the catalyst typically diminish their activity and alter their specificity, but they do not necessarily render catalysis impossible.

10.4.1.4 Probable Functionally Convergent Catalysis: Rhodium-Phosphine Hydroformylation Catalysts.

One of the most well-known multicentered organometallic catalysts is Stanley and co-worker's dinuclear rhodium-phosphine hydroformylation catalyst **25** (Scheme 10.9) [71,72].

1-Hexene is converted by **25** to a >25:1 mixture of linear:branched aldehydes without the need for excess phosphine and at a turnover frequency of $390\,h^{-1}\,Rh^{-1}$ (under 60 psi pressure at 80 °C) [71]. By contrast, the comparable $HRh(CO)(PPh_3)_3$ commercial catalyst requires a 319-fold excess of PPh_3 and then only achieves a 14:1 ratio of linear:branched aldehydes under the same circumstances [71]. Its turnover frequency is $875\,h^{-1}\,Rh^{-1}$.

What is particularly surprising about this system is the fact that **25** should be a much slower catalyst because of the more basic nature of its tetraphosphine ligand. It is certainly the case with comparable mononuclear, control catalysts. The dramatic rate enhancement is therefore caused by a bimetallic effect involving the two rhodium centers. This is confirmed by the poor catalytic rate of the analogous

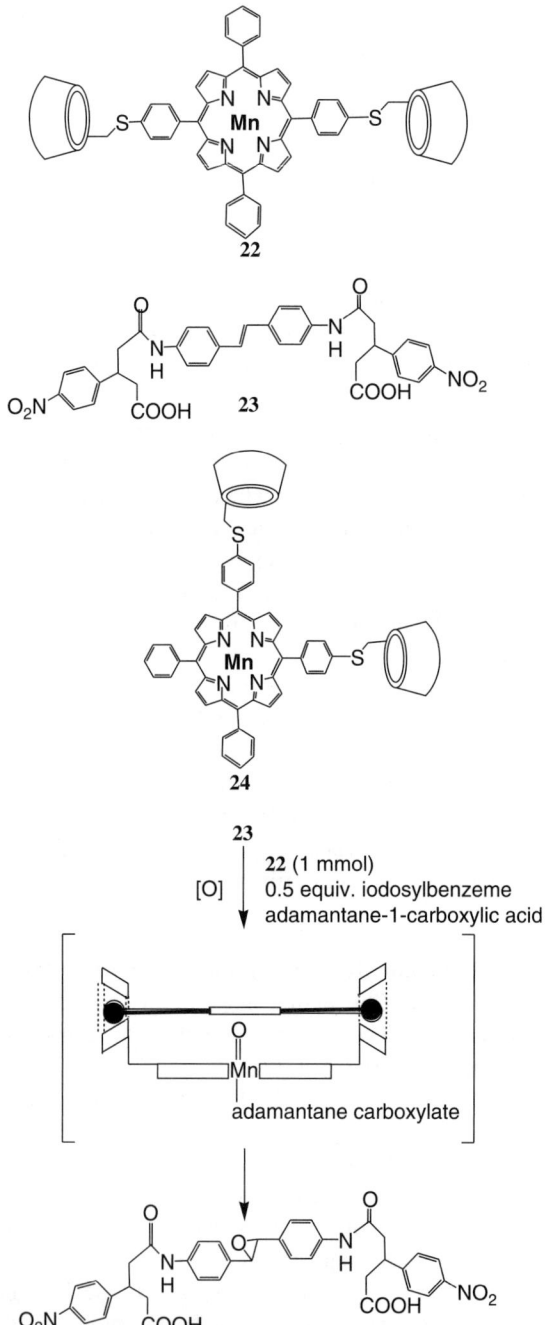

Figure 10.7. *Biomimetic epoxidation catalysis*: Mn porphyrin **22** oxidizes substrate **23** with high, but not perfect, selectivity. Porphyrin **24** shows little selectivity under comparable conditions.

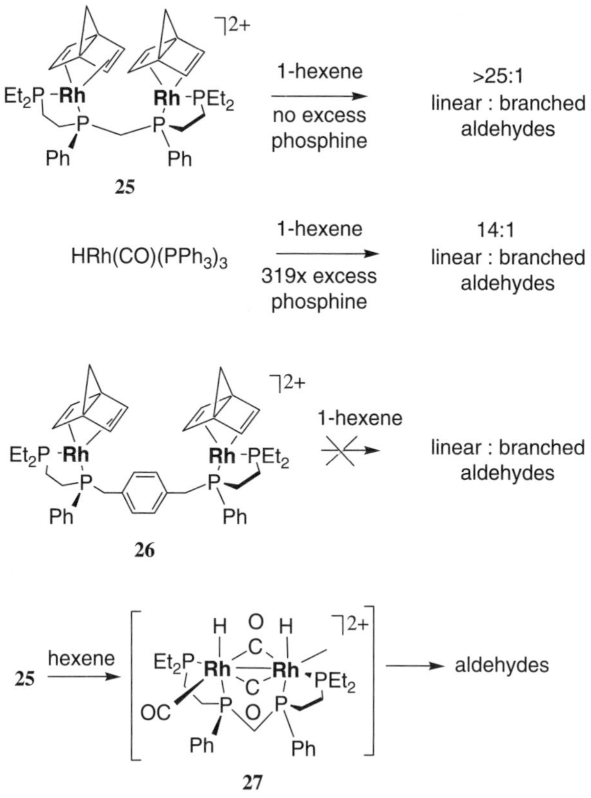

10.4 NONBIOLOGICAL, MULTICENTERED MOLECULAR CATALYSTS 249

Scheme 10.9.

p-xylene-bridged system **26**, in which Rh–Rh approach is strongly hindered. Early studies suggested that the key, rate-determining step in this catalysis involves an intramolecular hydride transfer from the one rhodium center to the other [71]. However, recent *in situ* FTIR and NMR studies suggest a more complicated mechanism [73]. The dirhodium(II) complex **27** that contains an unusual Rh(II)–Rh(II) bond is believed to be the catalyst, with linear aldehyde production favored by the crowded steric environment. Because the other ligands cannot easily bend away upon coordination of an alkene to **27**, the tetraphosphine ligand is sterically highly effective.

Several other binuclear rhodium complexes that catalyze hydroformylation have been described [74]. Dirhodium carboxylates have also been shown to catalyze certain carbenoid reactions with Michaelis–Menten kinetics, which is indicative of a time-dependent catalytic action [75].

Because little is known about the mechanism of catalysis in these species, firm conclusions cannot be drawn regarding the catalytic synergies achieved. However, the Michaelis–Menten kinetics observed in the dirhodium carboxylates and the fact that the spatial arrangement of the rhodium moieties is exceedingly important to their catalytic performance strongly suggest that the above dirhodium species display functional convergence. As such, they must be presumed to probably be time-dependent.

10.4.1.5 Possible Functionally Convergent Catalysis: Ruthenium-Based Water Oxidation Catalysts.
The oxidation of water to oxygen is of crucial importance in water electrolysis. This half-reaction is not only the rate-determining step in the process but also highly energy intensive.

A wide range of Ru–N complexes, including many mononuclear species, have now been shown to act as, or to form, *in situ*, active water oxidation catalysts. Comprehensive reviews of ruthenium-based water oxidation catalysts are available [32]. A common feature of many of these catalysts is the presence of bridging μ-oxo or μ-chloro moieties. The role of these species and the bridges is not yet entirely clear, but they are, very likely, involved in correctly arraying the ruthenium catalytic groups during the time that they bear activated reactants. If so, they display functionally convergent catalysis in their mode of action. A detailed listing of Ru- and Mn-oxo water oxidation catalysts is provided in Table 11.1 in Chapter 11.

Among the most widely studied intramolecular catalysts of water oxidation are μ-oxo-bridged, *N*-donor complexes, such as $[(bpy)_2(H_2O)Ru^{III}(\mu-O)Ru^{III}(H_2O)(bpy)_2]^{4+}$ **28** and the tri-ruthenium complex known as "ruthenium red" $[(NH_3)_5Ru^{III}(\mu-O)Ru^{IV}(NH_3)_4(\mu-O)Ru^{III}(NH_3)_5]^{6+}$ **29** (Scheme 10.10) (bpy = 2,2′-bipyridine) [76–78].

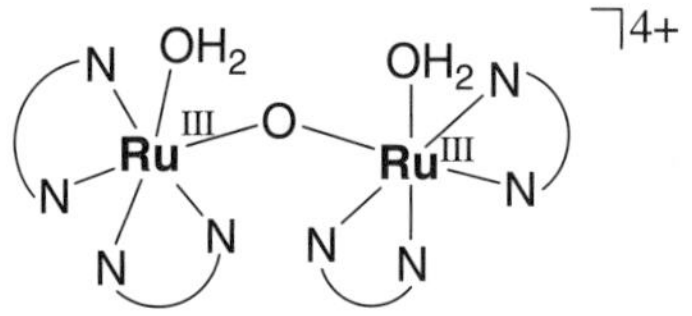

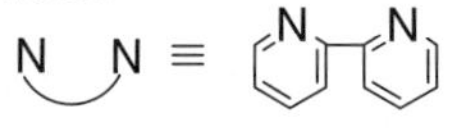

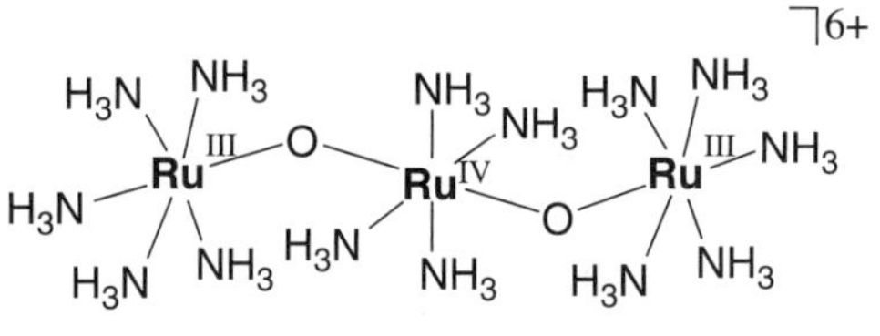

Scheme 10.10.

10.4 NONBIOLOGICAL, MULTICENTERED MOLECULAR CATALYSTS

The oxo-bridged diruthenium(III) complex **28** was first reported to act as a water oxidation catalyst by Meyer and coworkers in 1982 [76]. In these experiments, which involved oxidation by Ce(IV), only 25 turnovers were achieved on average before catalyst degradation and ligand exchange [76]; more turnovers could be obtained under other conditions [77]. This compound has nevertheless been extensively studied as a model molecular catalyst of water oxidation. Its mechanism has been subjected to particular scrutiny. Progress in this respect has been reviewed in detail recently [79]. Isotopic labeling studies provide what seem to be the most important clues regarding mechanism. When the terminal aquo-ligands in **28** were ^{18}O-labeled and oxidized with 4–5 equiv. of Ce(IV) in 1 M triflic acid, a mixture of $^{18}O_2$ (13%), $^{18}O^{16}O$ (64%), and $^{16}O_2$ (23%) was obtained [80]. Labeling of all oxygens in **28** with ^{18}O led, however, to the production of $^{18}O_2$ (60%) and $^{18}O^{16}O$ (40%) when oxidizing with 5–6 equiv. of Co(III) [81]. These results suggest that several mechanisms operate simultaneously if the catalytic process is intramolecular. Alternatively, the appearance of all three labeled forms can be explained, at least qualitatively, by a bimolecular mechanism [82,83]. In the former case, $^{16}O_2$ may be formed along the lines depicted in Figure 10.8(a) or (b), whereas the mechanism shown in

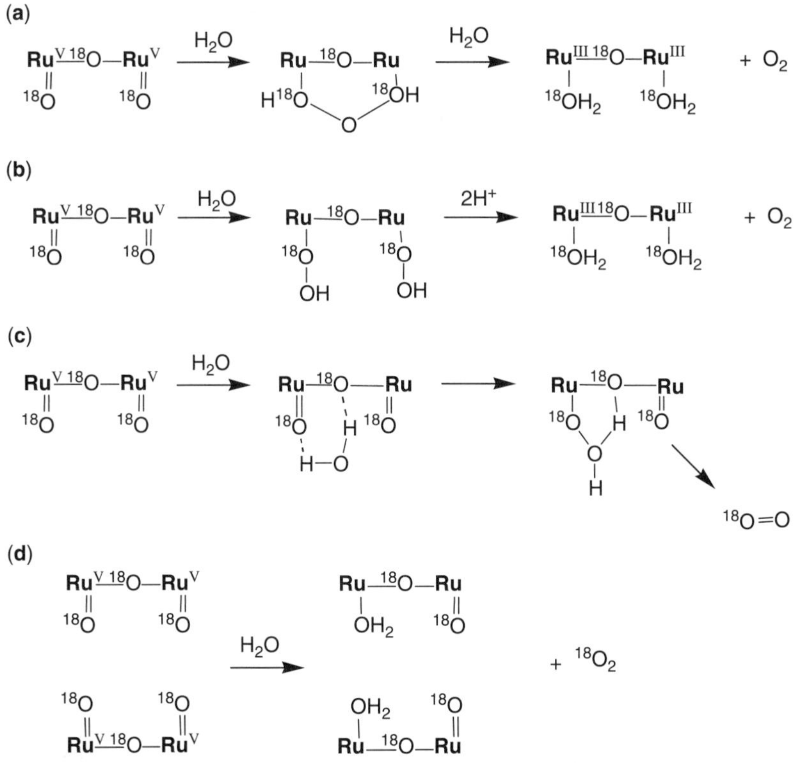

Figure 10.8. *Water oxidation by μ-oxo-bridged ruthenium complexes:* Possible mechanisms of water oxidation consistent with the data obtained. Mechanisms (a)–(c) involve intramolecular catalysis. Mechanism (d) involves intermolecular catalysis.

Figure 10.8(c) may lead to the $^{18}O^{16}O$. In the latter case, $^{18}O_2$ formation can be explained by the mechanism depicted in Figure 10.8(d), with the mixed and $^{16}O_2$ products possibly obtained by scrambling of the label in a cross-reaction between the Ru^VORu^V and the $Ru^{IV}ORu^{III}$ intermediates. Kinetic studies seem, unfortunately, to be unhelpful in clarifying this process as the step involving actual water oxidation is not rate limiting [82,83].

Ruthenium-red **29** is, by contrast, a highly active molecular catalyst of water oxidation. When adsorbed in high dispersion on a basal-plane pyrolytic graphite electrode coated with platinum black, it amplifies the rate of oxygen generation by ca. 8-fold when poised for 1 h at 1.3 V (vs. Ag/AgCl) in 0.1 M $NaClO_4$ [77]. Studies indicated an increase in the rate of gas generation with increased loading of **29**, up to a maximum of 9.3×10^{-9} mol of **29** per 6.0×10^{-6} mol of platinum black. Loadings in excess of this amount resulted in progressive deactivation of the catalyst caused by a competing bimolecular decomposition reaction. The critical intermolecular distance for decomposition was calculated to be 1.21–1.37 nm.

10.4.1.6 Functionally Complementary Catalysis: Intramolecular Epoxidation Catalysts. A range of multicentered epoxidation catalysts are known. Most of these catalysts involve intermolecular mechanisms, in which one metal ion binds and activates the epoxide and another binds and activates the substrate.

The dimeric complexes **30a–c** (Fig. 10.9) are intramolecular catalysts of this type, which efficiently and asymetrically ring-open epoxides [84]. The equivalent intermolecular mechanism usually accompanies this process. These catalysts have been discussed in a comparative light in several previous chapters.

The mechanism of this catalysis involves an epoxide bound to a chromium species that is ring-opened by an azide located on another chromium species. In the equivalent, untethered intermolecular catalyst **30d**, the rate of ring-opening declines rapidly as the concentration in solution of the corresponding free chromium species are decreased. However, when two such species are tethered to each other, as in **30a–c**, the reaction may proceed by an intramolecular pathway even at very low concentrations. A "head-to-tail" transition state seems to be favored for such a reaction.

Thus, when treated with cyclopentene oxide and TMS-N_3, **30a** was found to catalyze the asymmetric ring-opening reaction to completion over 24 h at 0.05 mol%. The corresponding untethered monomer was essentially inactive at this concentration. To achieve the same rate of reaction, 1 mol% of the monomer was required. Comparable enantiopurities (90–94% ee) were achieved.

Epoxidation catalysts of this type display functionally complementary catalysis, although a relatively minor level of convergence is usually present. This convergence is created by the tether in **30a–c**. Although small relative to that found in functionally convergent catalysts, this convergence nevertheless strongly affects the efficiency of the catalysis and explains the relative reaction rates of **30a–c**.

A key advantage of functionally complementary catalysts like **30a–c** is their structural and synthetic simplicity over true functionally convergent catalysts such as **8a–c** and **1**. This is, of course, highly beneficial from a practical and commercial point of view.

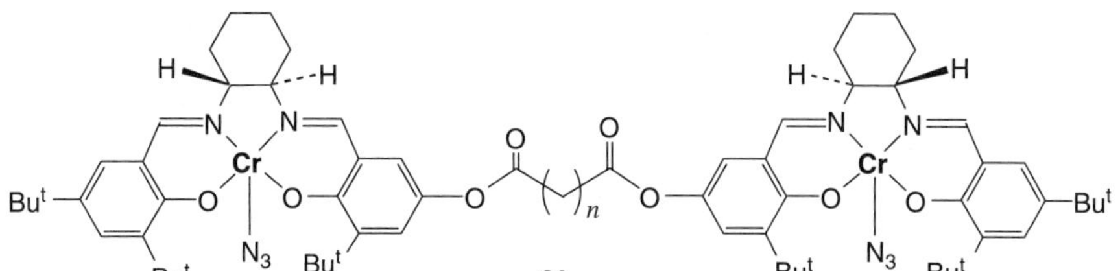

10.4 NONBIOLOGICAL, MULTICENTERED MOLECULAR CATALYSTS

30
a: $n = 5$ b: $n = 2$ c: $n = 10$

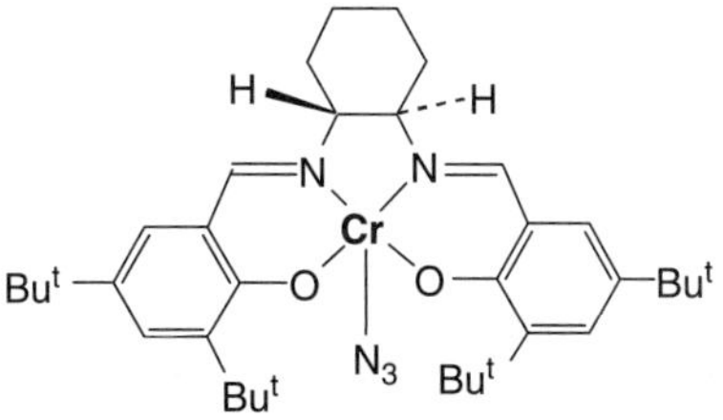

30d

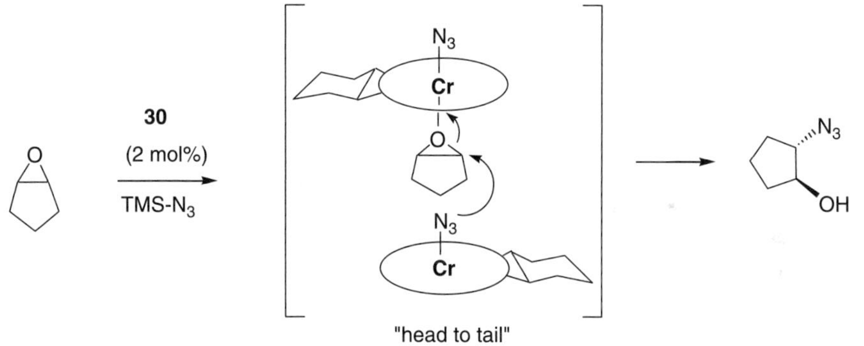

"head to tail"

Intramolecular reaction rate constant (k_{intra}):

30a: 42.9×10^{-2} min^{-1}, **30b**: 4.4×10^{-2} min^{-1}, **30c**: 3.8×10^{-2} min^{-1}

Figure 10.9. Functionally complementary asymmetric catalysis in an epoxide ring-opening reaction.

10.4.1.7 Metal Clusters in Multicentered Molecular Catalysis: Triruthenium Dodecacarbonyl Hydrogenation Catalysts.

Although not generally thought of as "biomimetic" in character, metal clusters are prospectively ready-made for multicentered molecular catalysis. Indeed, the field of metal clusters can be considered to overlap theoretically with enzymatic, homogeneous, and heterogeneous catalysis. On the one hand, supported bimetallic clusters of platinum mixed with rhenium or iridium are widely used in the petroleum reforming industry, where they have played a key role in the development and production of unleaded gasoline [85]. On the other hand, many enzyme active sites, such as the photosystem II water oxidizing complex, contain metal clusters [79].

The biomimetic aspects of metal clusters go further than that, however. Metal clusters may also act catalytically in accord with the general principles of enzymatic catalysis. Thus, metal clusters may 1) undertake multicentered binding and activation of a single reactant, 2) bind and activate reactants on adjacent or nearby metal atoms (that is, in close proximity to each other), 3) facilitate the formation of a transition state stabilized at multiple points and in three dimensions, and 4) induce cooperative influences among the metals involved, thereby assisting catalysis [86,87]. Metal clusters may also facilitate the migration of bound or activated reactants from one metal atom to another as is observed in some heterogeneous catalysts and, indeed, in certain enzymes [86,87].

Although widely studied and employed as catalysts, identifying the active species and thereby establishing the mechanism of catalysis by metal clusters is often highly problematic [86]. In this respect, they are also similar to heterogeneous and enzymatic catalysts. The polynuclear complex originally introduced or generated may not be the catalytically active species. Cluster fragmentation or degradation into new, unidentified, but catalytically active species is common. This phenomenon complicates attempts to understand the synergies in such catalysts since simple comparisons with the corresponding mononuclear complex cannot be readily made.

Phosphine (L) derivatives of triruthenium dodecacarbonyl clusters $[Ru_3(CO)_{12-x}(L)_x]$ have been extensively studied as hydrogenation catalysts. The mechanism of one such process has recently been elegantly and comprehensively established using parahydrogen (p-H_2) NMR methods [88]. The reaction of cluster **31** with p-H_2 in $CDCl_3$ containing a 100-fold excess of diphenylacetylene resulted in the initial formation of the hydride clusters **32**, **33**, and **34** in the ratios 58:9:1, respectively (Scheme 10.11). However, the NMR signal intensities of **32** and **33** rapidly increased until, after 30 s, they displayed comparable strengths. Studies indicated that the signal of **33** was amplified by its vigorous participation in the catalytic hydrogenation of the diphenylacetylene. Through the exhaustive use of NMR and comparative procedures, the mechanism of the catalysis was established in detail to be that shown in Figure 10.10. Cluster **33** reacts with diphenylacetylene and hydrogen, losing a CO ligand in the process, to form **35**, which is, in turn, cycled back to **33** by loss of the alkene and uptake of CO. The long-lived resting state **36** was also detected; this species catalyzes the reaction, albeit more slowly. Fragmentation of the cluster, via species **37**, was found to provide a minor competing catalytic pathway.

10.4.1.8 Statistical Approaches to Functionally Convergent Catalysis: Macromolecular Intramolecular Catalysts. An alternative to the use of small-molecule intramolecular catalysts that contain optimally arranged catalytic groups is to tether the catalytic groups to a reactive monomer, which is then polymerized into a linear or branched polymer. If correctly designed, the regularly repeating nature of the polymer and its flexibility may result in many of the catalytic groups being optimally arrayed with respect to each other. That is, there is a statistical probability that some proportion of the catalytic groups will be optimally organized to facilitate reactive collisions. Little control of the exact spatial organization is typically available in such systems, however, so that uniformly precise proximities and dispositions

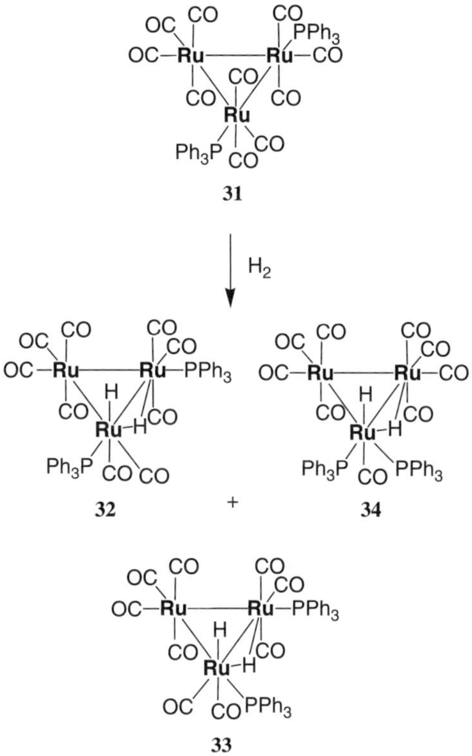

10.4 NONBIOLOGICAL, MULTICENTERED MOLECULAR CATALYSTS

Scheme 10.11.

cannot be achieved. A distribution of spatial arrangements and conformational movements will instead be obtained.

As noted, when short-lived activated reactants are employed, this means that at any one time a small, statistically relevant proportion of the catalytic groups present will be active and able to undertake functionally convergent, single-turnover catalysis. Many such single turnovers may produce high overall catalytic activity.

Polymer supports offer several important fundamental advantages in catalysis. As pointed out recently, enzymes are macromolecules and certain of their properties derive from this fact [89]. Synthetic polymers can similarly be used to generate highly desirable catalytic effects. These effects may be inherent to a particular chemical property of the polymer employed [89], such as an ability to act as an ion-transport media, or they may be incidental, being a function of the morphology or other bulk attribute of the polymer [90]. As noted in Chapter 2, polymer supports may, for example, act to [91]:

1. Promote coordinative unsaturation of the metal atoms/ions, thereby speeding up their catalytic reaction
2. Improve the selectivity of the catalysis, with an accompanying improvement in the product properties

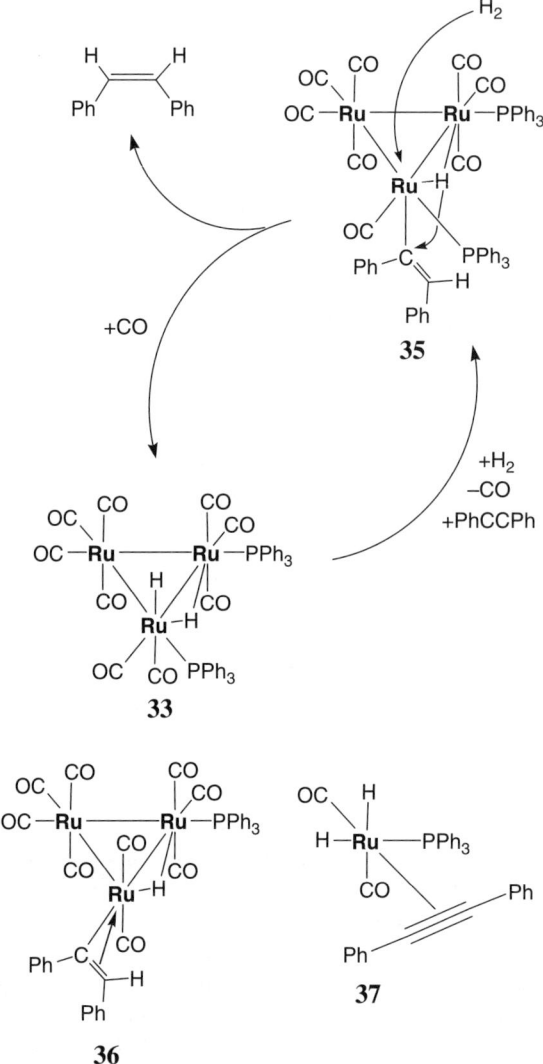

Figure 10.10. *Catalytic hydrogenation of an alkyne by a triruthenium dodecacarbonyl cluster*: The catalytic mechanism for the hydrogenation of diphenylacetylene was established by para-hydrogen NMR methods.

3. Retain the metal atoms/ions strongly, allowing maximum product generation by the catalyst
4. Allow multistep catalysis in which the product of one step is consumed in the next

From a purely practical point of view, polymer-supported molecular catalysts are also more easily handled in industrial applications than homogeneous catalysts in

10.4 NONBIOLOGICAL, MULTICENTERED MOLECULAR CATALYSTS

open solution. They can, for example, be physically withdrawn from product and reagent streams to avoid the catalyst becoming incorporated in the reaction products.

Polymers also display notable drawbacks. If establishing the mechanism in multicentered catalytic processes in homogeneous solution is problematic, then determining it in polymers is generally still more complicated. The main difficulty involves separating those properties that are inherent to the catalytic process from those that originate from the polymer. For example, whereas the catalytic step is typically rate-determining in open solution, hindered diffusion processes may well be rate-determining within a polymer. Heterogeneous polymer mixtures can, moreover, often not be studied using common analytical techniques, such as solution NMR.

Macromolecular, intramolecular catalysts that display functionally complementary mechanisms are known. For example, dendrimer **38** selectively catalyzes the ring-opening epoxidation of (S)-vinyl cyclohexyloxide, (S)-**39**, in the racemic mixture (±)-**39** (Scheme 10.12) [92]. This process is known to occur by a multicentered, intramolecular mechanism similar to that depicted in Figure 10.9.

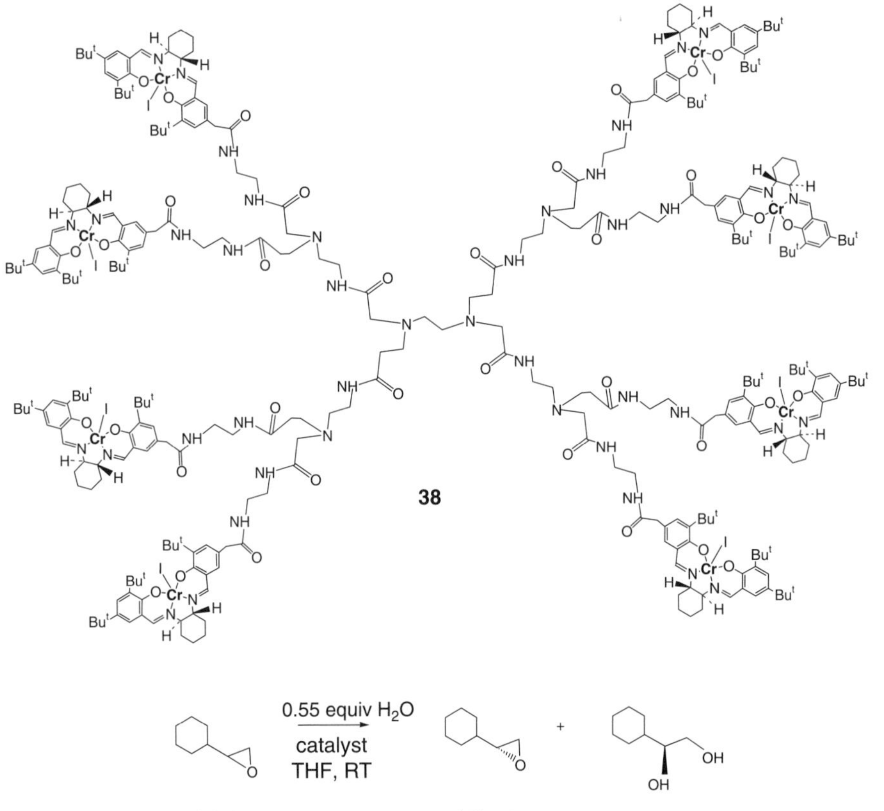

Scheme 10.12.

The effect of the minor convergence present in this process could be elegantly illustrated by comparing the hydrolytic kinetic resolution of **38** and the comparable mononuclear catalyst **30d** (Fig. 10.9) at similar catalyst concentrations/loadings. At 0.025 mol% loading, **30d** gave no measurable conversion after 40 h. However, a loading of 0.027 mol% of **38** gave 50% conversion with complete kinetic resolution (98%ee). This effect was suggested to originate in restricted conformations imposed by the dendrimer structure, thereby creating a greater effective molarity (i.e., increased convergence) [92]. Alternatively, the multimeric nature of the dendrimer may lead to higher order, cooperative electronic interactions.

To the best of our knowledge, true functionally convergent, multicentered catalysis within a nonbiological polymer has not been unequivocally demonstrated. However, a few reports describe catalytic processes that may possibly involve such processes. One example involves an intriguing report that details the preparation of a methyl acryl chloride polymer to which iron(III) tetra(*o*-aminophenyl)porphyrin **40** was attached using an amidization reaction (Scheme 10.13) [93]. The resulting porphyrin-tethered polymer was spin-coated onto a glassy carbon electrode and tested as an electrocatalyst of oxygen reduction. An unexpected result was obtained.

When the ratio of poly(methyl acryl chloride) to **40** in the synthetic step was such that the surface coverage of **40** was $(0.5-1.1) \times 10^{-9}$ mol cm^{-2}, four-electron reduction of oxygen to water was observed in 0.05 M H$_2$SO$_4$ solutions. However, two-electron reduction to hydrogen peroxide was observed when the surface coverage was decreased to $(1.0-1.5) \times 10^{-11}$ mol cm^{-2} by changing the ratio of

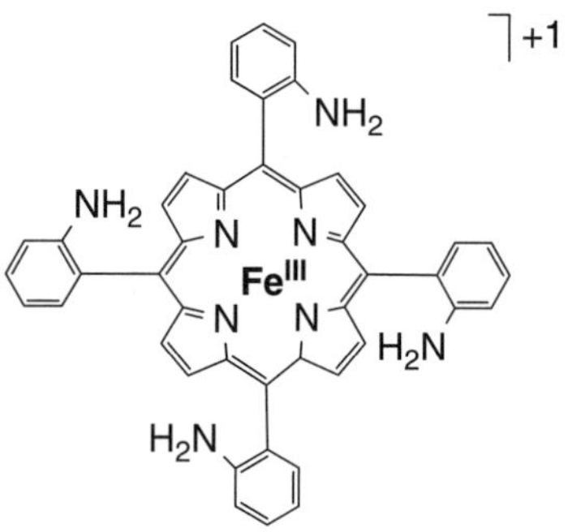

40

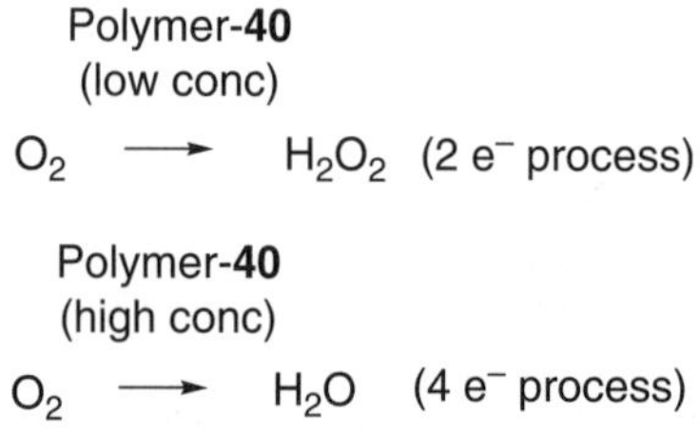

Polymer-**40**
(low conc)

O$_2$ ⟶ H$_2$O$_2$ (2 e$^-$ process)

Polymer-**40**
(high conc)

O$_2$ ⟶ H$_2$O (4 e$^-$ process)

Scheme 10.13.

10.4 NONBIOLOGICAL, MULTICENTERED MOLECULAR CATALYSTS

poly(methyl acryl chloride) to **40** in the synthetic step under otherwise invariant conditions [92].

Several explanations are, of course, conceivable for this phenomenon, including the possibility of single-centered catalysis caused by cooperative electronic effects. However, it is not immediately obvious how such an effect could originate given that the polymer should have little electronic influence on the porphyrin system and that this should, in any case, be much the same at the two different concentrations studied. Instead, a substantial number of monomers may well become adventitiously proximate for convergent catalysis at high concentrations.

A similar effect was observed for cobalt phthalocyanine **41** (Scheme 10.14). Under normal conditions, **41** catalyzes the conversion of dioxygen to exclusively H_2O_2 [94]. However, polymeric cobalt phthalocyanine **42** displays mixed product selectivity, generating about 50% H_2O, along with H_2O_2 under certain conditions [95]. The phthalocyanine structures in **41** and **42** are somewhat different, so that a simple comparison of this type may, of course, not be valid.

10.4.2 Intermolecular Catalysts

10.4.2.1 Functionally Complementary Catalysis. Intermolecular catalysis involving two or more catalytic groups that are not connected to each other, is known in, for example, the epoxidation catalyst **30d** (Fig. 10.9). As noted, catalysis in

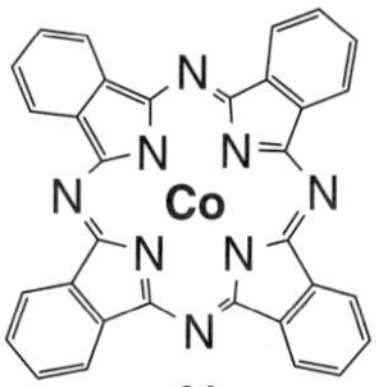

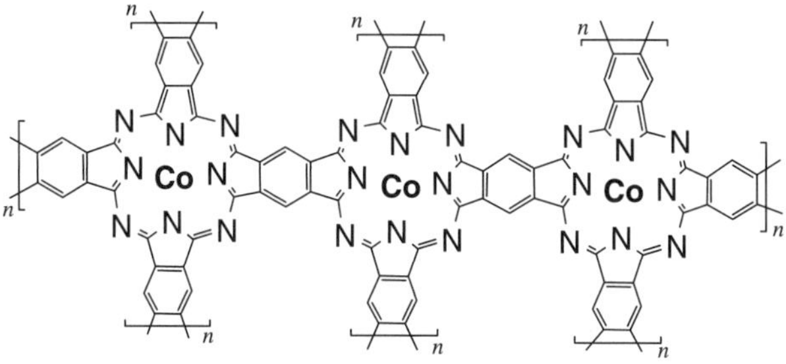

$$O_2 \xrightarrow{41} H_2O_2 \quad (2\ e^-\ \text{process})$$

$$O_2 \xrightarrow{42} H_2O \quad (4\ e^-\ \text{process})$$

Scheme 10.14.

260   A LITERATURE SURVEY OF MULTICENTERED HOMOGENEOUS CATALYSIS

such cases typically requires intermediates that are bound and activated for substantial proportions of the time (see Section 6.3 in Chapter 6). A particular spatial arrangement of the catalytic groups need not, therefore, be maintained to achieve catalysis.

10.4.2.2 Statistical Approaches to Functionally Convergent Catalysis: Concentration Effects in Intermolecular Catalysts. In the realm of small-molecule activation, cooperative intermolecular catalysis has been claimed in the reactions of a range of monomeric ruthenium amine complexes, such as **43** and **44** (Scheme 10.15) [32,96–98]. These molecules catalyze water oxidation in both open solution and when immobilized in various porous polymers. Kinetic studies reveal their catalytic activities to be second order in the concentration of the monomers, prompting the suggestion of an intermolecular mechanism, which would presumably be conceptually akin to that depicted in Figure 10.8(d).

Such a mechanism has, however, not been unambiguously established. It is possible, and perhaps even very probable, that the slow step in the catalysis involves formation of a bimetallic species that is the true catalyst. Given the activities of comparable μ-oxo species in the same reaction (Section 10.4.1.5), it is possible that such species are formed *in situ*.

Metalloporphyrins and pthalocyanines have provided several other interesting examples that may possibly involve intermolecular multicentered catalysis. Certain mononuclear metalloporphyrins and phthalocynanines adsorbed onto graphite can catalyze the four-electron reduction of O_2 to H_2O in the same way that the cofacial metallodiporphyrins discussed in Section 10.4.1.1 do. This has led to suggestions that certain of these species are effectively adsorbed in a side-on, pairwise arrangement or become adventitiously eclipsed and thereby capable of intermolecular catalysis [94].

An example in this respect is the tetrasulfonated iron phthalocyanine **45** (Scheme 10.16). At low concentrations, the electrocatalytic reduction of dioxygen by **45** has been reported to involve a two-electron process giving H_2O_2. As the concentration of **45** is increased however, this rapidly gives way to a four-electron

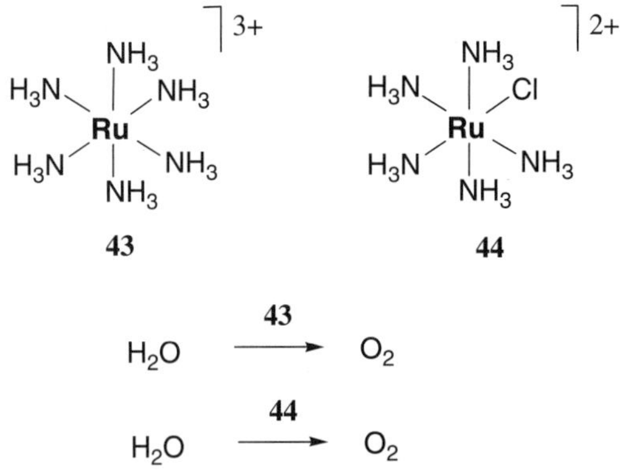

Scheme 10.15.

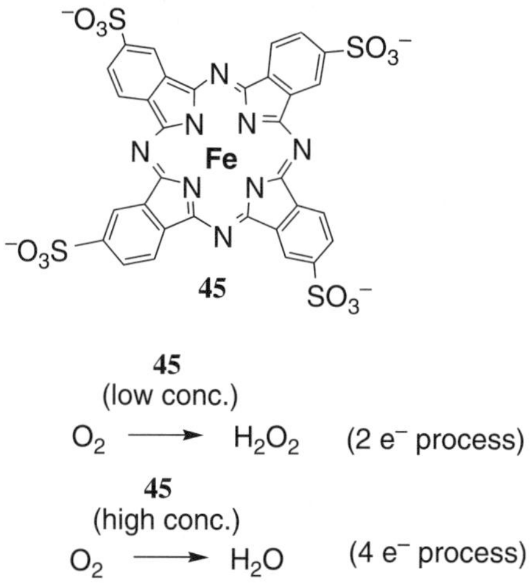

10.4 NONBIOLOGICAL, MULTICENTERED MOLECULAR CATALYSTS

Scheme 10.16.

45 (low conc.)
$O_2 \longrightarrow H_2O_2$ (2 e⁻ process)

45 (high conc.)
$O_2 \longrightarrow H_2O$ (4 e⁻ process)

reduction to H_2O [94,99,100]. The fact that some supported metalloporphyrin and phthalocyanine catalysts retain their catalytic properties and, indeed, improve their durability after heat treatment to 800–1000 °C, despite the destruction of the organic ligands, adds another level of complexity to this fascinating situation [94].

These systems are extremely rare examples of molecular catalytic systems that have the same properties as their structurally analogous heterogeneous catalysts. The fact of identical catalytic properties suggests, albeit inconclusively, that the key role of the ligand in such catalysts is to space correctly the metal ions for transition state formation.

10.4.2.3 Statistical Approaches to Functionally Convergent Catalysis: Self-Assembled, Supramolecular Catalysts. Perhaps the most promising approach to intermolecular catalysis involves the use of supramolecular interactions in a "self-assembled" species to bring separate catalytic groups into optimum spatial dispositions. Compounds displaying a wide variety of structural motifs can now be prepared using such influences [101].

One remarkable report in this respect has described a multicomponent metal cluster that self-assembles under turnover conditions [102]. The species $\alpha\text{-}[(Co^{II})PW_{11}O_{39}]^{5-}$ is a catalyst of the oxidation of 1,2-diphenylethene by PhIO. It is self-assembled *in situ* in the presence of the reagents and provides an immediate catalytic effect. The components of the catalyst (Co^{2+}, $H_2PO_4^-$, WO_4^{2-}, and H^+) do not produce a catalytic effect either individually or in various incomplete combinations. Thus, the cluster spontaneously self-assembles during catalysis, meaning that it cannot degrade because of fragmentation. The mechanism of its catalysis has unfortunately not been reported so that it is not clear whether it is single-centered or multicentered.

10.4.3 Footnote: Unexpected Mechanistic Changes in Multicentered Catalysts

As an important postscript to this chapter, it is worth noting that small changes to multicentered molecular catalysts, particularly intramolecular catalysts, can create substantial changes in their mechanisms, including a reversion to complete or simultaneous single-centered action. Examples already touched on include **22** (Fig. 10.7) and **6** (Scheme 10.5), both of which display simultaneous multicentered and single-centered catalysis, albeit in very different proportions (Sections 10.4.1.1 and 10.4.1.3). In these cases, the single-centered and multicentered mechanisms generate products that are quite different, so that their dual modes of action are obvious. However, when a single-centered catalytic process generates a product that is expected from a multicentered process, then this can be extremely misleading to researchers.

An example of such an occurrence is Anson and coworker's catalyst **46** (Scheme 10.17) [30,103]. As mentioned, cofacial diporphyrins like **3** and **11** (Scheme 10.3) successfully convert O_2 to H_2O only when adsorbed on edge-plane graphite electrodes. This conversion is thought to be caused by the rapidity with which electrons can be shuttled from the graphite to the active Co^{II} center. Catalyst **46** was therefore prepared with the intention that the four peripheral Ru^{II} centers could intramolecularly, and therefore rapidly, provide the Co^{II} center with four electrons for the reduction. In this way, the reaction was expected to be complete before electron transfer would be required from the electrode surface, thereby allowing the use of other electrode substrates.

As it turned out, **46** did, indeed, catalyze the formation of H_2O from O_2 in, for example, a Nafion layer on a graphite electrode [103], but not for the reasons expected [30]. Instead studies showed that the only role played by the peripheral ruthenium ions was to back-bond to the porphyrin, thereby increasing the extent of d-electron

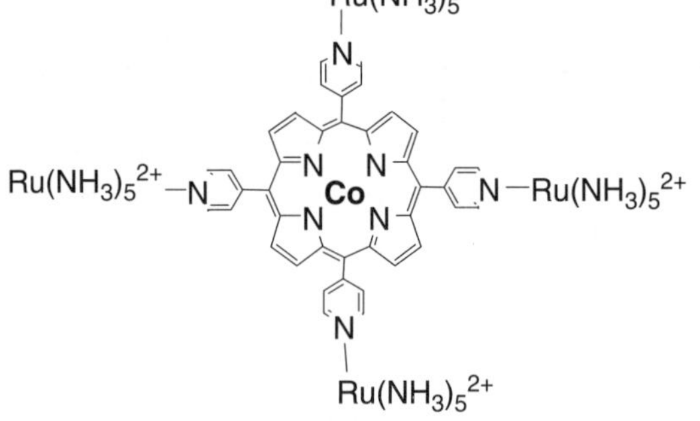

$$O_2 \xrightarrow{46} H_2O \quad (4\ e^-\ \text{process})$$

(single-centered catalysis)

Scheme 10.17.

transfer from the Co-porphyrin to the O_2 and its intermediates [30]. This electronic effect most likely increased the strength of the catalyst–reactant interaction, causing the peroxide intermediate to be activated for more sustained periods. The activation presumably involved an increase in the electron density of the O–O antibonding molecular orbitals. Dissociation of H_2O_2 prior to O–O cleavage was consequently prevented.

ACKNOWLEDGMENTS

The authors thank for their comments: Fred Menger (Emory University), Wolf Sasse (CSIRO) and Peter Osvath (CSIRO). GFS also thanks Bruce King (University of Georgia) who, rather than publishing an embryonic version of this manuscript that had been accepted for a special edition in 2003, agreed to allow the proper and full development of this topic.

REFERENCES

1. Breslow, R. *Chem. Soc. Rev.* **1972**, *1*, 553.
2. Breslow, R. In *Bioinorganic Chemistry, Advances in Chemistry, Vol. 100*, American Chemical Society, 1971.
3. Breslow, R. *Acc. Chem. Res.* **1980**, *13*, 170.
4. Breslow, R. *Acc. Chem. Res.* **1995**, *28*, 146, and references therein.
5. Breslow, R.; Dong, S. D. *Chem. Rev.* **1998**, *98*, 1997.
6. Breslow, R. *Chem. Rec.* **2001**, *1*, 3.
7. Bender, M. In *Enzyme Mechanisms*, Eds. Page, M. I., Williams, A., The Royal Society of Chemistry, 1987.
8. Saenger, W. In *Structural and Functional Aspects of Enzyme Catalysis*, Eds. Eggerer, H., Huber, R., Springer-Verlag, 1981.
9. Behr, J. P.; Lehn, J. M. In *Structural and Functional Aspects of Enzyme Catalysis*; Eds. Eggerer, H., Huber, R., Springer-Verlag, 1981.
10. (a) Stoddart, J. F. In *Enzyme Mechanisms*, Eds. Page, M. I., Williams, A., The Royal Society of Chemistry, 1987; (b) Klotz, I. In *Enzyme Mechanisms*, Eds. Page, M. I., Williams, A., The Royal Society of Chemistry, 1987.
11. Interfacial catalysis; see, for example: (a) Cleij, M. C.; Scrimin, P.; Tecilla, P.; Tonellato, U. *Langmuir* **1996**, *12*, 2956; (b) Kunitake, T.; Shinkai, S. *Adv. in Phys. Org. Chem.* **1980**, *17*, 435.
12. Breslow, R.; Overman, L. E. *J. Am. Chem. Soc.* **1970**, *92*, 1075.
13. Breslow, R. *Chem. Br.* **1983**, *19*, 126.
14. Masel, R. I. *Chemical Kinetics and Catalysis*; Wiley-Interscience, 2001, p. 707–742.
15. Collman, J. P.; Wagenknecht, P. S.; Hutchison, J. E. *Angew. Chem. Int. Ed. Engl.* **1994**, *33*, 1537, and references therein.
16. Palmer, T. In *Understanding Enzymes (4th ed.)*, Prentice Hall/Ellis Horwood, 1995.

17. Humphrey, A. J.; Turner, N. J. In *Enzyme Chemistry: Impact and Applications (3rd ed.)*, Eds. Suckling, C. J., Gibson, C. L., Pitt, A. R., Blackie Academic & Professional, 1998.
18. van den Beuken, E. K.; Feringa, B. L. *Tetrahedron* **1998**, *54*, 12985.
19. Sanders, J. K. M. *Chem. Eur. J.* **1998**, *4*, 1378.
20. Rowlands, G. *Tetrahedron* **2001**, *57*, 1865.
21. Shibasaki, M.; Sasai, H.; Arai, T. *Angew. Chem. Int. Ed. Engl.* **1997**, *36*, 1236.
22. Kirby, A. J. In *Enzyme Mechanisms*, Eds. Page, M. I., Williams, A., The Royal Society of Chemistry, 1987.
23. Menger, F. M. *Acc. Chem. Res.* **1985**, *18*, 128, and references therein.
24. Menger, F. M. *Acc. Chem. Res.* **1993**, *26*, 206.
25. Fersht, A. In *Enzyme Structure and Mechanism (2nd ed.)*, W. H. Freeman, 1984,
26. Page, M. I. In *Enzyme Mechanisms*, Eds. Page, M. I., Williams, A., The Royal Society of Chemistry, 1987.
27. McCollum, D. G.; Bosnich, B. *Inorg. Chim. Acta* **1998**, *270*, 13.
28. Bosnich, B. *Inorg. Chem.* **1999**, *38*, 2554, and references therein.
29. Mueller-Westerhoff, U. T.; Rheingold, A. L.; Swiegers, G. F. *Angew. Chem. Int. Ed. Engl.* **1992**, *31*, 1352.
30. Anson, F. C.; Shi, C.; Steiger, B. *Acc. Chem. Res.* **1997**, *30*, 437, and references therein.
31. Halpern, J. *Inorg. Chim. Acta* **1982**, *62*, 31, and references therein.
32. Yagi, M.; Kaneko, M. *Chem. Rev.* **2001**, *101*, 21, and references therein.
33. Collman, J. P.; Elliott, C. M.; Halbert, T. R.; Tovrog, B. S. *Proc. Natl. Acad. Sci. USA* **1977**, *74*, 18.
34. Collman, J. P.; Marrocco, M.; Denisevich, P.; Koval, C.; Anson, F. C. *J. Electroanal. Chem. Interfac. Electrochem.* **1979**, *101*, 117.
35. Collman, J. P.; Denisevich, P.; Konai, Y.; Marrocco, M.; Koval, C.; Anson, F. C. *J. Am. Chem. Soc.* **1980**, *102*, 6027.
36. Durand, R. R. J.; Benscome, C. S.; Collmann, J. P.; Anson, F. C. *J. Am. Chem. Soc.* **1983**, *105*, 2710.
37. (a) Collman, J. P.; Hendricks, N. H.; Kim, K.; Bencosme, C. S. *J. Chem. Soc., Chem. Commun.* **1987**, 1537; (b) Liu, H. Y.; Abdalmuhdi, I.; Chang, C. K.; Anson, F. C. *J. Phys. Chem.* **1985**, *89*, 665; (c) Collman, J. P.; Kim, K. *J. Am. Chem. Soc.* **1986**, *108*, 7847.
38. Guilard, R.; Lopez, M.-A.; Tabard, A.; Richard, P.; Lecomte, C.; Brandes, S.; Hutchison, J. E.; Collman, J. P. *J. Am. Chem. Soc.* **1992**, *114*, 9877.
39. Collman, J. P.; Hutchison, J. E.; Lopez, M.-A.; Tabard, A.; Guilard, R.; Seok, W. K.; Ibers, J. A.; L'Her, M. *J. Am. Chem. Soc.* **1992**, *114*, 9869.
40. Collman, J. P.; Chng, L. L.; Tyvoll, D. A. *Inorg. Chem.* **1995**, *34*, 1311.
41. Chang, C. J.; Deng, Y.; Shi, C.; Chang, C. K.; Anson, F. C.; Nocera, D. G. *Chem. Commun.* **2000**, *15*, 1355.
42. (a) Naruta, Y.; Sasayama, M.-A.; Sasaki, T. *Angew. Chem. Int. Ed. Engl.* **1994**, *33*, 1839; (b) Shimazaki, Y.; Nagano, T.; Takesue, H.; Ye, B. Y.; Tani, F.; Naruta, Y. *J. Inorg. Biochem.* **2003**, *96*, 227.
43. Ricard, D.; Andrioletti, B.; L'Her, M.; Boitrel, B. *Chem. Commun.* **1999**, 1523.
44. Collman, J. P.; Fu, L.; Herrmann, P. C.; Zhang, X. *Science* **1997**, *275*, 949.

45. Collman, J. P.; Schwenninger, R.; Rapta, M.; Bröring, M.; Fu, L. *Chem. Commun.* **1999**, 137.
46. Collman, J. P.; Fu, L.; Herrmann, P. C.; Wang, Z.; Rapta, M.; Bröring, M.; Schwenninger, R.; Boitrel, B. *Angew. Chem. Int. Ed. Engl.* **1998**, *37*, 3397.
47. Collman, J. P.; Rapta, M.; Bröring, M.; Raptova, L.; Schwenninger, R.; Boitrel, B.; Fu, L.; L'Her, M. *J. Am. Chem. Soc.* **1999**, *121*, 1387.
48. Collman, J. P. *Inorg. Chem.* **1997**, *36*, 5145.
49. Collman, J. P.; Ha, Y.; Wagenknecht, P. S.; Lopez, M.-A.; Guilard, R. *J. Am. Chem. Soc.* **1993**, *115*, 9080.
50. Collman, J. P.; Hutchinson, J. E.; Wagenknecht, P. S.; Lewis, N. S.; Lopez, M. A.; Guilard, R. *J. Am. Chem. Soc.* **1990**, *112*, 8206.
51. Corey, E. J.; Noe, M. C. *J. Am. Chem. Soc.* **1996**, *118*, 319.
52. Bitterwolf, T. E.; Ling, A. C. *J. Organomet. Chem.* **1973**, *57*, C15.
53. Mueller-Westerhoff, U. T.; Haas, T. J.; Swiegers, G. F.; Leipert, T. K. *J. Organomet. Chem.* **1994**, *472*, 229.
54. Hillman, M.; Michaile, S.; Feldberg, S. W.; Eisch, J. J. *Organomet.* **1985**, *4*, 1258.
55. Michaile, S.; Hillman, M. *Organomet.* **1988**, *7*, 1059.
56. Waleh, A.; Loew, G. H.; Mueller-Westerhoff, U. T. *Inorg. Chem.* **1984**, *23*, 2859.
57. Karlsson, A.; Broo, A.; Ahlberg, P. *Can. J. Chem.* **1999**, *77*, 628.
58. Mueller-Westerhoff, U. T.; Swiegers, G. F. *Chem. Lett.* **1994**, 67.
59. Diaz, A. F.; Mueller-Westerhoff, U. T.; Nazzal, A.; Tanner, M. *J. Organomet. Chem.* **1982**, *236*, C45.
60. Rheingold, A. L.; Mueller-Westerhoff, U. T.; Swiegers, G. F.; Haas, T. J. *Organomet.* **1992**, *11*, 3411.
61. Karlsson, A.; Hilmersson, G.; Ahlberg, P. *J. Phys. Org. Chem.* **1997**, *10*, 590.
62. Cassens, A.; Eilbracht, P.; Nazzal, A.; Prössdorf, W.; Mueller-Westerhoff, U. T. *J. Am. Chem. Soc.* **1981**, *103*, 6367.
63. Mueller-Westerhoff, U. T. *Angew. Chem. Int. Ed. Engl.* **1986**, *25*, 702, and references therein.
64. Mueller-Westerhoff, U. T.; Nazzal, A. *J. Am. Chem. Soc.* **1984**, *106*, 5381.
65. Mueller-Westerhoff, U. T.; Nazzal, A.; Tanner, M. *J. Organomet. Chem.* **1982**, *236*, C41.
66. Molenveld, P.; Engbersen, J. F. J.; Reinhoudt, D. N. *Chem. Soc. Rev.* **2000**, *29*, 75, and references therein.
67. Steinhagen, H.; Helmchen, G. *Angew. Chem. Int. Ed. Engl.* **1996**, *35*, 2339.
68. Kirby, A. J. *Angew. Chem. Int. Ed.* **1996**, *35*, 707.
69. Sawamura, M.; Nagata, H.; Sakamoto, H.; Ito, Y. *J. Am. Chem. Soc.* **1992**, *114*, 2586.
70. Breslow, R.; Zhang, X.; Xu, R.; Maletic, M.; Merger, R. *J. Am. Chem. Soc.* **1996**, *118*, 11678.
71. Laneman, S. A.; Stanley, G. G. In *Homogeneous Transition Metal Catalyzed Reactions: Developed from a Symposium*, Eds. Moser, W. R., Slocum, D. W., American Chemical Society, 1992.
72. Laneman, S. A.; Fronczek, F. R.; Stanley, G. G. *J. Am. Chem. Soc.* **1988**, *110*, 5585.

73. Matthews, R. C.; Howell, D. K.; Peng, W.-J.; Train, S. G.; Treleaven, W. D.; Stanley, G. G. *Angew. Chem. Int. Ed. Engl.* **1996**, *35*, 2253.
74. (a) Escaffre, P.; Thorez, A.; Kalck, P. *J. Chem. Soc., Chem. Commun.* **1987**, 146. (b) Jenck, J.; Kalck, P.; Pinelli, E.; Siani, M.; Thorez, A. *J. Chem. Soc., Chem. Commun.* **1988**, 1428.
75. Pirrung, M. C.; Liu, H.; Morehead, A. T. *J. Am. Chem. Soc.* **2002**, *124*, 1014.
76. Gersten, S. W.; Samuels, G. J.; Meyer, T. J. *J. Am. Chem. Soc.* **1982**, *104*, 4029.
77. Ogino, I.; Nagoshi, K.; Yagi, M.; Kaneko, M. *J. Chem. Soc., Faraday Trans.* **1996**, *92*, 3431.
78. Gilbert, J. A.; Eggleston, D. S.; Murphy, W. R.; Geselowitz, D. A.; Gersten, S. W.; Hodgson, D. J.; Meyer, T. *J. Am. Chem. Soc.* **1985**, *107*, 3855.
79. Ruettinger, W.; Dismukes, G. C. *Chem. Rev.* **1997**, *97*, 1, and references therein.
80. Geselowitz, D.; Meyer, T. *Inorg. Chem.* **1990**, *29*, 3894.
81. Hurst, J.; Zhou, J.; Lei, Y. *Inorg. Chem.* **1992**, *31*, 1010.
82. Binstead, R. A.; Chronister, C. W.; Ni, J.; Hartshorn, C. M.; Meyer, T. J. *J. Am. Chem. Soc.* **2000**, *122*, 8464.
83. Chronister, C. W.; Binstead, R. A.; Ni, J.; Meyer, T. J. *Inorg. Chem.* **1997**, *36*, 3814.
84. (a) Konsler, R. G.; Karl, J.; Jacobsen, E. N. *J. Am. Chem. Soc.* **1998**, *120*, 10780; (b) Jacobsen, E. N. *Acc. Chem. Res.* **2000**, *33*, 412.
85. Adams, R. D. *J. Organomet. Chem.* **2000**, *600*, 1.
86. Norton, J. R. In *Fundamental Research in Homogeneous Catalysis*, Eds. Tsutsui, M., Ugo, R., Plenum Press, 1977.
87. Basset, J. M.; Besson, B.; Choplin, A.; Hugues, F. In *Fundamental Research in Homogeneous Catalysis, Vol. 4*, Eds. Graziani, M., Giongo, M., Plenum Press, 1982.
88. Blazina, D.; Duckett, S. B.; Dyson, P. J.; Lohman, J. A. B. *Chem. Eur. J.* **2003**, *9*, 1046.
89. Liu, L.; Breslow, R. *J. Am. Chem. Soc.* **2002**, *124*, 4978.
90. Twyman, L. J.; King, A. S. H.; Martin, I. K. *Chem. Soc. Rev.* **2002**, *31*, 69.
91. Hirai, H.; Toshima, N. In *Tailored metal catalysts*; Iwasawa, Y., Ed. D. Reidel Publishing: Dordrecht, Holland, 1986.
92. Breinbauer, R.; Jacobsen, E. N. *Angew. Chem. Int. Ed. Engl.* **2000**, *39*, 3604.
93. Bettelheim, A.; Chan, R. J. H.; Kuwana, T. *J. Electroanal. Chem.* **1979**, *99*, 391.
94. Vasudevan, P.; Santosh; Mann, N.; Tyagi, S. *Transition Met. Chem.* **1990**, *15*, 81 and references therein.
95. Behret, H.; Binder, H.; Sandstede, G.; Scherer, G. G. *J. Electroanal. Chem.* **1981**, *117*, 29.
96. Yagi, M.; Sukegawa, N.; Kasamastu, M.; Kaneko, M. *J. Phys. Chem. B.* **1999**, *103*, 2151.
97. Yagi, M.; Nagoshi, K.; Kaneko, M. *J. Phys. Chem. B.* **1997**, *101*, 5143.
98. Yagi, M.; Kasamastu, M.; Kaneko, M. *J. Mol. Catal. A: Chem.* **2000**, *151*, 29.
99. Zecevic, S.; Simic-Glavaski, B.; Yeager, E. *J. Electroanal. Chem.* **1985**, *196*, 339.
100. Brezina, M.; Khalil, W.; Koryta, J.; Musilova, M. *J. Electroanal. Chem.* **1977**, *77*, 237.
101. Swiegers, G. F.; Malefetse, T. J. *Chem. Rev.* **2000**, *100*, 3483.
102. Hill, C. L.; Zhang, X. *Nature* **1995**, *373*, 324.
103. Shi, C.; Anson, F. C. *J. Am. Chem. Soc.* **1991**, *113*, 9564.

11

TIME-DEPENDENT ("MECHANICAL"), NONBIOLOGICAL CATALYSIS. 1. A FULLY FUNCTIONAL MIMIC OF THE WATER-OXIDIZING CENTER (WOC) IN PHOTOSYSTEM II (PSII)

ROBIN BRIMBLECOMBE, G. CHARLES DISMUKES, GREG A. FELTON, LEONE SPICCIA, AND GERHARD F. SWIEGERS

11.1 INTRODUCTION

Hydrogen (H_2) has long been considered an ideal fuel for the future. When burned in the presence of oxygen (O_2), it produces water (H_2O) as the only waste product. Hydrogen, therefore, potentially offers a clean, nonpolluting alternative to fossil fuels, which generate pollutants like carbon dioxide (CO_2), hydrocarbons, and various nitrogen oxides (NO_x). Carbon dioxide is a greenhouse gas that is currently believed to be the source of a recent trend toward rising temperatures globally.

Hydrogen has the added advantage that its reaction with oxygen may be made to take place in a solid-state fuel cell, which harnesses the resulting energy not as heat or pressure but as an electrical current. Fuel cells offer greater inherent energy efficiency than simple combustion of the type employed in, for example, internal combustion engines.

Mechanical Catalysis: Methods of Enzymatic, Homogeneous, and Heterogeneous Catalysis,
Edited by Gerhard F. Swiegers
Copyright © 2008 John Wiley & Sons, Inc.

Because of these factors and various other politico-economic imperatives, several nations, including the United States, have set themselves the strategic goal of developing a "hydrogen economy" to replace the existing fossil fuel economy.

A key issue in this respect is the source of the hydrogen to be used in such an economy. Where would it come from? This is a contentious matter because the cheapest source of hydrogen gas is, currently, fossil fuels. Hydrogen is readily obtained as a by-product in the processing of oil and natural gas. Hydrogen generated in this way offers the most economically viable approach to creating an embryonic hydrogen economy. However, its use lies at odds with the strategic intention of replacing fossil fuels.

An ideal source of hydrogen is water (H_2O), which is not only readily available in human society but can also be electrochemically split into hydrogen (H_2) and oxygen (O_2) as shown in Equation (11.1) [1].

$$2\,H_2O \longrightarrow O_2 + 2\,H_2 \qquad \Delta H^0 = 286\ \text{kJ/mol} \qquad (11.1)$$

The energy source for such water-splitting would, ideally, be sunlight [2–4]. Hydrogen made from water using sunlight prospectively offers an abundant, renewable, clean, and borderless energy resource, whose effects could be profound and lasting [2–4]. However, no technically practical and economic catalytic system exists to facilitate this reaction. The potential of solar-produced hydrogen has, consequently, never been realized [2–4].

The difficulties involved in sustainably undertaking solar-powered water-splitting can be understood when considering the process that must be undergone [1]. In an electrochemical half-cell, the water-splitting reaction comprises two half-reactions:

$$2\,H_2O \longrightarrow O_2 + 4\,H^+ + 4\,e^- \qquad E^{0'} = 0.82\ \text{V at pH 7} \qquad (11.2)$$
$$2\,H^+ + 2\,e^- \longrightarrow H_2 \qquad (11.3)$$

The overall reaction depicted in Equation (11.1) is thermodynamically highly demanding ($\Delta H^0 = 286\ \text{kJ/mol}$). In its uncatalyzed form, this reaction requires an input temperature of 2500 °C in order to break the four strong O–H bonds (enthalpy $\Delta H^0 = 494\ \text{kJ/mol}$ each) [1].

This energy barrier may potentially be lowered by coupling these dissociation steps with the step of O–O bond formation, which is favorable and energy releasing ($\Delta H^0 = -494\ \text{kJ/mol}$). However, to do this, one needs a catalyst that is capable of rearranging these five bonds in a coordinated sequence without releasing the reactive intermediates [1]. Achieving such mechanistic complexity and high-energy input is a daunting challenge and the major limitation preventing the development of practical light-driven water-splitting devices. Meeting this challenge is considered to be a "holy grail" of chemistry [4].

The only molecular catalyst that is known to be capable of sustainably photooxidizing water using visible light is the *Water-Oxidizing Complex* (WOC) of *Photosystem II* (PSII), which is found in various photosynthetic organisms in nature. This catalyst converts an excited state of chlorophyll (P680*) into a cation

11.1 INTRODUCTION

radical by charge separation that is then used to extract electrons from an inorganic core (at 1.0 V vs. Ag/AgCl) [5,6].

The composition of the catalytic core of the PSII-WOC has been deduced from spectroscopic and physicochemical studies. It consists of a $Mn_4Ca_1O_x$ cluster bridged by oxygen atoms derived from water molecules [5–7]. This structure is highly conserved, being found in essentially an identical state in numerous photosynthetic organisms.

The absence of evolutionary diversity in the amino acids and inorganic core, ($Mn_4Ca_1O_xCl_y$) of the photosystem II water-oxidizing complex (PSII-WOC) indicates that combinatorial biosynthesis in natural habitats on earth that allow oxygenic photosynthesis, has produced a *single catalyst* capable of oxidizing water ($2 H_2O \rightarrow O_2 + 4 H^+ + 4 e^-$). In effect, therefore, natural selection over several billion years seems to have evolved a single type of enzyme that catalytically photo-oxidizes water. This realization has important implications for catalyst research and implies a necessity to incorporate nature's chemical principles to develop manmade catalysts for use in robust photoelectrochemical cells.

An atomic model of the resting oxidation state of the PSII-WOC core isolated from the cyanobacterium *Thermosynechococcus sp.* has recently been derived from single-crystal X-ray diffraction (XRD) data [7]. Although the low resolution (3.2 to 3.6 Å) leads to differences in data interpretation, one model [8,9] indicates an inorganic core comprising Mn_4Ca cluster, described as "capped tetrahedral" with four metal atoms in a symmetrical trigonal prism or heterocubane arrangement of Mn_3Ca, connected to a single Mn atom (Fig. 11.1) [6,10]. Four oxygens have been

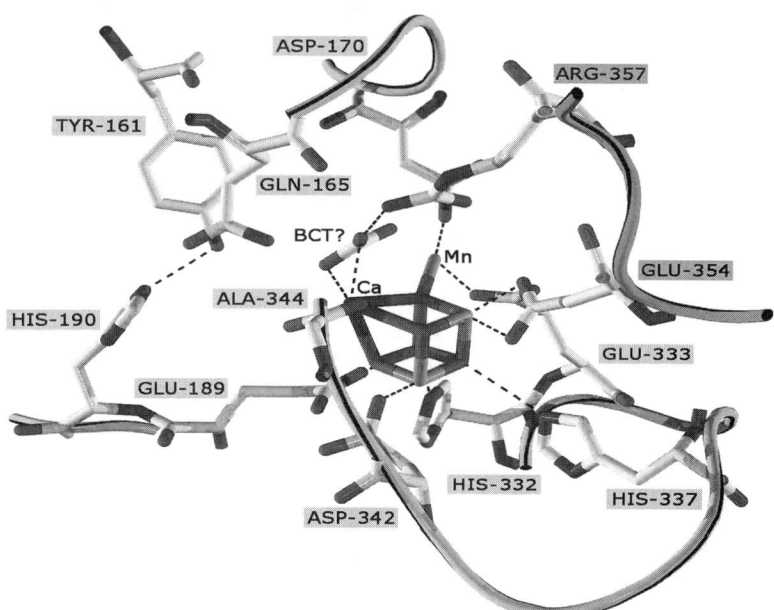

Figure 11.1. XRD model of Photosystem II water-oxidizing complex (PSII-WOC) [6].

proposed to link the tetrahedral metal array, with one μ_4-oxo bridging to the fourth Mn atom. Two oxophilic ligands may be water or hydroxide ions based on H-bonding patterns to amino acids [6]. The resulting $CaMn_4O_4$ cluster is compatible with Mn EXAFS and EPR studies [11].

This new structural information, coupled with electronic structure analyses, has been used to propose a pathway by which the WOC core achieves O–O bond formation and catalytic turnover. The proposed mechanism addresses the role of the Mn atoms [12], the bridging oxophilic ligands, the Ca^{2+} ion, and proton removal from the catalytic site through amino acid side chains and transfer to a (bi)carbonate ion which may be present in the active site [8].

In previous chapters we have noted that many enzymes seem to employ time-dependent ("mechanical") actions during catalysis. By this is meant that, after binding their reactants and polarizing their electronic structures to make them conducive for reaction (*"activation"*), the enzymes guide the bound reactants into collision with each other along pathways and with orientations that are ideal for reaction. That is, they employ concerted actions, involving *coordinated interplays* between the different catalytic groups so as to ensure that the bound reactant functionalities are always brought into reactive contact, or pulled apart, in a very specific and highly repeatable way. As noted in Section 8.5 of Chapter 8, such actions are more correctly termed *convergent* since the catalytic groups are disposed to act in a synchronized way; that is, their movements *converge* to create a concerted action.

The effect of convergent actions is to minimize the energy barrier, known as the *activation energy* or E_A, that must be overcome during a collision between the reactants. As a result, a high likelihood exists that such collisions will be successful, leading to products. The catalytic rate k is then, effectively, dependent on the *collision frequency* A rather than on the *activation energy* E_A in the Arrhenius equation (11.4), which describes the rate at which reactant collisions generate products:

$$k = A \exp\left(\frac{-E_a}{RT}\right) \quad (11.4)$$

We have termed such catalysis *time-dependent* because the *collision frequency* involves, effectively, a quantification of time. The catalytic action may be considered fundamentally mechanical in character, because the reaction occurs at the rate at which the catalyst brings the reactants into mechanical contact with each other.

By contrast, a catalyst whose rate is determined by the activation energy E_A is termed *energy-dependent* in its action. In such a case, the catalytic action is fundamentally thermodynamic in character, since the reaction rate is determined by the rate at which the energy barrier to the reaction is overcome in collisions between the reactants.

In Section 10.2 of Chapter 10, we discussed the criteria for rationally designing *time-dependent* ("mechanical") molecular catalysts. We noted that such catalysts must display two key attributes:

(1) The catalyst must flex, conformationally, about a certain optimum structure, which is, ideally, complementary to the transition state of the reaction to be catalyzed.

11.1 INTRODUCTION

(2) The interactions between the catalyst and the reactants must be as dynamic as possible, while still allowing for the reactants to be sufficiently activated (electronically polarized) for reaction during the time they are bound to the catalyst.

Attribute (1) is necessary to ensure that any attached reactants will be carried into collision with each other along the optimum approach trajectories. In other words, attribute (1) ensures the greatest possible thermodynamic efficiency in the collision leading up to reaction.

Attribute (2) is then, further, necessary to ensure coordinated actions, as well as rapid uptake of reactants, turnover, and ejection of the products. In particular, the catalyst must polarize the electronic structure of the attached reactants sufficiently that they will react with each other when brought into collision along the optimum pathways.

How, though, does one design a catalyst that concurrently meets both of the above requirements? In Section 10.2 in Chapter 10, we noted that a time-dependent catalyst will generate a catalytic effect only if *each* of the above attributes is *simultaneously* near-optimum.

This process creates a significant problem: How can one discover and optimize new time-dependent catalysts if they only generate a catalytic effect when their key features are already near-ideal? In such a case, it is not possible to optimize each variable separately in a trial-and-error approach that slowly but inexorably leads to new catalysts. Instead, both variables must be made simultaneously optimum, drastically expanding to near impossibility, the number of variants that must be tried. Catalyst discovery is thereby rendered an almost arbitrary, hit-and-miss affair.

One possible approach to this problem that we discussed in Section 10.2.3 of Chapter 10 is to create a *starting point* by mimicking a known time-dependent catalyst of the reaction, such as an appropriate enzyme.

The idea is that, if the enzyme in question employs a time-dependent action, its active site must already have a near-optimum structure [attribute (1)]. Moreover, its catalytic groups must offer sufficient dynamism and activation to bring about the reaction [attribute (2)].

Thus, a species that models and incorporates these aspects as closely as possible may, conceivably, have both of attribute (1) and (2) sufficiently optimum to generate a catalytic effect. Even if such a catalytic effect is only very weak, the model species will then, nevertheless, offer a genuine and useful starting point from which to work toward a more efficient time-dependent catalyst using systematic trial-and-error. In other words, if the model species generates even a weak catalytic effect that is similar to that of the enzyme, it may become possible to optimize methodically and separately each of attributes (1) and (2) in this initial catalyst using a trial-and-error approach.

In the case of water-oxidation catalysis, it is clear that the water-oxidizing center in Photosystem II (PSII-WOC) can sustainably oxidize water. If one assumes that this is because it meets the above two criteria, then to mimic this catalyst, one needs to find a model complex that has a similar shape and composition, and which dynamically interacts with its reactants.

A wide variety of manganese complexes have been developed as models of the PSII-WOC [13,14]. One group involves a class of tetramanganese-oxo *cubane* molecules whose structures are similar in form and motif to that in Figure 11.1 [11,12,15]. These complexes **1** have the formula $Mn_4O_4L_6$ (where $L^- = (p\text{-}RC_6H_5)_2PO_2^-$; R = H (**1a**), OMe (**1b**)). They possess a "cubical" $Mn_4O_4^{6+}$ core that assembles spontaneously from simple Mn^{2+} and MnO_4^- (permanganate) salts in the presence of phosphinate ligands [11,12,15].

Figure 11.2 compares in schematic form the proposed structure of the active site in the PSII-WOC and the cubical structure of the cubanes **1a-b**. As can be seen, there is a clear structural similarity.

What is particularly significant about this class of model complex is that the components of the cubanes also interact highly dynamically with both each other and with water molecules, undergoing ready self-assembly in solution [11,12,15].

Thus, the cubanes **1a-b** seem, at face value, to achieve both of the key attributes described above. As such, they conceivably offer a model that could potentially serve as an initial starting point for the systematic development of an efficient, nonbiological, time-dependent, water-oxidation catalyst (or photocatalyst).

In this chapter, we examine the methoxy-substituted cubane **1b** as a possible water-oxidation photocatalyst. We show that, when it is ion-exchanged into a thin Nafion membrane deposited on a suitable electrode, it or its reactive intermediate catalyzes the conversion of water into oxygen when illuminated with light. Moreover, it opens the possibility of interacting electrochemically with the Ru-bipyridine and other dyes used in *dye-sensitized solar cells*, thereby potentially allowing **1b** to be employed in the fabrication of a fully functional, *water-splitting*,

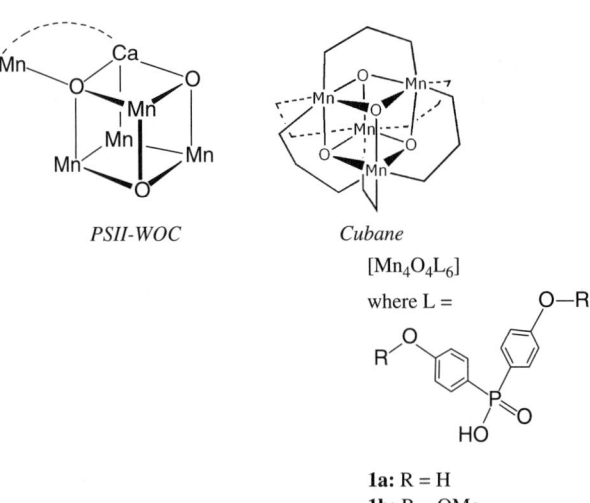

Figure 11.2. (Left): Schematic of the proposed structure of the Mn_3CaO_4Mn active site of the PSII-WOC. (Right): Structure of the synthetic molecule $Mn_4O_4L_6$ **1** [where $L^- = (p\text{-}RC_6H_5)_2PO_2^-$; R = H (**1a**), OMe (**1b**)] and its central $Mn_4O_4^{6+}$ cubane core.

dye-sensitized solar cell. The operation of such a cell would mimic, in principle, the PSII-WOC. The characteristics of the catalysis and its implications for catalyst development are discussed.

11.2 THE PHYSICAL AND CHEMICAL PROPERTIES OF THE *CUBANES* 1a-b

11.2.1 Chemical Structures

The cubanes **1a-b** were first synthesized in the group of R. Charles Dismukes at Princeton University [15]. X-ray crystal structures indicated a cubical core arrangement of Mn and O atoms, in which the phosphinate ligands bridge pairs of Mn atoms, one to each of the six faces of a cube. An XRD structure of **1a** is depicted in Figure 11.3 [15]. As can be seen, the core is surrounded by 12 hydrophobic phenyl rings that pack together, contributing to the driving force needed to hold together this otherwise unstable core. The bidentate phosphinates are essential to forming the $Mn_4O_6^{6+}$ core, which has unusually long Mn–O bonds that are weaker and, consequently, more reactive than in other Mn-oxo complexes.

Two variations of the $[Mn_4O_4L_6]$ molecule have been studied: **1a**, where L = diphenyl phosphinate [15]; and **1b**, where L = bis(*p*-methoxyphenyl)phosphinate [16].

The parent **1a** has tetrahedral symmetry in its core indicating delocalized valence electrons among the four identical Mn sites, in both the solid state and the solution phases [15].

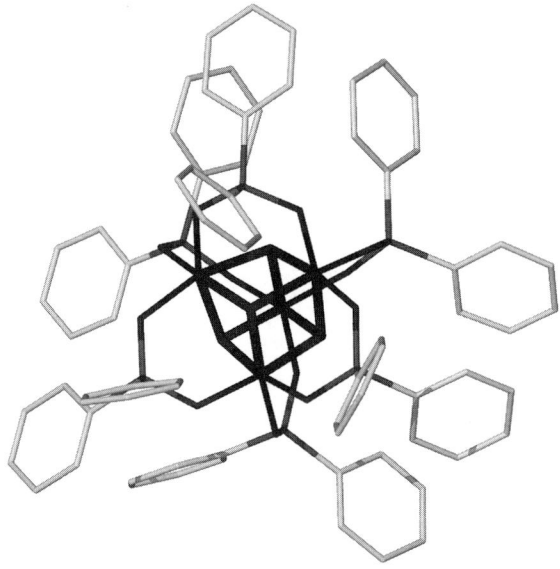

Figure 11.3. XRD-derived structure of the cubane $(Ph_2PO_2)_6Mn_4O_4$ **1a** [15].

The derivative **1b** displays Jahn Teller distortion, giving it C_{2v} point symmetry with pairs of discrete Mn(III) and Mn(IV) sites [16]. The ligand binding energy of **1b** is 14.7 kcal/mol lower than for **1a**. This distortion is likely caused by the greater electron donating properties of the methoxy-group, which induces electron repulsion in the core.

The crystal structure of the chemically oxidized cubane **1a**$^+$ reveals a one electron transition $[Mn_4^{(2III, 2IV)}O_4L_6] + HClO_4 \rightarrow [Mn_4^{(III, 3IV)}O_4L_6]^+(ClO_4^-) + e$, and displays stronger metal-to-ligand binding.

Finally, it should perhaps be explicitly noted that the Mn–O bonds involve coordinate, not covalent, bonds. Coordinate bonds are more directional and stronger (bond energies ca. 10–30 kcal mol^{-1}) than phenyl-pairing interactions and the weak interactions of biology (bond energies ca. 0.6–7 kcal mol^{-1}). They are nevertheless noncovalent in character and, therefore, like the phenyl-pairing interactions, inherently more dynamic than covalent bonds. This dynamism gives the cubanes some interesting chemical properties.

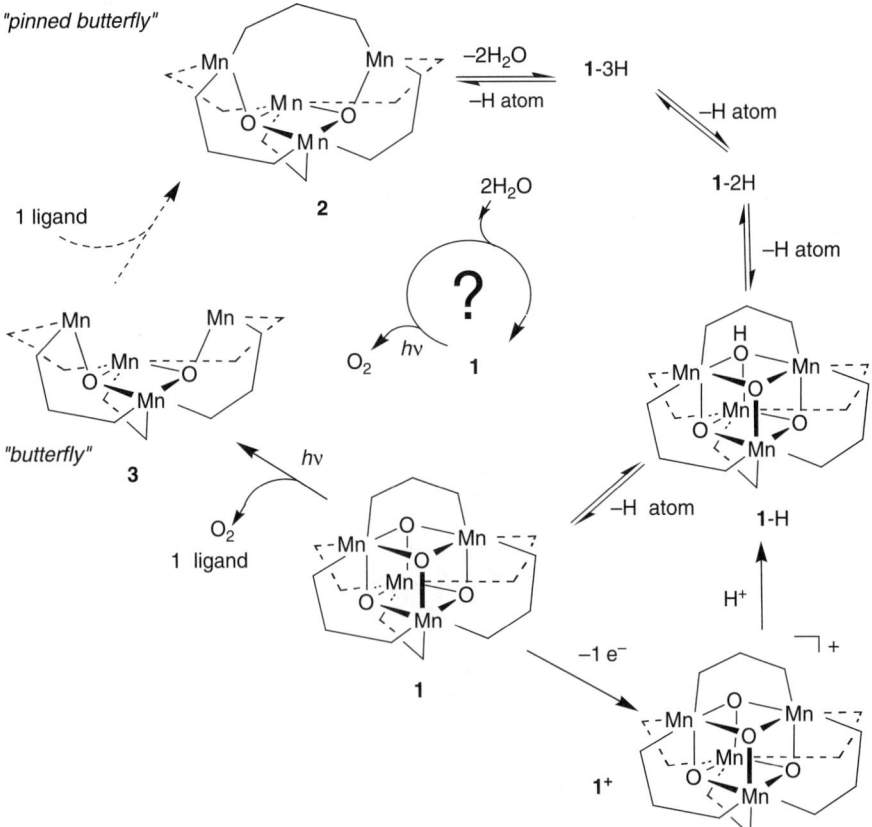

Figure 11.4. Reaction pathways and a possible photocatalytic cycle for cubanes of type **1a-b** (represented here as **1**).

11.2.2 Stepwise Hydride Abstraction, Leading to Water Release

A feature of the chemistry of the cubanes **1** is their ability to abstract H atoms (H$^+$ + e$^-$) from substrates such as the CH$_3$O–H bond of methanol (bond strength = 440 kJ/mol) (Fig. 11.4) [16–19]. The one-electron oxidized analogs (**1**$^+$) abstract hydrides from amines (N–H bond strength $\geq$ 530 kJ/mol). Thus, **1**$^+$ is thermodynamically capable of direct cleavage of the HO–H bond of water. Four such H atoms may be abstracted in total by **1**, at which stage the cubane releases two water molecules, leaving the so-called *pinned butterfly* complex **2**. The right-hand side of Figure 11.4 shows this sequence in detail. As can be seen, two of the four oxygen atoms in the cubane core of **1** undergo ready reduction by proton-coupled electron transfer (PCET) reactions to release two H$_2$O molecules into solution.

11.2.3 Dioxygen Generation

As shown at the bottom left of Figure 11.4, a unique feature of the cubane Mn$_4$O$_6^{6+}$ is its ability to evolve an O$_2$ molecule from the core oxygen atoms, following photo-induced release of a single phosphinate ligand anion in the gas phase [20,21]. Ultraviolet excitation achieves this release with 60–100% quantum yield in the gas phase depending on the phosphinate derivative. The sole other photochemical product is the remaining *butterfly* complex, [Mn$_4$O$_2$(Ph$_2$PO$_2$)$_5$]$^+$ **3**, which comprises all of the remaining elements of the cubane.

The overall gas phase reaction, shown in Equation (11.5), is predicted to be favorable ($\Delta H^\circ = -25$kJ/mol) on the basis of experimental bond enthalpies.

$$[Mn_4O_4(Ph_2PO_2)_6] \longrightarrow O_2 + Ph_2PO_2^- + [Mn_4O_2(Ph_2PO_2)_5]^+ \quad (11.5)$$

A low quantum yield is observed for this reaction in condensed phases because of the energy demands involved in O$_2$ release and rapid ligand rebinding. That is, dioxygen release is not observed when the cubane is dissolved in liquid solution and irradiated.

High-level density functional theory (DFT) calculations confirm the weak exothermicity [22], and they predict a large energy requirement for O–O bond formation by the intact cubane. This energy requirement originates from the inflexibility of the complex, which does not allow for ready collision of two corner oxygens during conformational flexing. However, the energy requirement is substantially lowered (96–126 kJ/mol) upon release of a phosphinate ligand, which provides for greater conformational flexability, thereby allowing the reaction to occur in the gas phase.

The likely mechanism for the formation of dioxygen according to DFT is depicted in Figure 11.5 [22]. Irradiation causes the release of a single ligand by photo-promotion of ligand electrons into an anti-bonding orbital. The remaining complex is sufficiently flexible that, at the temperatures of the gas phase, it can undergo substantial conformational flexing. This is predicted by DFT to involve, first, a lengthening of the Mn–Mn distance at the uncapped end of the molecule, followed by a shortening of the O–O distance, resulting in a collision between the

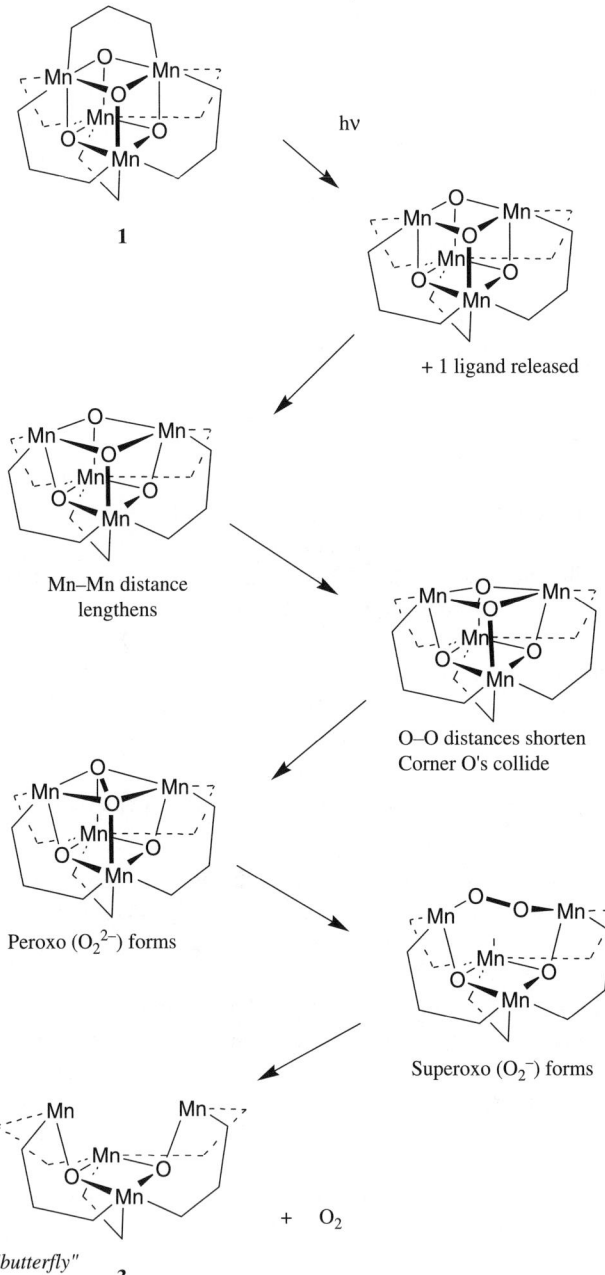

Figure 11.5. Likely mechanism for light-induced formation of O_2 by **1a-b** (represented here as **1**) (complex charges not shown).

two, free corners oxygens. Given the exceedingly high quantum yields for dioxygen formation, this collision must have a high likelihood of success, leading to, first, peroxo and to, then, superoxo formation. Finally, the dioxygen molecule is released, leaving only the *butterfly* complex **3**.

The energy required to bring about dioxygen release seems to be needed to drive the conformational changes that take place. That is, the Mn ions seem to activate the corner oxygens sufficiently strongly that they react with each other when propelled into collision during the conformational flexing in the gas phase. The pathways and trajectories followed by these corner oxygens in the moments leading up to the collision seem to be optimum or near to optimum from a thermodynamic viewpoint. This must be the case because if multiple collisions were required between the corner oxygens before reaction, then 1) the actual energy requirement in the gas phase would be higher than the calculated one, and 2) there would be a nonzero chance of the released ligand rebinding before reaction, with an accompanying decrease in the quantum yield. The latter is precisely what is observed in condensed, liquid phases, indicating that there is then insufficient energy to bring about the required conformational flexing of the cubane framework before ligand rebinding.

11.2.4 A Possible Catalytic Cycle

The above properties of the cubane suggest a prospective catalytic cycle. The cycle involves the following steps, which are also depicted in Figure 11.4. After irradiation of the cubane **1**, with subsequent dioxygen release, the *butterfly* complex **3** is formed. Rebinding of the released phosphinate ligand with **3** should generate **2**. Under suitable conditions, it should, moreover, be possible for **2** to take up two water molecules and be oxidized along the reverse pathway $1H_3 \rightarrow 1H_2 \rightarrow 1H_1 \rightarrow 1$, thereby regenerating **1**. This sets up another cycle, which commences with **1** being again converted to **3**, with accompanying photochemical generation of O_2.

Thus, there is in the cubanes **1a-b**, the possibility of a closed catalytic cycle that contains steps not entirely dissimilar to those seen in the PSII-WOC. The cycle is shown as a circular arrow, framing a question mark, in the center of Figure 11.4. All indications are that such a cycle would involve a time-dependent catalytic action.

Unfortunately, it has, thus far, proved impossible to realize such a catalytic cycle, partly because the ligand released during O_2 production is either lost and not available, or it rebinds before O_2 release. In the liquid phase, O_2 is not released. The hydrophobicity of **1** makes it, in any case, extremely poorly soluble in solvents containing the water reactant.

11.2.5 Other Reactions

When dissolved in CH_2Cl_2, the cubanes can act as powerful catalysts for the oxidation of a range of organic substrates, including thioethers, hydrocarbons, alkenes, benyzl alcohol, and benzaldehyde [23]. The proposed mechanisms for these oxidations involve the disruption of the binding of a phosphinate ligand, facilitating the temporary coordination of reagents to a Mn ion within the cubane core.

Investigation of the reactivity and catalytic potential of the cubane has, however, been limited to organic systems because of the poor solubility of **1** in anything other than organic solvents. This organic solubility derives from the hydrophobic shell that the phenyl phosphinate ligands form around the cubane core.

11.2.6 Summary

As noted, the chemical reactivity of the cubane model complexes is such that they are exceptionally strong oxidizing agents, with the core oxygen atoms capable of oxygenation reactions and H atom abstraction reactions. Moreover, cubanes have been shown by mass spectrometric analysis to be able to release molecular oxygen in the gas phase. Light excitation causes photodissociation of a phosphinate ligand facilitating the collision and subsequent release of two core oxygen atoms as O_2. This generation of molecular oxygen is not observed in condensed phases. Cubanes have also been shown to be capable of releasing two water molecules by PCET of two core oxygen atoms.

Based on these properties, it is proposed that the cubanes could undertake a possible catalytic cycle in that, after formation of O_2, they should be able to bind and oxidize two water molecules, releasing their protons and then reincorporating them into the core to restore the cubane and thereby create a cycle in which oxidation of water can be repeatedly facilitated.

Unfortunately, it has not proved possible to investigate the manganese-oxo cubanes as sustainable homogeneous catalysts for water oxidation because of their insolubility in water. This property originates from the hydrophobic outer shell formed by the 12 phenyl rings of the phosphinate ligands.

Moreover, it is not clear that the energy requirement for the reaction can be overcome in homogeneous solutions containing water. Such a reaction depends on dynamic disassembly of at least one ligand in the complex when illuminated by light. Such a disassembly may not be favored in aqueous solution.

11.3 NAFION PROVIDES A MEANS OF SOLUBILIZING AND IMMOBILIZING HYDROPHOBIC METAL COMPLEXES

Nafion is a perfluorosulfonate cation exchange polymer that has been widely used in a range of electrochemical applications. The tetrafluoroethylene (Teflon) backbone provides a robust, inert, semirigid support for the proton exchanging sulphonic acid groups. Nafion is therefore well known as a proton exchange membrane, which can be placed between the two electrodes in fuel cells to facilitate the selective exchange of protons while preventing electron conduction. Nafion is also commonly used as a coating material for electrochemical sensors and for chemically modifying electrodes [24].

Investigations of the substructure of Nafion coatings on solid surfaces have revealed that the polymer layers contain hydrophilic channels through the otherwise hydrophobic membrane; these channels allow the diffusion of small molecules such as water throughout the membrane [25]. The size and distribution of these pores is highly dependent on the preparation method used to make the membrane. The pore sizes in Nafion membranes cast from solution are influenced by the rate of evaporation of the casting solvent and the presence of water. The hydrophilic pores in Nafion membranes cast from aqueous solutions have been found to be approximately 20–30 nm in size [25]. These channels are permeable to cations but not anions. Lying between these channels are hydrophobic pockets within which organic soluble species, such as the phosphinate ligands and cubanes, may be solubilized. Within these pockets, high concentrations of organic soluble species may potentially be trapped.

Nafion has, consequently, been used to support various electrocatalysts, including water oxidation catalysts such as, for example, pentaammine-chlororuthenium(III) $[Ru(NH_3)_5Cl]^{2+}$ [26]. A recent comparison of this catalyst in homogenous aqueous solution and in heterogeneous Nafion membrane demonstrated a greater activity for the membrane-supported complex [26]. This was attributed to favorable interactions of the catalysts with water and to the suppression of a bimolecular decomposition step caused by the immobilization of the complexes in the membrane.

11.4 PHOTOELECTROCHEMICAL CELLS AND DYE-SENSITIZED SOLAR CELLS FOR WATER-SPLITTING

Photoelectrochemical (PEC) cells use light energy to drive redox reactions. Like normal electrochemical cells, they contain a cathode and an anode separated by electrolyte-containing reactants that are oxidized or reduced at the respective electrodes. In photoelectrochemical cells, at least one of the half reactions is driven by solar energy; that is, sunlight is converted into chemical energy.

A common example involves the use of a semiconductor photocatalyst, such as TiO_2, to oxidize water, with the released electrons migrating along an external circuit to the counter electrode, where they reduce protons. How does this work?

As schematically depicted in Figure 11.6, absorption of a light photon by the titania semiconductor causes an electron to be promoted to the conduction band. The "hole" that is thereby created migrates to the semiconductor surface, where it facilitates the conversion of water into dioxygen. The promoted electron travels via the external circuit to the other electrode, where it facilitates proton reduction. Dioxygen (O_2) is therefore produced at the titania and hydrogen (H_2) at the counter electrode. In effect, water (H_2O) is converted to H_2 and O_2 by the effect of light.

Unfortunately, this process is not very efficient in the case of TiO_2 because titania has a relatively large bandgap of 3.0–3.2 eV [4,27]. This limits its light absorption to the ultraviolet (UV) range of the spectrum, which comprises only 4% of the solar energy spectrum. This is a major limiting factor in the efficiency of current photoelectrochemical cells.

To improve the situation, one could use small bandgap semiconductors that absorb visible light but still oxidize water at potentials close to the thermodynamic limit. For example, whereas materials like TiO_2 and WO_3 are wide bandgap

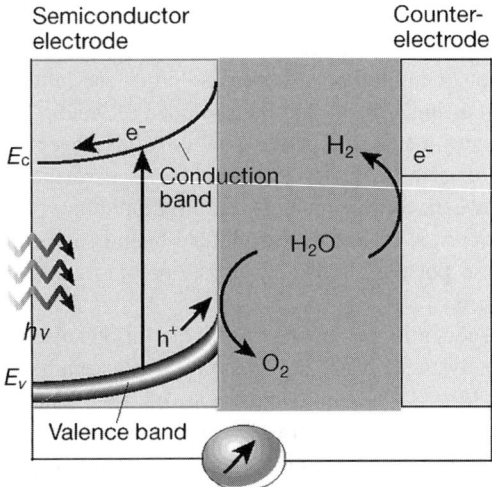

Figure 11.6. Schematic illustration of the design of a photoelectrochemical cell for water-splitting.

semiconductors that photooxidize water to O_2, their thermodynamic efficiencies are poor because of the mismatch in the required and actual electrochemical potentials (>3 V for WO_3 vs. 1.23 V for water-splitting) and their poor solar spectral absorption [4,27]. Small bandgap semiconductors like CdSe, CdTe, and Si have better solar efficiencies but typically photocorrode under electrolysis conditions. Application of such materials consequently require both the covalent attachment of suitable catalysts for water-splitting and chemical modification of the semiconductor surface to suppress photocorrosion [28]. Several doped binary and other semiconductors that are capable of sustained water-oxidation catalysis using visible light are, nevertheless, known [29,30].

Another alternative is to improve the light absorbing efficiency of TiO_2 by applying a layer of a *photoactive dye* with improved visible light absorption properties [27]. For example, many Ru(II)-diimine complexes (e.g., Ru-bipyridines) have strong absorption maxima in the visible spectrum. The introduction of carboxylate groups to the dyes allow them to be tethered securely and bound strongly to the titania surface. Absorption of solar radiation by these dyes results in the promotion of an electron from the Ru(II) metal into a conduction state, allowing it to be injected into the conduction band of TiO_2. The resulting "hole" causes oxidation of a redox active species in the electrolyte. The promoted electrons travel via an external circuit to undertake reduction of a redox species at the other electrode. Thus, solar irradiation is converted into an electrical current. This *sensitizing* of the TiO_2 facilitates access to the visible region solar spectrum, which makes up the majority of energy available in the solar radiation [27].

The most successful *dye-sensitized solar cell* (DSSC) of this type is the so-called *Graetzel cell*, which employs a $3I^-/I_3^-$ couple as the redox active couple in an acetonitrile electrolyte [Fig. 11.7(a)] [27]. Illumination causes the excitation of a dye

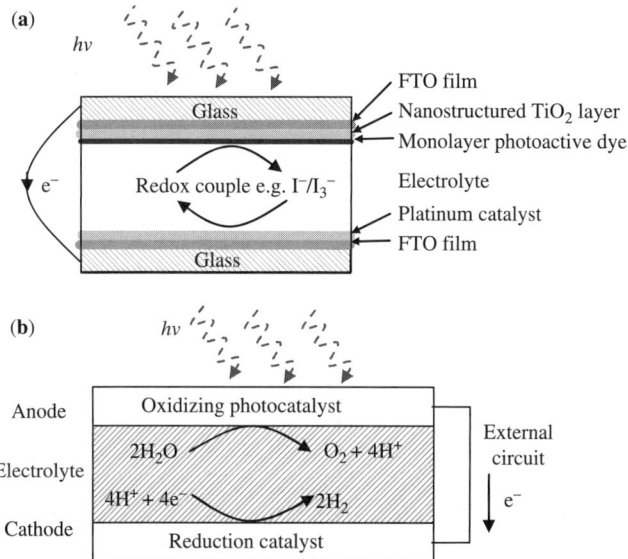

Figure 11.7. Schematic illustration of: (a) a *Graetzel* dye-sensitized solar cell, (b) an equivalent, water-splitting cell.

electron into the conduction band of the TiO_2, thereby creating a voltage between the two electrodes of the cell. Electrons flow from the TiO_2 electrode to a Pt counter electrode via an external circuit. Some of the resulting potential energy can be used to do work. At the counter electrode, the electrons reduce I_3^- to I^-, which then diffuses in solution to the dye sensitized TiO_2 where it is converted back to I_3^- by the oxidized dye, thereby completing the circuit.

Unfortunately the oxidizing potentials of the oxidized Ru(II) dyes and other photoactive dyes, such as the porphyrins, are not high enough to oxidize water efficiently. In addition, the oxidation of water can produce highly reactive intermediates that typically destroy the photoactive dye. Thus, the application of dye-sensitized TiO_2 for photooxidation of water has achieved little success. What is needed is an intermediate catalyst that decreases the overpotential of water oxidation and thereby makes available the possibility of water oxidation. Such a catalyst should allow for the fabrication of a *water-splitting, dye-sensitized solar cell* of the type illustrated in Figure 11.7(b).

With efficiencies >11% (solar to electricity), dye-sensitized solar cells are starting to challenge silicon photovoltaic cells in terms of their overall cost–benefit efficiency [27]. The chemical inertness of the nanostructured electrodes in acid/base conditions and their resistance to photocorrosion make these materials ideal for the development of water oxidation catalysts with high efficiency and low photodecomposition. All that is required is the discovery of a suitable intermediate catalyst that can mediate the oxidation of water by the dyes employed.

More efficient solar powered hydrogen production can be achieved by *tandem cell* arrangements, which use photovoltaic cells in tandem with an electrolyzer, yielding

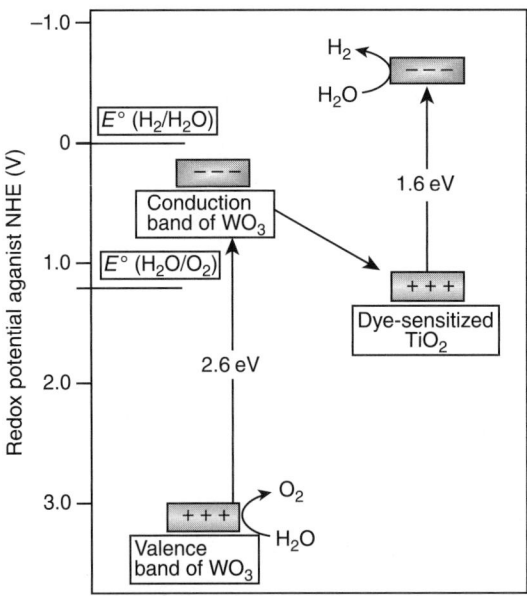

Figure 11.8. Thermodynamic potentials for water-splitting in a tandem PE Cell [27].

hydrogen generating efficiencies of up to 12–20% [31]. However, cells of this type typically use single-crystal silicon photovoltaic technology, which in conjunction with the electrolyzer, are expensive to produce.

An example of a tandem cell developed by Graetzel is shown in Figure 11.8 [31]. The photocathode reduces water to H_2 on a Pt nanoparticle electrode integrated into a mesoporous TiO_2 semiconductor, to which is attached a visible-absorbing dye that captures more solar energy. The photoanode employs a WO_3 semiconductor to oxidize water to O_2. The resulting, cheap device displays 10% power efficiency. The interfaces maintain electrical and photoconductivity between all regions of the semiconductor materials despite their large surface areas.

11.5 PHOTOCATALYTIC WATER OXIDATION BY *CUBANE* 1B DOPED INTO A NAFION SUPPORT

11.5.1 Solution Electrochemistry

The solution electrochemistry of the cubanes has been extensively studied. Reversible potentials (E_f^{ox}), derived from the average of oxidation (E_p^{ox}) and reduction (E_p^{red}) real potentials [($E_p^{ox} + E_p^{red}$)/2], have been found to be:

1a: 690 ± 3 mV, and
1b: 542 ± 8 mV

(vs. Fc/Fc$^+$ in CH_2Cl_2 (0.1M ^{t}Bu$_4$NPF$_6$)), for the oxidation process:

$$Mn_4O_4L_6 \longleftrightarrow [Mn_4O_4L_6]^+ + e^- \quad (11.6)$$

This corresponds to a difference in E_f^{ox} for the two compounds of 148 mV. The values of E_f^{ox} were almost unaffected by scan rate.

These data are in agreement with a recent study that reported a difference in the reversible E_f^{ox} of 150 mV [16]. An earlier study indicated a difference of 109 mV, with E_f^{ox}'s of 813 mV (**1a**) and 704 mV (**1b**) (vs. Ag/Ag$^+$(0.1M AgNO$_3$ in CH_3CN)) [16]. The ΔE_p values in this case were, however, very large (e.g., ΔE_p for **1b** was found to be 297 mV at 100 mV/s, compared with 95 mV in the current study). It suggests that the uncompensated resistance was much larger, which would have introduced considerable uncertainty [32].

The peak separation (ΔE_p) for **1a** indicates a reversible process, which is similar to the known reversible one electron oxidation of ferrocene/ferrocenium (Fc/Fc$^+$) [84 mV peak separation at 10 mV/s in CH_2Cl_2 (0.1M ^{t}Bu$_4$NPF$_6$)]. The oxidation of **1b** is quasi-reversible.

11.5.2 Electrochemistry of 1b Doped into a Nafion Membrane

To overcome the insolubility of the cubanes in hydrophilic solutions, **1b**$^+$ was ion-exchanged into a spin-coated Nafion membrane (ca. 7 μm thick) deposited onto a glassy carbon disk electrode (3 mm diameter) from a saturated, 3 mM acetonitrile solution. Cyclic voltammetry of the doped, membrane-coated electrode immersed in aqueous electrolyte displays an oxidation at $E_{ox}^f = 0.84$ V (vs. Ag/AgCl) that seems, at face value, to be similar to the **1b** ↔ **1b**$^+$ transition described earlier. The oxidation couple is not observed for undoped membranes.

The **1b** that is likely generated in the Nafion membrane upon cycling to more negative potentials seems to be stable and retained. These observations suggest that **1b**$^+$ is taken up into the Nafion by cation exchange with the protons of the sulphonic acid functional groups. Reduction of **1b**$^+$ then likely generates **1b**, which seems to be trapped within the hydrophobic pockets of the Nafion channels.

11.5.3 Electrocatalytic Effects Are Observed Under CV Conditions

A notable feature of the electrochemistry of the **1b**$^+$-doped Nafion system is the fact that, as the potential is swept up toward the water oxidation potential (1.4–1.6 V), the current increases at a significantly steeper gradient than the current of the control undoped Nafion-coated electrode. It is also true relative to the comparable uncoated electrode.

These experiments suggest that the **1b**$^+$-doped Nafion may increase the rate of water oxidation compared with an undoped Nafion coated or an uncoated glassy carbon electrode. Thus, it is possible that **1b**$^+$ may be able to catalyze the oxidation of water when supported in a Nafion membrane. As **1b** and **1b**$^+$ are both completely insoluble in water, the Nafion membrane could conceivably provide an interface for the cubane to interact with the water reactant. In such a case, the cubane would, presumably, be held in the hydrophobic regions of the Nafion membrane, at whose interfaces it would be exposed to water molecules that diffuse into the sulphonic acid/sulphonate lined channels.

To investigate the origin of this phenomenon, we studied the properties of the **1b**$^+$-doped Nafion system under potentiostatic conditions at 1.00 V versus Ag/AgCl. This potential was selected because the bare glassy carbon electrode does not catalyze water oxidation under these circumstances. Moreover, the **1b** ↔ **1b**$^+$ oxidation is complete at 1.00 V. Coincidentally, 1.00 V is also the potential at which the PSII-WOC catalyzes water oxidation (see Section 11.1).

11.5.4 A Photo-Electrocatalytic Effect Is Observed at 1.00 V (vs. Ag/AgCl)

As depicted by the solid line in Figure 11.9, exposure of a **1b**$^+$-doped Nafion-coated electrode to light (arrow in Fig. 11.9), while poised at 1.00 V and immersed in aqueous 0.1 M Na_2SO_4, causes a dramatic increase in current. Removal of the light causes an immediate decrease in the current back to the baseline (broken line). Re-exposure to light immediately regenerates the current.

Key Point. **Thus, a significant photocurrent is generated in this system upon illumination with light. By contrast, the comparable undoped Nafion-coated electrode does not display a clear photocurrent upon illumination (see the broken line in Fig. 11.9). The glassy carbon electrode on its own, without a Nafion coating, generates no current whatsoever under these circumstances at potentials negative of 1.20 V.**

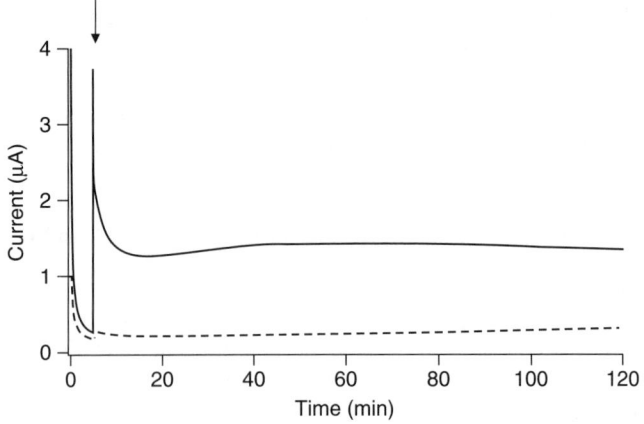

Figure 11.9. Photocurrent of doped Nafion on glassy carbon: Amperogram at 1.00 V (vs. Ag/AgCl) obtained at 22 °C for **1b**$^+$ doped into Nafion (———) and undoped Nafion (- - - -) deposited on a 3-mm-diameter glassy carbon disk electrode in H_2O (0.1M Na_2SO_4), light source exposed at the 5-minute time point. Arrow indicates the onset of illumination of the electrode.

11.5 PHOTOCATALYTIC WATER OXIDATION

The chronoamperogram of the photocurrent generated by the **1b**$^+$-doped Nafion-coated electrode shown in Figure 11.9 (solid line) shows that, after ca. 10 min of continuous light exposure, the current stabilizes at a level 6-fold greater than is observed for an equivalent, control glassy carbon electrode coated with Nafion only (broken line in Fig. 11.9).

*Key Point. Over a period of 2 h, the photocurrent generated from the **1b**$^+$-doped Nafion-coated electrode is stable at over six times the current of the undoped Nafion-coated electrode. In fact, over 15 h of continuous illumination, the photocurrent was found to degrade slowly to ca. four times the control current, where it remained invariant for an additional 50 h of testing.*

By measuring the electronic absorption of the doping solution before and after doping, the amount of cubane that had been ion-exchanged into the Nafion membrane used in the experiment displayed in Figure 11.9 was estimated at 6 ± 2 ng. This is an exceedingly tiny quantity of material.

As the cubane is only ion-exchanged into the Nafion membrane in the positively charged, oxidized form (**1b**$^+$), one could reasonably expect that most of the cubane within the Nafion should be in the oxidized state. Any neutral cubane should be readily reoxidized at 1.00 V.

The light is therefore either generating a new, previously unobserved oxidation of the cubane (**1b**$^+ \rightarrow$ **1b**$^{n+1}$ + ne$^-$), or it is activating a chemical reaction between the cubane and the working solution. Given the tiny quantity of cubane taken up by the Nafion and the consistency of the current over periods of up to 65 h of testing, it is not reasonable to conclude that the current is caused by a simple, irreversible oxidation of the cubane. It must therefore be caused by a reaction with a large quantity of an available chemical feedstock. But, what reaction?

If the photocurrent originates from a simple reaction of the cubane with another species, the current would rapidly decrease toward zero within seconds. Instead, it is sustained over long time periods. Thus, the cubane must be *mediating* a reaction with another species that is present in sufficient quantities to provide the large number of electrons traveling through the external circuit. Since this reaction does not seem to occur in the absence of **1b**, the cubane or its product upon ion-exchange into Nafion, must be operating as an electrocatalytic relay.

11.5.5 If the Photocurrent Is Caused by Water Oxidation Catalysis, This Involves a Decrease in the Overpotential of 0.4 V

A photocurrent can be discerned in the **1b**$^+$-doped Nafion-coated glassy carbon electrode at potentials positive of ca. 0.80 V, which corresponds to the earliest current for the oxidation peak at 0.84 V. The photocurrent grows with increasing applied potential up to 1.00 V and thereafter reaches saturation. This arguably corresponds to a point at which the oxidation peak at 0.84 V is complete.

The photocurrent is therefore intimately connected with the oxidation at 0.84 V. Moreover, if the current is caused by water oxidation catalysis, then there is a decrease in the overpotential upon irradiation, of ca. 0.40 V. To the best of our knowledge, the largest decline in overpotential for a manmade molecular catalyst of water oxidation on a carbon electrode is provided by *Ru-red*, which exhibits a decline in overpotential of ca. 0.15 V under comparable circumstances ($Ru\text{-}red = [(NH_3)_5Ru^{III}(\mu\text{-}O)Ru^{IV}(NH_3)_4(\mu\text{-}O)Ru^{III}(NH_3)_5]^{6+}$) [33]. This decline occurs, however, without the influence of any illumination.

11.5.6 The Photocurrent Is Observed only in the Presence of Water. The System Saturates at Low Water Content, Consistent with a Time-Dependent Catalytic Action

A **1b**$^+$-doped Nafion-coated electrode displays little to no photocurrent when it is immersed in pure acetonitrile (0.1 M Bu$_4$NPF$_6$). However, when the acetonitrile contains 10% water (v/v), the same electrode displays a notable photocurrent upon exposure to light. The equivalent undoped Nafion-coated electrode exhibits no significant photocurrent in either pure acetonitrile nor acetonitrile containing 10% water. Thus, the photocurrent is only observed in the presence of the cubane and water.

The *net photocurrent* (that is, the measured current minus the background current) was examined as a function of increasing water content within an acetonitrile electrolyte. Figure 11.10 depicts the resulting data. As shown, an unambiguous increase in photocurrent is observed up to 8% (v/v) water in the acetonitrile. As the photocurrent is stable for hours on end and there is only a tiny amount of cubane present in the Nafion membrane, the water must be acting as the source of electrons in the photocurrent.

A noteworthy feature of Figure 11.10 is the fact that the photocurrent saturates at the relatively low water content of 8%. If the electrocatalysis does involve water oxidation, then the saturation is unlikely to be caused by to a mass transport limitation since Nafion layers are known to sustain substantially higher absolute rates of water

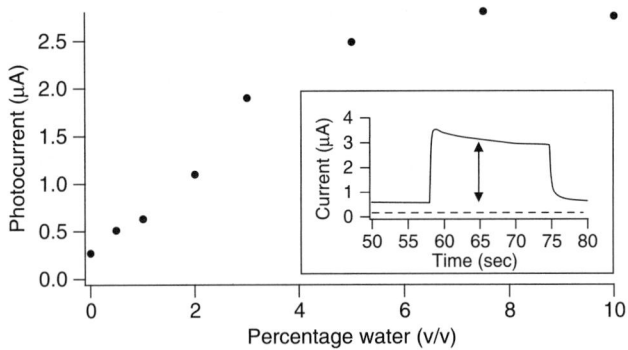

Figure 11.10. Photocurrent of **1b**-doped Nafion membrane on a 3-mm-diameter glassy carbon disk electrode immersed in a solution of acetonitrile (0.1 M Bu$_4$NPF$_6$) containing increasing percentages (v/v) of water. The insert displays photocurrent for 10% (v/v) water solution (———) and 0% water solution (- - - -).

oxidation by immobilized molecular catalysts [30]. As mentioned in Section 5.4 of Chapter 5, saturation kinetics is, however, a characteristic of time-dependent catalysis, being seen in, for example, the Michaelis–Menten kinetics that characterizes many enzymes. The existence of saturation in this system is therefore consistent with, and suggestive of, a time-dependent catalytic action, although it cannot be considered definitive proof at this stage, given the current paucity of information about this system.

11.5.7 The pH Dependence of the Photocurrent Is Consistent with Water Oxidation

The photocurrent generated by $1b^+$-doped Nafion displays a strong relationship with the pH of the aqueous electrolyte. At an applied potential of 1.00 V (vs. Ag/AgCl), a $1b^+$-doped Nafion-coated electrode displays a linearly increasing photocurrent with increasing pH.

A well-documented feature of water oxidation is that its electrochemical potential varies by 59 mV per pH unit [12]. This trait derives from the fact that the reaction involves one H^+ per e^-. An increase in the pH of an electrochemical water oxidation reaction is, consequently, expected to generate a linearly increasing current under otherwise invariant conditions. This is precisely what is observed, adding support to the possibility that the observed photocurrent originates from the oxidation of water.

11.5.8 Bulk Water Is a Reactant and Oxygen Is Generated

Isotopic enrichment and mass spectrometry studies were carried out to determine the reactants and products of the photo-electrocatalysis noted above.

An undoped Nafion-coated electrode was exposed to light at 1.00 V (vs. Ag/AgCl) for 1 h in a 50% $^{20}H_2O$ solution containing 0.1 M Na_2SO_4. The treated solution was passed through a MIMS. No increase in $^{36}O_2$ above the natural concentration in untreated water was detected.

As indicated in Figure 11.11, an equivalent experiment using a $1b^+$-doped Nafion-coated electrode immersed in the 50% $^{20}H_2O$ solution with 0.1 M Na_2SO_4 resulted, however, in the formation of a detectable increase in the amount of $^{36}O_2$. This result is only possible if bulk water is a reactant in the electrocatalysis and dioxygen, O_2, is a product.

The experiment therefore provides unambiguous evidence that the cubane-doped Nafion catalyzes the oxidation of water to dioxygen when exposed to light with an external potential of 1.00 V. Bulk water is unequivocally a source of the dioxygen.

11.5.9 The Quantity of Gas Generated Matches the Current Obtained. Notable Turnover Frequencies Are Implied

A custom-build gas-collecting cell was used to investigate the oxygen-producing capability of the cubane in Nafion. The gas cell was completely filled with H_2O

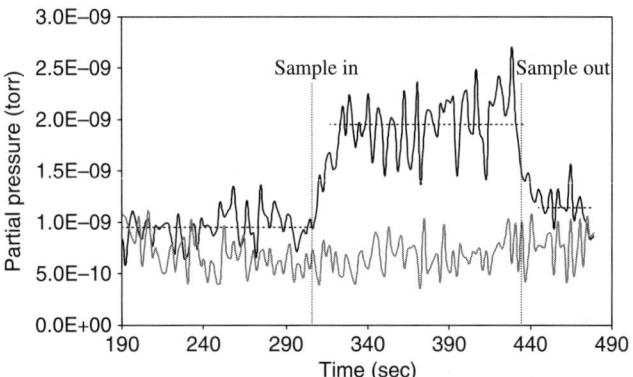

Figure 11.11. Membrane-Inlet Mass Spectrometry (MIMS) trace of a solution of 50% $^{20}H_2O$ with 0.1 M Na_2SO_4 exposed to a **1b**$^+$-doped Nafion electrode and light for one hour and passed through a MIMS tracing amu 36 (top trace with dashed lines show average partial pressure). The lower, gray line is the same solution exposed to an undoped Nafion-coated electrode with light for one hour. Untreated water was passed though the machine before and after the sample as a baseline.

(0.1 M Na_2SO_4), and a 100 × 150 mm Pt plate electrode was used as the working electrode. After 7.5 h of light excitation at 1.00 V (vs. Ag/AgCl), an electrode coated in undoped Nafion produced no observable gas bubbles. However, when the working electrode was coated with **1b**$^+$-doped Nafion, gas bubbles were observed to evolve at the electrode surface; 32 µl of gas was collected in the gas collection chamber.

At 25 °C and 1 atm, 32 µl of gas is equivalent to 1.31 µmol or 42 µg of O_2. The total charge recovered from the experiment was 0.488 Coulombs, which is equivalent to 5.1 µmol of electrons. Assuming that the gas produced is pure oxygen, the amount of gas generated from this charge would be 1.27 µmol, or 41 µg of O_2. Thus, the amount of gas collected from the experiment was, effectively, equivalent to the charge generated.

By measuring the electronic absorption of the doping solution before and after doping, the amount of **1b**$^+$ that was ion-exchanged into the Nafion membrane on the Pt plate was estimated at 15 ± 5 nmol. If it is assumed that all of this participates in the catalysis, then the average turnover frequncy per putative cubane or its active intermediate was ((1.31 µmol/15 nmol)/7.5) = 11.6 molecules of O_2 h^{-1}. Given the thickness of the Nafion (7 µm), it must be considered highly unlikely that all of the dopant cubane (or their active intermediate) participate in the catalysis.

To obtain an alternative estimation of the turnover frequency, bulk electrolysis of the observed **1b** → **1b**$^+$ oxidation was used to estimate that 0.272 nmol of **1b**$^+$ was ion-exchanged into a typical Nafion membrane. Such a membrane readily generated a net charge of 0.163 C under sustained illumination over 65 h in aqueous 0.1 M Na_2SO_4 at 1.00 V. This equates to 1566 turnovers per **1b**$^+$ ion-exchanged into the Nafion, with an average turnover frequency of 24.1 molecules of O_2 cubane^{-1} h^{-1}.

11.5 PHOTOCATALYTIC WATER OXIDATION

TABLE 11.1. Turnover Frequencies (kO_2) of Known, Active, Molecular Water Oxidation Catalysts

	k (molecules $O_2 \cdot$ catalyst^{-1} h^{-1})			
System[a]	Nafion Membrane	Homogeneous System	Type	Ref.[b]
$[(NH_3)_3Ru^{III}(\mu\text{-}Cl)_3Ru^{II}(NH_3)_3]^{2+}$	227	201	4 e$^-$	
$[(NH_3)_5Ru^{III}(\mu\text{-}O)Ru^{IV}(NH_3)_4$ $(\mu\text{-}O)Ru^{III}(NH_3)_5]^{6+}$ (*Ru-red*)	162	184	4 e$^-$	
cis-$[Ru^{III}(NH_3)_4Cl_2]^+$	50	7	4 e$^-$	
$[(tpy)Ru^{II}(bpp)(H_2O)_2Ru(tpy)]^{3+}$		50	4 e$^-$	34
$[(NH_3)_3Ru^{III}(\mu\text{-}O)Ru^{III}(NH_3)_3]^{4+}$	47	47	4 e$^-$	
[Mn$_4$O$_4$L$_6$]$^+$ (*Cubane* 1b)[c]	**24**		**4 e$^-$**	
$[Ir(bphp)_2(H_2O)_2]$		16[d]	4 e$^-$	35
$[Ru^{III}(NH_3)_5Cl]^{2+}$	10	1	2 e$^-$	
$[(bpy)_2(H_2O)Ru^{III}$ $(\mu\text{-}O)Ru^{III}(bpy)_2(H_2O)]^{4+}$	9	14	4 e$^-$	
$[1,2\text{-}C_6H_4\text{-}(TPP\text{-}Mn)_2]^{2+}$		6[e]	4 e$^-$	36
$[(bnppp)Ru_2(\mu\text{-}Cl)(CH_3C_5H_4N)_4]^{3+}$		2.8[f]	4 e$^-$	37
$[Mn_2^{IV}(H_2O)_2(mcbpen)_2)]^{2+}$		1.5[g]	4 e$^-$	38
$[(bpy)(H_2O)Mn(\mu\text{-}O)_2Mn(H_2O)(bpy)]^{3+}$		1[h]	4 e$^-$	39
$[(tpy)(H_2O)Mn(\mu\text{-}O)_2Mn(H_2O)(tpy)]^{3+}$		1[i]	4 e$^-$	40
$[Ru^{III}(en)_3]^{3+}$	0.3		2 e$^-$	
$[Ru^{III}(NH_3)_6]^{3+}$	0.1	0.05	2 e$^-$	

[a] tpy = 2,2':6',2''-terpyridine; bpy = 2,2'-bipyridine; bpp = bipyridylpyrazine; TPP = triphenylporphyrin; bnppp = 3,6-bis-[6'-(1'',8''-napthyrid-2''-yl)-pyrid-2'-yl]pyridazine; bphp = 5-methyl,4''-fluoro,2-phenylpyridine.
[b] Data drawn from Reference 30 unless stated otherwise.
[c] This study; L = (p-MeO-C$_6$H$_5$)$_2$PO$_2^-$.
[d] Over 7 days.
[e] Over 4 h at 2.0 V vs. Ag/Ag$^+$. Turnover frequency at 0.00 V vs. Ag/Ag$^+$ = 0.
[f] Over 3200 turnovers in total.
[g] Initial rate; estimated from Clark electrode data over first 230 seconds (ca. 2 × 10^{-5} mol/L O$_2$ generated using 0.2 mM catalyst).
[h] Over 1 h; initial rate = 7.2 h^{-1}.
[i] Over 4 h; initial rate = 12 h^{-1}.

These data compare favorably with known, active molecular catalysts of water oxidation, many of which display very limited net turnover capacities caused by parallel, decomposition side-reactions (Table 11.1) [30,34–40].

Key Point. The 1b$^+$-doped Nafion system seems to comprise the most active known Mn-based molecular catalyst of water oxidation. To the best of our knowledge, no nonbiological Mn molecular water oxidation catalyst has been shown to remain constantly active for more than 65 h of continuous operation.

11.5.10 Photocurrent as a Function of the Illumination Wavelength

Long-wavelength, high-pass filters were used to control the wavelength of the light employed to illuminate the $1b^+$-Nafion-coated electrodes. The photocurrents obtained were found to be strongly dependent on the illumination wavelength. The largest photocurrents were obtained with UV light having wavelengths below 395 nm (70% of total current measured with unfiltered light). The use of visible wavelengths of >455 nm yielded 35% of the photocurrent obtained with white light, whereas wavelengths >495 nm gave 20% of the photocurrent.

Thus, the cubane or its active intermediate within the Nafion sustains water oxidation catalysis when excited with light from the blue-green spectral region. To the best of our knowledge, such behavior is unprecedented for a manmade molecular catalyst. Water oxidation catalysis in the visible region has previously been reported for doped binary and other semiconductors [29,30].

11.5.11 The Photoaction Spectrum of the Catalysis Corresponds to the Main LMCT Absorption Peak of 1b

Monochromated light was used to determine the peak wavelength for photoexcitation. The *net photocurrent* (total photocurrent minus the background current) was measured for a range of excitation wavelengths, using a Nafion-coated Pt electrode that was either undoped (control) or doped with $1b^+$ as per the standard technique employed. Figure 11.12 depicts the resulting data.

The control undoped Nafion-coated electrode showed relatively low photoexcitation with a small excitation peak at 350 nm [Fig. 11.12(b)].

The $1b^+$-doped Nafion-coated electrode, however, displayed significant photocurrent from 325 nm to 525 nm, with peak response at 360 nm and a prominent shoulder at 425 nm [Fig. 11.12(a)].

The peak at 360 nm corresponds to the main ligand-to-metal charge transfer (LMCT) absorption observed in the electronic spectra of the cubane in solution [Fig. 11.12(d)]. Furthermore, the peak excitation corresponds to the maximum difference in absorption between $1b^+$ and $1b$, suggesting that the energy provided by this light corresponds to the peak charge transfer of the oxidized cubane ($1b^+$). This adds support to the hypothesis that it is the oxidized form of the cubane that photocatalyzes the oxidation of water.

The light source used in these experiments was a Newport-Oriel Instuments, Monochromator Model 74000, Arc Lamp-200W Hg(Xe)OF model 6292, at a single wavelength. This light source has an irradiation peak at ca. 440 nm that may contribute to the shoulder peak.

A 1-mm-diameter Pt electrode was used for these experiments, and the intensity of the monochromated light was significantly less than the white light used in previous experiments, explaining the dramatically reduced photocurrent displayed in Figure 11.12 compared with earlier experiments. The difference in photocurrent between the undoped Nafion control and the $1b^+$-doped Nafion was proportional to previous experiments.

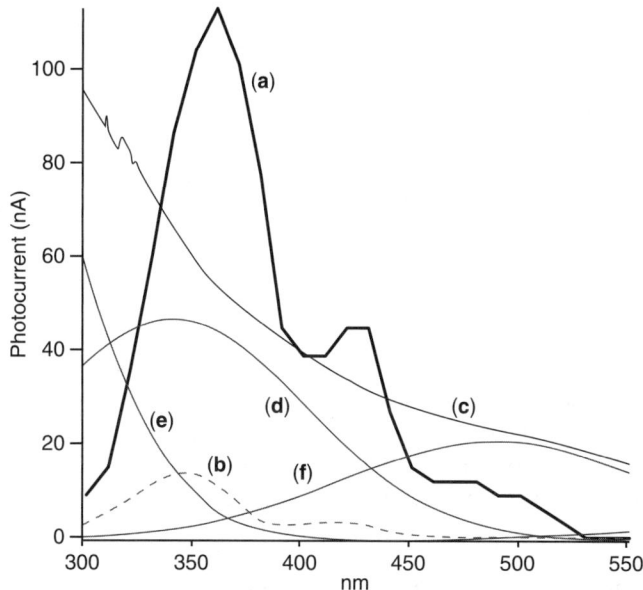

Figure 11.12. Plot of wavelength dependence of the photocurrent from controlled potential electrolysis at 1.00 V (vs. Ag/AgCl) obtained at 22 °C for (a) **1b**$^+$ ion exchanged into Nafion (thick black line) and (b) undoped Nafion only (dashed line) on a 3-mm-diameter glassy carbon electrode in H$_2$O (0.1 M Na$_2$SO$_4$) using monochromated (10 nm bandwidth) wavelengths of light from 300 nm to 550 nm (Xe light source). Overlaid is (c) the electronic absorption spectrum of **1b**$^+$ (thin solid line) in 0.5 mM in CH$_2$Cl$_2$, 22 °C, and (d)/(e)/(f) its component absorption peaks (thin solid lines) obtained by least-squares deconvoluton into Gaussian bands. The peak photocurrent action corresponds to a charge transfer absorption observed in the electronic spectrum.

11.6 THE CHALLENGE OF DYE-SENSITIZED WATER-SPLITTING

The cubane **1b**$^+$ in Nafion or its active intermediate, therefore, absorbs most strongly in the UV region of the solar spectrum. As noted in Section 11.4, the UV region makes up only 4% of the total solar spectrum, so that catalysis by the above system can only ever be relatively inefficient from the point of view of harnessing sunlight.

One potential solution to this problem that was alluded to in Section 11.4 is to extend the light absorption into the visible region by electrochemical coupling with suitable dyes that absorb well into the visible region. In absorbing a photon of light, such dyes may promote electrons into the conduction band of an attached electrode or semiconductor. The use of such dyes (or *sensitizers*) has dramatically expanded the absorption range of solar electrochemical cells, such as the *Gratezel cell* described in Section 11.4 and depicted in Figure 11.7. If such dyes can be made to couple electrochemically with the cubane **1b**$^+$, then they will potentially also expand the absorption range of the cubane and therefore improve the solar efficiency of its water oxidation catalysis. Additionally, the cubane could then potentially be used as a catalytic water-oxidizing relay in a dye-sensitized electrochemical cell. In effect, one would then have a *Water-Splitting, Dye-Sensitized Solar Cell*.

As noted in Section 11.4, a key disadvantage of *Graetzel*-type titania-[Ru(bipy)$_3$]$^{2+}$ electrochemical cells is that the oxidation potential of [Ru(bipy)$_3$]$^{2+}$ renders it unable to oxidize water (bipy = 2,2'-bipyridyl). Such cells cannot therefore be used for solar water-splitting. The incorporation of a cubane as a water-oxidizing relay has the potential to overcome this limitation, providing such cells with two energy outputs: electrical current *and* hydrogen gas generation. Present-day *Graetzel* cells generate only electrical current. They must, additionally, use an acetonitrile electrolyte that is extremely problematic from a fabrication point of view (it readily evaporates, destroying the electrical performance). A modified, water-splitting Graetzel cell would, by contrast, use water as a less problematic electrolyte.

Because of intellectual property and other constraints in place at the time of writing, we cannot report on the above possibilities at present. Future publications will, however, detail extensive work in this respect.

11.7 THE MECHANISM OF THE CATALYSIS

At this stage, we cannot definitively identify the catalytically active species within the Nafion coating doped with **1b**$^+$. Nor can we unequivocally pronounce upon the actual mechanism of the catalysis. However, it should be noted that other Mn ions, which are similarly deposited within Nafion films, do not yield comparable photocatalytic properties.

Thus, Nafion-coated anodes doped with Mn^{2+}, [(bipy)$_2$Mn(μ-O)$_2$Mn(bipy)$_2$]$^{3+}$ (bipy = 2,2'-bipyridine), or colloidal MnO$_2$ produce no photocatalytic effect. Additionally, the photocurrent of Nafion films that had been subjected to ion-exchange with **1b**$^+$ declines to that of undoped Nafion membranes, when the catalytically active ingredient is leached out by extended immersion/ion-exchange in concentrated Cu^{2+} CH$_3$CN/CH$_2$Cl$_2$ solutions.

The photocurrent is, moreover, observed with a wide range of electrolytes (Na$_2$SO$_4$, NaF, Bu$_4$NPF$_6$, Bu$_4$NClO$_4$) and for **1b**$^+$-doped Nafion membranes deposited on various conductive surfaces (glassy carbon, Pt, F-doped SnO$_2$ coated glass). No photocurrent is observed for undoped electrodes or for Nafion doped with salts, like NaClO$_4$ or sodium diphenylphosphinate. Thus, the photocurrent seems to require the presence of **1b**$^+$, Nafion, and water.

In the current study, Nafion has been used to overcome the hydrophobic nature of the cubane catalyst, which prevents its dissolution in pure water. It is possible that, during the doping process in acetonitrile solution, the hydrophobic cubane cations **1b**$^+$ are drawn into the water channels of the membrane, attracted by the sulfonate anions. Subsequent immersion of the electrode in an aqueous electrolyte may then cause the cubane cations to migrate into the hydrophobic pockets of the Nafion polymer. Within these hydrophobic domains, one may speculate that illumination of the charge transfer absorption band results in photodissociation of one of the phosphinate ligands from the cubane core, facilitating the release of dioxygen and the formation of the *butterfly* cubane according to the gas-phase mechanism shown in Figure 11.5. Although this explanation seems to be reasonable, with a clear precedent in the gas phase, physical confirmation of this mechanism is not as yet available.

Other possible influences are open for additional investigation. For example, it is conceivable that the sulfonic acid group may donate a proton to the more basic phosphinate, which causes dissociation of a single phosphinate ligand, thereby increasing the opportunity for O_2 evolution. The sulfonate anions of the Nafion may also provide a means of slowing the rebinding of the phosphinate ligand in their normal bridging geometry after the photodissociation process, which was previously shown to inhibit O_2 evolution in condensed phases.

What is certainly clear is that the Nafion seems to be a very friendly home for the cubane molecule or its active intermediate, when compared with bulk solution. It seems to prevent exposed Mn ions from undergoing deactivation by ligand exchange reactions with excess water molecules.

The necessity of using Nafion suggests that the combination of hydrophobic and hydrophilic domains and the proton exchange sites of the Nafion create an environment around the cubane that is analogous to the active site of the PSII-WOC enzyme. Reactive intermediates may be held within the catalytic site by a combination of hydrophilic, hydrophobic, and electrostatic forces that facilitate the interaction between the reactants and the catalyst, thereby prolonging the lifetime of intermediates and increasing their opportunity to react.

11.8 CONCLUSIONS

In conclusion, we have shown that a Nafion membrane doped with extremely small quantities of a cubical model complex **1b**$^+$ seems to demonstrate a significant photocatalytic effect in the oxidation of water. The active species and mechanism of the catalysis is unknown at present. Future work will examine these aspects in detail, as well as, *inter alia*: 1) the influence of the illumination on the observed decline in overpotential, 2) the oxidation states of the Mn atoms, 3) the role of the Nafion in facilitating cubane–water interaction and ligand dissociation/binding, and 4) the use of the cubane as a potential water-oxidation relay in a *Water-Splitting, Dye-Sensitized Solar Cell*.

For the cubane-doped Nafion system reported here, both atoms in the O_2 derive from water molecules and no sacrificial reactant is required to drive the process. The catalyst is regenerated by an external electrical potential that could conceivably be replaced by a solar cell, which is analogous to the oxidizing potential provided by chlorophyll in photosystem II. We have therefore, arguably, demonstrated a functional, synthetic model of the PSII-WOC.

Additional studies may provide deeper insight into the mechanism of action. We hope that a detailed understanding of these aspects will make it possible to develop other bioinspired molecular catalysts with improved efficiency for abiotic water oxidation.

Although we have not demonstrated that **1b**$^+$ undertakes time-dependent, mechanical catalysis of water oxidation, the evidence available at present is consistent with such a possibility. Moreover, an assumption that a time-dependent catalytic action occurs in PSII-WOC has led us to develop an entirely new and efficient water oxidation photoelectrocatalyst.

REFERENCES

1. Bockris, J. O. M.; Dandapani, B.; Cocke, D.; Ghoroghchian, J. *Int. J. Hydrogen Energy* **1985**, *10*, 179.
2. Dresselhaus, M. *Basic Research Needs for Solar Energy Utilization*, DOE Office of Basic Energy Sciences (OBES), 2003.
3. Lewis, N.; Crabtree, G. DOE Office of Basic Energy Sciences (OBES), 2005.
4. Bard, A. J.; Fox, M. A. *Acc. Chem. Res.* **1995**, *28*, 141.
5. Messinger, J. *Phys. Chem. Chem. Phys.* **2004**, *6*, 4764.
6. Dasgupta, J.; van Willigen, R. T.; Dismukes, G. C. *Phys. Chem. Chem. Phys* **2004**, *6*, 4793.
7. Ferreira, K. N.; Iverson, T. M.; Maghlaoui, K.; Barber, J. Iwata, S. *Science* **2004**, *303*, 1831.
8. Kamiya, N.; Shen, J. R. *Proc. Nat. Acad. Sci. USA* **2003**, *100*, 98.
9. Biesiadka, J.; Loll, B.; Kern, J.; Irrgang, K.-D.; Zouni, A. *Phys. Chem. Chem. Phys.* **2004**, 4733.
10. Sauer, K. *Biochim. Biophys Acta-Bioenerg.* **2004**, *1655*, 140.
11. Carrell, T. G.; Tyryshkin, A. M.; Dismukes, G. C. *J. Biol. Inorg. Chem.* **2002**, *7*, 2.
12. Ruettinger, W.; Dismukes, G. C. *Chem. Rev.* **1997**, *97*, 1, and references therein.
13. McEvoy, J. P., Gascon, J. A., Batista, V. S.; Brudvig, G. W. *Photochem. Photobiol. Sci.* **2005**, *4*, 940.
14. R. Manchanda, G. B., Crabtree, R. H. *Coord. Chem. Rev.* **1995**, *144*, 1.
15. Rüttinger, W. F.; Campana, C.; Dismukes, G. C. *J. Am. Chem. Soc.* **1997**, *119*, 6670.
16. Wu, J.-Z.; Sellitto, E.; Yap, G. P. A.; Sheats, J.; Dismukes, G. C. *Inorg. Chem.* **2004**, *43*, 5795.
17. Ruettinger, W. F.; Dismukes, G. C. *Inorg. Chem.* **2000**, *39*, 1021.
18. Ruettinger, W. F.; Ho, D. M.; Dismukes, G. C. *Inorg. Chem.* **1999**, *38*, 1036.
19. (a) Carrell, T. G.; Bourles, E.; Lin, M.; Dismukes, G. C. *Inorg. Chem.* **2003**, *42*, 2849; (b) Maniero, M.; Ruettinger, W. F.; Bourles, E.; McLendon, G.; Dismukes, G. C. *Proc. Nat. Acad. Sci. USA* **2003**, *100*, 3703.
20. Ruettinger, W.; Yagi, M.; Wolf, K.; Bernasek, S.; Dismukes, G. C. *J. Am. Chem. Soc.* **2000**, *122*, 10353.
21. Yagi, M.; Wolf, K. V.; Baesjou, P. J.; Bernasek, S. L.; Dismukes, G. C. *Angew. Chem. Int. Ed. Engl.* **2001**, *40*, 2925.
22. Dismukes, G. C. **2008**. Unpublished results.
23. Carrell, T. G.; Cohen, S.; Dismukes, G. C. *J. Mol. Catal. A Chem.* **2002**, *187*, 3.
24. Hoyer, B.; Jensen, N.; Busch, L. P. *Electroanalysis* **2001**, *13*, 843.

25. Gargas, D.; Bussian, D.; Burratto, S. *Nano Lett.* **2005**, *5*, 2184.
26. Yagi, M.; Kentaro, M. K.; Nagoshi, *J. Phys. Chem. B* **1997**, *101*, 5143.
27. Grätzel, M. *Nature* **2001**, *414*, 338.
28. Buriak, J. *Chem. Commun.* **1999**, 1051.
29. See, for example: Ishikawa, A.; Takata, T.; Kondo, J. N.; Hoa, M.; Kobayashi, H.; Domen, K. *J. Am. Chem. Soc.* **2002**, *124*, 13457, and references therein.
30. Yagi, M.; Kaneko, M. *Chem. Rev.* **2001**, *101*, 21, and references therein.
31. Hagfeldt, A.; Gratzel, M. *Acc. Chem. Res.* **2000**, *33*, 269.
32. Wu, J.-Z.; De Angelis, F.; Carrell, T. G.; Yap, G. P. A.; Sheats, J.; Car, R.; Dismukes, G. C. *Inorg. Chem.* **2006**, *45*, 189.
33. Yagi, M.; Ogino, I.; Miura, A.; Kurimura, Y.; Kaneko, M. *Chem. Lett.* **1995**, 863.
34. Sens, C.; Romero, I.; Rodriguez, M.; Llobert, A.; Parella, T.; Benet-Bucholz, J. *J. Am. Chem. Soc.* **2004**, *126*, 7798.
35. McDaniel, N. D.; Coughlin, F. J.; Tinker, L. L.; Bernhard, S. *J. Am. Chem. Soc.* In press.
36. Naruta, Y.; Sasayama, M.-A.; Saski, T. Angew. *Chem. Int. Ed. Engl.* **1994**, *33*, 1839.
37. Zong, R.; Thummel, R. P. *J. Am. Chem. Soc.* **2005**, *127*, 12803.
38. Poulson, A. K.; Rompel, A; McKenzie, C. J. *Angew. Chem. Int. Ed. Engl.* **2005**, *44*, 6916.
39. Ramaj, R.; Kira, A.; Kaneko, M. *Angew. Chem. Int. Ed. Engl.* **1986**, *25*, 825.
40. Limberg, J.; Vrettos, J. S.; Liable-Sands, L. M.; Rheingold, A. L.; Crabtree, R. H.; Brudvig, G. W. *Science*, **1999**, *283*, 1524.

12

TIME-DEPENDENT ("MECHANICAL"), NONBIOLOGICAL CATALYSIS. 2. HIGHLY EFFICIENT, "BIOMIMETIC" HYDROGEN-GENERATING ELECTROCATALYSTS

Jun Chen, Junhua Huang, Gerhard F. Swiegers, Chee O. Too, and Gordon G. Wallace

12.1 INTRODUCTION

The two-electron reduction of inorganic acids (H^+) to hydrogen gas (H_2) is efficiently catalyzed by the hydrogenase enzymes in biology [1]. However, the best nonbiological, heterogeneous catalyst is Pt metal, which serves as the universal standard for electrochemical potentials in the Normal Hydrogen Electrode (NHE). A longstanding research goal in the catalysis of multielectron redox processes involving small gaseous molecules like O_2, H_2, and N_2 has been to discover molecular catalysts whose activity exceeds that of the best available heterogeneous catalyst and approaches that of the relevant enzymes [2].

The formation of H_2 from H^+ involves several steps, including the uni-atomic reduction step ($H^+ + e^- \rightarrow H^\bullet$) in which atomic hydrogen is formed. On metal surfaces, this step is highly unfavorable ($E°$ ca. -2.10 V for Pt) [2]. Two $H^\bullet$ species must also find each other on the metal surface in order to form H_2. An overpotential is, therefore, required to drive the reaction.

Mechanical Catalysis: Methods of Enzymatic, Homogeneous, and Heterogeneous Catalysis,
Edited by Gerhard F. Swiegers
Copyright © 2008 John Wiley & Sons, Inc.

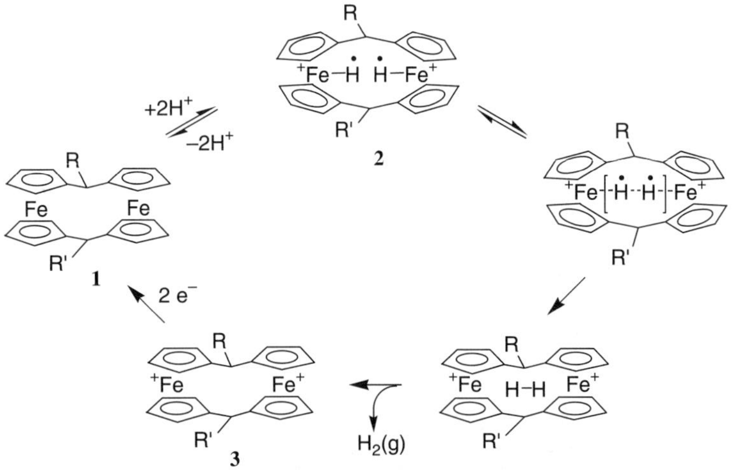

a: R = R' = H; **b**: R = Me, R' = H; **c**: R = Me, R' = Me-p(styrene)

Scheme 12.1.

Molecular species in which stabilized H• atoms are formed in close proximity to each other prospectively overcome these limitations. One class of such catalysts is the [1.1] *ferrocenophanes* **1a–c**, which facilitate the formation of H_2 from H^+ by a homolytic combination of two reduced protons that are formed on the ferrocene iron atoms during protonation (Scheme 12.1) [3,4] (see Chapters 3–6). The bis(ferrocenium) ions **3a–c** can be regenerated *in situ* using a sacrificial reductant to close the catalytic cycle or by direct electron transfer from an electrode [3]. We have previously discussed the ferrocenophane catalysts in Section 5.8 of Chapter 5 and Section 10.4.1.2 in Chapter 10.

INDO-SCF calculations indicate that the key catalytic intermediate **2** involves two $^+$Fe—H• moieties in which the positive charge of the protons effectively reside on the Fe atoms, with the hydrogens in an activated, atomic form (H•) [5,6]. These species undergo exceedingly rapid Fe—H ↔ Cp C—H exchange, with lifetimes too short to be observed on the NMR timescale, even at −122 °C [4,6].

Because of the very brief lifetime of the activated H• reactants, the rate of formation of H_2 by **3a–c** depends on the frequency with which the H• reactants can be brought into collision with each other. That is, it depends on the so-called *catalyst-mediated collision frequency*. The brevity of the Fe–H binding interactions means that there is very little time for the reactant functionalities to collide with each other on each occasion that they are bound to the catalyst. Moreover, such collisions can only occur when Fe—H binding on the two ferrocenes is synchronized. That is, it can only occur when both Fe catalytic groups simultaneously bear activated H• reactants.

Catalysis in such systems therefore only occurs if the participating ferrocenes flex about a spatial arrangement that is complementary to the structure of the H···H

12.1 INTRODUCTION

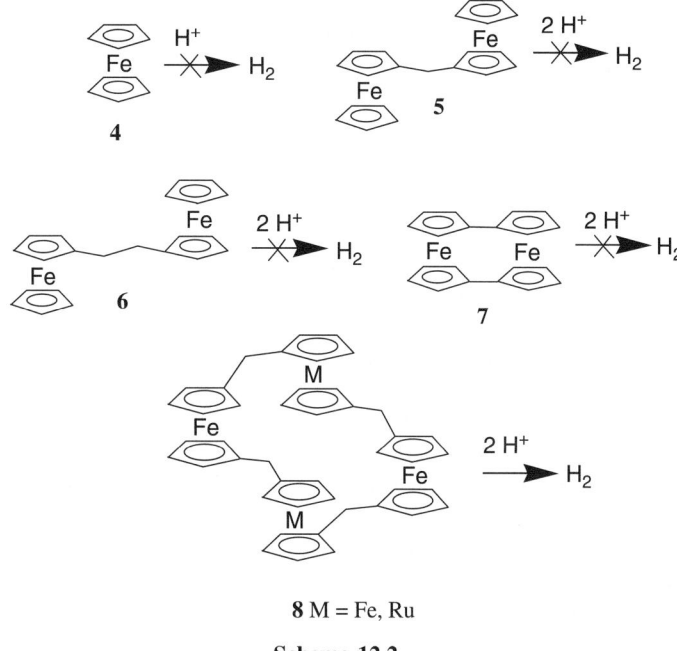

Scheme 12.2.

transition state [7]. Free ferrocene **4**, diferrocenylmethane **5**, diferrocenylethane **6**, and [0.0]ferrocenophane **7** do not, consequently, catalyze the reaction in open solution (Scheme 12.2). The tetramers **8** do catalyze the reaction, albeit poorly, because flexing allows opposing ferrocenes to mediate collisions, albeit less efficiently, and for a lesser, but still significant, proportion of the time. [4,8]

As noted in Chapters 3–6, catalysis that is governed by the collision frequency of the bound reactants is termed *time-dependent* or *mechanical* in character. This terminology reflects the fact that the collision frequency is, effectively, a quantification of time (s^{-1}). A collision is also a mechanical event; that is, it involves mechanical contact between two chemical entities.

The most prominent class of catalyst within which to find time-dependence is the catalysts of biology, known as enzymes. Many enzymes seem to employ a time-dependent action (see Chapters 5 and 6).

Most nonbiological catalyzed reactions are not controlled by the frequency with which the reactants collide. Rather, they depend on the so-called *activation energy* E_a of the transition state and on the rate at which it is overcome during collisions. Catalytic reactions of this type are termed *energy-dependent* or *thermodynamic* and are governed by the entity of the transition state.

There is great interest in studying nonbiological, time-dependent catalysts because of their inherent *biomimetic* or *bioinspired* qualities. Since **1a-c** already employ time-dependent catalytic actions, they and their derivatives are ideal species with which to examine the nature of, and practical utility of, time-dependent, nonbiological catalysis.

Conducting electroactive polymers (CEPs) offer a useful means of immobilizing and continuously regenerating redox catalysts if they are attached to the polymer or present within it as a counter-ion [8]. Additionally, they allow the study of ion-paired or polymer-tethered *monomeric* catalytic groups as a means of creating *macromolecular, intramolecular* time-dependent catalysts of the type described in Section 10.4.1.8 in Chapter 10. Such species prospectively offer a practical means of preparing highly efficient time-dependent catalysts by a *combinatorial approach*.

In Section 10.2.4 of Chapter 10, we discussed the use of a combinatorial approach as a means of discovering and creating new time-dependent catalysts. Such an approach was noted to have several potentially significant advantages over a conventional trial-and-error strategy. The key benefit was that one did not need to know, or iteratively search for, the optimum structural arrangement of the catalytic groups. Instead, a *library* of all available structural arrangements would be simultaneously present in the experiment. If the active arrangement was populated to even a small extent, this population should generate a substantial catalytic effect. Because of the character of time-dependent catalysis, the remaining structural arrangements would be passive onlookers that did not produce any sort of catalysis.

We termed a combinatorial experiment of this type a *Statistical Proximity* catalyst, since it would employ the same *proximity effect* that is present in many enzymes, but this would be created by the *statistical* influence of high local concentrations of monomeric catalytic groups.

Such an approach—if it is feasible—could drastically simplify the preparation of time-dependent homogeneous catalysts. Instead of synthesizing complicated molecules in which the catalytic groups are precisely connected to each other by suitably constrained but flexible covalent interconnections (e.g., **1a–c**), one could, at least in theory, achieve the same effect by drastically concentrating the equivalent monomeric catalytic groups (e.g., monomeric ferrocenes) within a fixed volume. A "statistical proximity" catalyst that avoids the need for multistep organic synthesis, but achieves the same overall catalytic effect, has important practical and commercial implications.

Given the capacity of the proximate ferrocene groups in **1a–c** to catalyze hydrogen formation from acids, we were interested in examining the effect of tethering [1.1] ferrocenophanes and related monomeric species to a conducting polymer. We also wanted to know whether a conducting polymer could be induced to incorporate unusually high local concentrations of a monomeric ferrocene species and, if so, whether it would be an efficient combinatorial, "statistical proximity" catalyst.

A series of platinum electrodes were therefore coated with polypyrrole (PPy) containing various ion-paired or covalently tethered ferrocene groups. Comparative testing of the hydrogen-generating capacity of the modified electrode was carried out in 1 M strong acids of the type employed in the Normal Hydrogen Electrode (NHE). As a control, the platinum electrodes were also coated with polypyrrole that was free of ferrocene and tested under analogous conditions.

A communication detailing early results of these experiments has been published [9], as have synthetic and characterization details of the various monomers and polymers employed [10].

12.2 MONOMER AND POLYMER PREPARATION

The present work will revisit some of these results and foreshadow, in summary form, future publications that will describe in greater detail the studies undertaken.

Chapters 1–7 in this series dealt with the theory of time-dependent catalysis. Chapters 8–10 dealt with the rational design of time-dependent catalysts. The remaining chapters detail practical examinations and illustrations of the concepts described. The present work falls within these remaining chapters and is intended to investigate the general utility and applicability of time-dependent catalysis, particularly in respect of catalyst discovery using a combinatorial, "statistical proximity" approach.

12.2 MONOMER AND POLYMER PREPARATION

The monomers **9–15** (Scheme 12.3) were prepared and characterized by multistep synthetic procedures described elsewhere [10].

As shown, species **9–11** and **14–15** involve single ferrocene groups. Monomeric ferrocene is known to be entirely inactive as a hydrogen generation catalyst in open solution. Species **9–11** and **14–15**, therefore, provide individual ferrocene catalytic groups that should be inactive unless they happen to become suitably proximate to other individual ferrocene groups during polymerization.

Species **12–13** contain [1.1]ferrocenophane units that have been tethered to pyrrole monomers. These moieties are known to be time-dependent catalysts of hydrogen evolution.

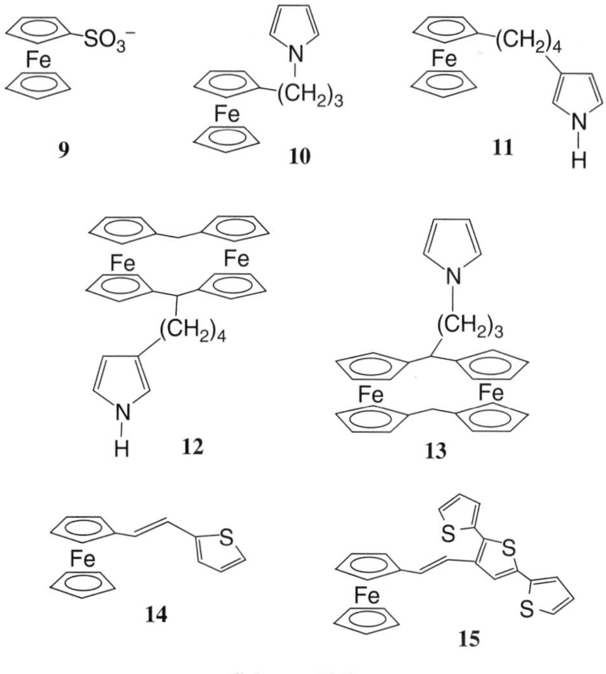

Scheme 12.3.

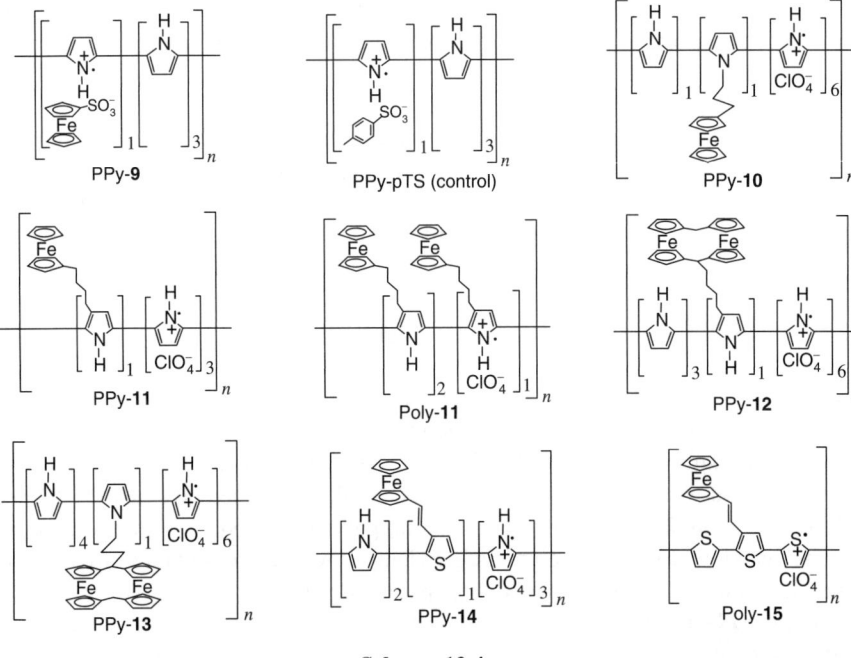

Scheme 12.4.

The base monomer is also varied in this series, with pyrrole units in **10–13**, whereas in **14–15**, thiophene and terthiophene units are employed.

Electropolymerization of the monomers was undertaken to generate a wide range of homo- and copolymers [10]. A selection of such polymers was characterized and found to have the structures shown in Scheme 12.4 [10]. A control coating, PPy-pTS, involving polypyrrole doped with *p*-toluene sulfonate was also prepared and characterized.

12.3 CATALYTIC EXPERIMENTS

12.3.1 PPy-9 and PPy-12 Display Anodic Shifts in the Most Positive Potential for Hydrogen Generation

In the first set of experiments, comparisons were made between a coating containing ion-paired ferrocene monomers, PPy-**9**, and the equivalent [1.1]ferrocenophane-substituted polypyrrole, PPy-**12**. These coatings were separately electrodeposited on a platinum disk electrode by potentiostatic growth as described elsewhere [9]. PPy-pTS was similarly prepared as a control coating. The electrocatalytic properties of the polymers on the Pt electrode were then examined.

In the representative results described below, all coatings were deposited on a platinum disk electrode of electrochemical area $0.0177\,\text{cm}^2$ to a charge of 100 mC. This was found to provide the greatest catalytic effect for PPy-**9**, which forms a

12.3 CATALYTIC EXPERIMENTS

uniform layer, 5 μm thick, under these conditions. A custom-designed cell was used in the experiments to ensure precise invariance in the positions of the electrodes and for collection and measurement of the gases produced. Because of the 1 M strong acid electrolyte, even substantial variations in the spatial positions of the electrodes relative to each other did not dramatically alter the currents obtained. To ensure that the $2H^+ \rightarrow H_2$ catalysis at the cathode was rate limiting, a large platinum mesh was employed as the anode.

Initial studies examined the most positive potential at which hydrogen generation was observed (the hydrogen decomposition potential).

When the bare platinum electrode was swept from positive to negative potential, a current from hydrogen formation commenced at -0.20 V [Fig. 12.1(a)] [11]. The PPy-pTS coated electrode began generating hydrogen from -0.24 V [Fig. 12.1(b)].

When coated with PPy-**9** or PPy-**12**, however, ferrocene reduction peaks were observed [points F in Fig. 12.1(c)–(d)] followed by distinctly increasing currents from 0.07 V (PPy-**9**) or 0.02 V (PPy-**12**). An unusual feature of these currents was the presence of maxima at -0.20 V [points G in Fig. 12.1(c)–(d)]. These currents and maxima did not seem to be caused by the polypyrrole backbone, whose electrochemical response in this region was minuscule [cf. PPy-pTS in Fig. 12.1(b)].

To understand these results, we compared them with the voltammetric profile previously observed for catalytic hydrogen generation by polystyrene-bound [1.1] ferrocenophane **1c** in 1 M acid (Fig. 12.1(e), Scheme 12.5) [11].

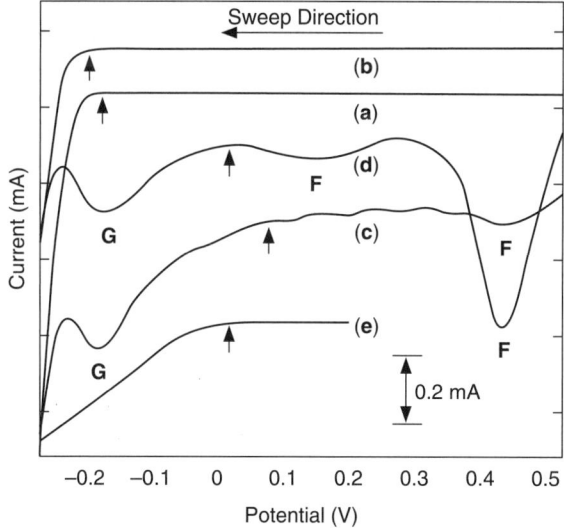

Figure 12.1. Current–voltage plot (first scan, vs. Ag/AgCl [3 M NaCl]) under invariant conditions in 1 M H_2SO_4 of a platinum electrode: (a) before coating and after coating with (b) PPy-pTS, (c) PPy-**9**, and (d) PPy-**12**. Curve (e) indicates the comparative area-equivalent response (vs. Ag/AgCl) of **1c** during catalytic hydrogen generation in 1 M $HClO_4$ as displayed in Reference 13. Arrows mark the commencement of hydrogen generation in each system as they are swept from positive to negative potential.

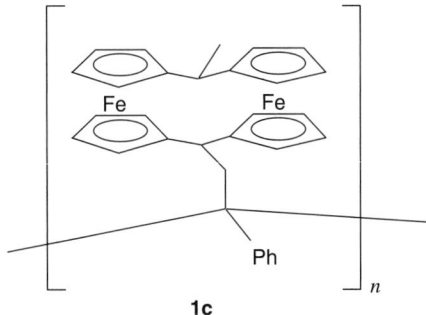

Scheme 12.5.

As shown in Figure 12.1(e), the current associated with hydrogen generation by **1c** commences at the same potential as that observed for PPy-**12**. This, along with the response of the bare platinum and the PPy-pTS-coated electrode, and the absence of other possible processes, suggests that the currents at potentials negative of the arrows in Figure 12.1(c)–(d) are also from hydrogen formation. Moreover, the mechanism of the catalysis by PPy-**12** is suggested to be the same as that in **1a–c**.

A substantial anodic shift, therefore, seems to occur in the hydrogen decomposition potential of platinum when it is coated with PPy-**9** or PPy-**12**. This shift is greater for PPy-**9** than for PPy-**12** presumably because $FcSO_3^-$ is fully reduced at more positive potentials than **1a–d**.

The maxima at points G in Figure 12.1(c)–(d) are not observed for **1d** and must therefore be associated with the polypyrrole. The peak currents were linearly dependent upon scan rate [12].[1] The current function also increased with scan rate [12]. The maxima are therefore, most likely, from adsorption of hydrogen by the polymer; this was presumed to block bulk transport of the reactants or the products, thereby starving the catalytic process [13]. Additional evidence of an adsorption process is the fact that the peaks are seen essentially only on the first scan and that no peak of even remotely comparable size is observed on any reverse scan.

The adsorption is substantial at potentials positive of −0.24 V, being >100-fold larger than the corresponding hydrogen adsorption peaks on bare platinum [13]. The adsorption effects disappear at potentials negative of −0.24 V, when polypyrrole is progressively reduced to its neutral state.

12.3.2 PPy-9 and PPy-12 Increase the Rate of Hydrogen Generation on Pt by ca. 7-Fold after 12 h at −0.44 V

Additional tests examined the rate of hydrogen generation by PPy-**9** and PPy-**12** under potentiostatic conditions. To avoid complications from the adsorption effects, these experiments were conducted at −0.44 V, which is 0.20 V cathodic of −0.24 V.

---

[1] Current function = peak current/(scan rate)$^{1/2}$.

12.3 CATALYTIC EXPERIMENTS

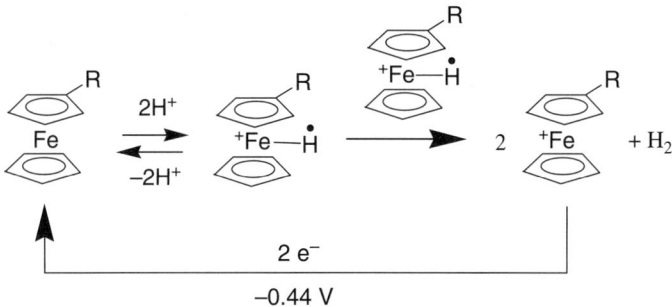

R = SO$_3^-$ or R = remainder of PPy-**12**

Scheme 12.6.

Since polypyrrole is likely to be substantially in its reduced, neutral state at -0.44 V, the FcSO$_3^-$ anions must be physically trapped rather than ion-paired within the PPy-**9** coating during these experiments. They consequently cannot be present in substantial quantities at the polymer surface.

At -0.44 V, ferrocene sulfonate is in its reduced, anionic form. Unsubstituted ferrocene is in its reduced, neutral form at this potential. Both of these species are transformed into their oxidized forms only at potentials that are more positive than $+0.45$ V (see points F in Fig. 12.1). Thus, any oxidized forms of ferrocene that are generated during catalysis of the type illustrated in Scheme 12.6 should be immediately transformed back to the reduced state at an applied potential of -0.44 V, thereby closing the catalytic cycle.

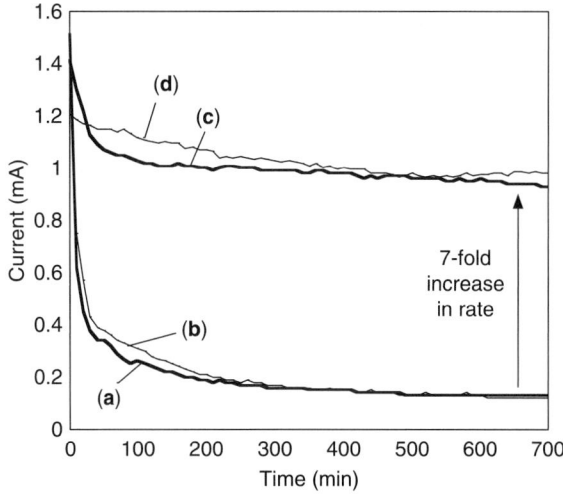

Figure 12.2. Current-time plot [vs. Ag/AgCl (3 M NaCl)] under invariant conditions of a platinum electrode poised at -0.44 V in 1 M H$_2$SO$_4$: (a) before coating and after coating with (b) PPy-pTS, (c) PPy-**9**, and (d) PPy-**12**.

As shown in Figure 12.2(a), the current at the uncoated platinum electrode declined sharply in the initial period of operation at −0.44 V. This is typical of freshly cleaned platinum, even in ultra-pure water (<10 ppb organics), and it seems to originate from selective poisoning of the most active sites on the Pt surface [14]. After 12 h, a steady current of 0.130 mA was obtained.

Under analogous conditions, the same electrode produced a 7-fold greater current after 12 h when coated with PPy-**9** (0.930 mA) or PPy-**12** (0.980 mA) [Fig. 12.2(c)–(d)].

By contrast, coating with PPy-pTS resulted in a current of only 0.140 mA after 12 h [Fig. 12.2(b)]. Separate two-electrode experiments indicated these results to be general for PPy-**9** versus bare Pt to potentials of −3 V at 20 °C and 80 °C (Fig. 12.3).

This divergence is not explained by different coating densities or quantities, since varying the PPy-pTS coating charge from 20 to 100 mC did not substantially change its rate of hydrogen generation after 12 h.

The volumes of hydrogen gas collected in these and other systems employing larger electrodes were also typically within 10% of the quantity expected from the cumulative charge passing through the electrode. GC-MS of the gas indicated it to be pure hydrogen.

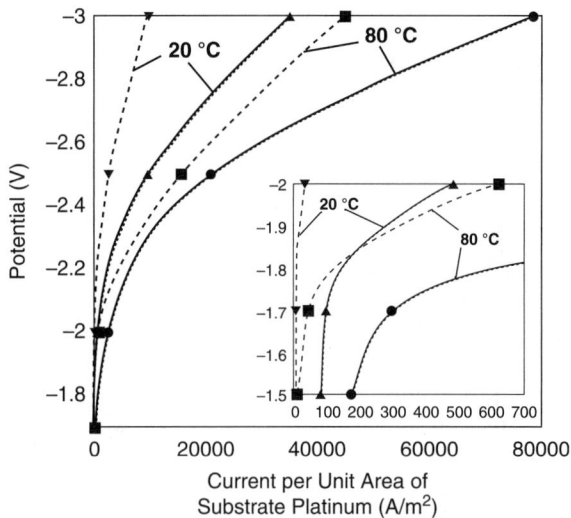

Figure 12.3. Polarization curves for water electrolysis in 1 M H_2SO_4 (two-electrode system) after 1 h of operation at 20 °C or 80 °C by a platinum disk cathode before coating (dotted lines) and after coating with PPy-**9** (solid lines). The inset graph details the data at the bottom left of the main graph. The coating was deposited galvanostatically (1 mA cm^{-2} for 30 min). Electrochemical area of uncoated electrode: 0.0170 cm^2; coated electrode: 0.0580 cm^2.

12.3.3 PPy-9 and PPy-12 Increase the Rate of Hydrogen Generation on Pt per Unit Area by ca. 3.5-Fold

Studies also examined the effect of the area of the coating on the rate of catalysis. Cyclic voltammetry data (10 mV s^{-1} in 0.1 M K$_4$Fe(CN)$_6$/1.0 M NaNO$_3$) was applied to the standard equation (12.1) [15],

$$i_p = 0.4463 nFAC_o^*(nF/RT)^{1/2} v^{1/2} D_o^{1/2} \qquad (12.1)$$

which indicated the electrochemical areas of the electrodes to be:

0.0177 cm^2 (bare Pt)
0.0355 cm^2 (Pt/PPy-**9**)
0.0364 cm^2 (Pt/PPy-**12**)
0.0209 cm^2 (Pt/PPy-pTS)

[where i_p = peak current (amps), n = electrons per molecule oxidized or reduced, F = Faraday's constant (C), A = area (cm^2), C_o^* = bulk concentration (mol cm^{-3}), R = gas constant (J mol^{-1} K^{-1}), T = temperature (K), v = scan rate (V s^{-1}), and D_o = diffusion coefficient (cm^2 s^{-1})] [15].

The current densities during the operation of the electrodes could therefore be calculated to be as shown in Figure 12.4. After 12 h of operation, they were:

7.33 mA cm^{-2} (bare platinum)
26.20 mA cm^{-2} (Pt/PPy-**9**)

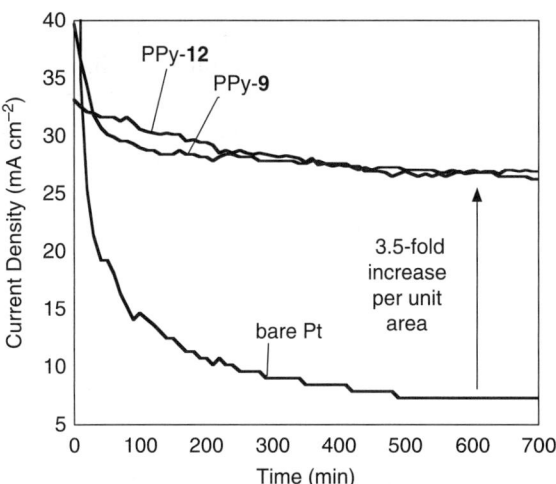

Figure 12.4. Current per unit electrochemical area-time plot [vs. Ag/AgCl (3 M NaCl)] under invariant conditions of a platinum electrode poised at −0.44 V in 1 M H$_2$SO$_4$.

26.93 mA cm^{-2} (Pt/PPy-**12**)
6.76 mA cm^{-2} (Pt/PPy-pTS)

The ferrocene-containing modified electrodes consequently displayed greater rates of hydrogen generation per unit electrochemical area. Moreover, increasing the electrochemical area of the platinum electrode by coating it with PPy-pTS seemed to have a negligible effect on its overall rate of catalysis.

PPy-**9** and PPy-**12** are, consequently and unambiguously, vigorous catalysts of $2H^+ \rightarrow H_2$.

12.3.4 The Mechanism of Catalysis in PPy-9. Is PPy-9 a Combinatorial ("Statistical Proximity") Catalyst?

As PPy-**9** is a powerful hydrogen generation catalyst that is far more easily, cheaply, and conveniently prepared than PPy-**12**, we wondered whether PPy-**9** was a combinatorial *statistical proximity* catalyst? That is, does the mechanism of action of PPy-**9** involve a homolytic combination of activated H$^{\bullet}$ species similar to that in **1a** (Scheme 12.1), but where a statistically significant proportion of the ferrocene species are adventitiously correctly arrayed at any one time?

To this end, the average "concentration" of ferrocene sulfonate within five different samples of PPy-**9** was measured. It was found to be 2.02–2.30 M by a determination of the weight and dimensions of uniformly deposited coatings of PPy-**9**, whose weight percentage of Fe was known from elemental analysis. This is substantially more than the saturation concentration of NH_4^+ $FcSO_3^-$ in 1 M H_2SO_4 open solution, which is 0.41 M.

At an average concentration of 2.02 M, the average Fe–Fe distance in PPy-**9** is calculated to be 9.3 Å. This is more than the 3.4–4.8 Å in **1a** [4]. However, a significant number of ferrocene sulfonate ions must clearly still be suitably closely arrayed for catalysis. When protonated, such $FcSO_3^-$ ions will be neutral, thereby facilitating a closer approach to each other than the protonated cations in **1a**. The greater basicity of anionic $FcSO_3^-$ relative to neutral ferrocene in **1a** could conceivably also make a difference.

It was, consequently, not impossible that PPy-**9** was a statistical proximity catalyst.

To assess the role of the ferrocene groups in PPy-**9**, we examined and measured, over time, the retention of ferrocene sulfonate by PPy-**9** after immersion in a solution of 1 M sulfuric acid at −0.44 V.

As shown in Figure 12.5, most ferrocene sulfonate is lost within the first few minutes by ion-exchange with the acidic solution.

However, after 12 h, ca. 12–15% of the original ferrocene solfonate remains within the polymer. These molecules must be trapped within the deepest layers of the polymer, near to the Pt surface.

Thus, either the ferrocene immediately at the Pt surface provides the catalytic effect or some other mechanism is at work.

To assess the role of the ferrocene sulfonate, a series of de-doping trials were undertaken.

12.3 CATALYTIC EXPERIMENTS

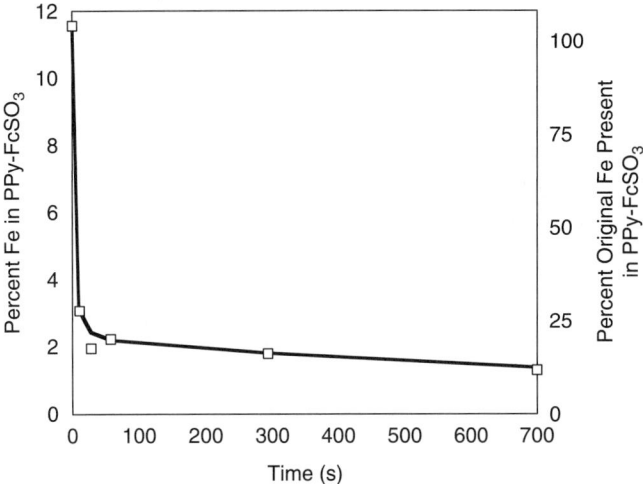

Figure 12.5. Retention of ferrocene sulfonate dopant by PPy-**9** on Pt, over time, after immersion in 1 M sulfuric acid while poised at −0.44 V. After 12 h, ca. 12–15% of the original FcSO$_3^-$ remains.

In these studies, PPy-**9**/Pt coatings were prepared and left to generate hydrogen at −0.44 V for 1 h. They were then subjected to one or more multiples of ten potentiodynamic sweeps to +0.8 V and back at 100 mV/s. The potential of −0.44 V was then reestablished and the rate of hydrogen generation was monitored for 1 h.

The effect of these sweeps is to momentarily and partially reoxidize polypyrrole, drawing in anions from solution. When the polymer is then re-reduced, an excess of anions temporarily exists within the polymer, causing de-doping of the ferrocene sulfonate.

The results of these de-doping experiments are shown in Figure 12.6 (top graph). As shown, a consistent decrease in the catalytic rate of hydrogen production by PPy-**9** is observed with increasing de-doping cycles. This appears to be matched by an accompanying loss of ferrocene, as evidenced by a corresponding decrease in the ferrocene oxidation and reduction peak currents (Figure 12.6, bottom graphs).

The de-doping experiments consequently support the involvement of the ferrocene sulfonate dopant in the catalytic process.

However, the situation proved to be more complicated than these experiments suggested.

12.3.5 Polypyrrole Is Likely Involved in the Catalytic Cycle

Subsequent work demonstrated that when a control coating of PPy-pTS/Pt was subjected to the equivalent "de-doping" procedure that had been applied to PPy-**9**, its current from hydrogen generation at −0.44 V *increased*. This result was found to be caused by an *increase* in its electrochemical area. Thus, the channels and

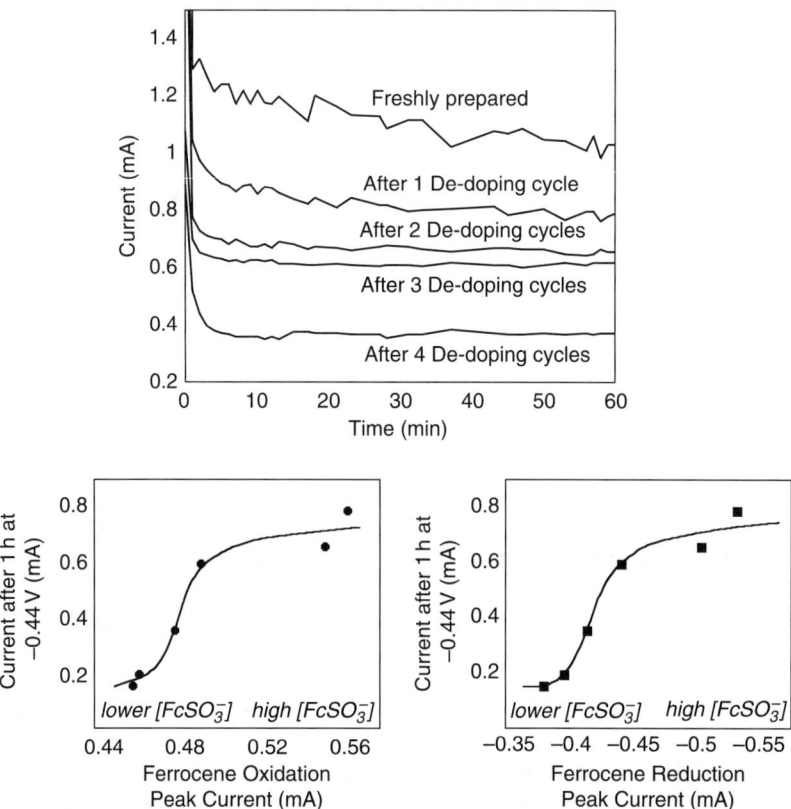

Figure 12.6. Rate of hydrogen production by PPy-**9** on Pt, over time, when poised at −0.44 V in 1 M H_2SO_4, before and after de-doping cycles (top graph). Each de-doping cycle consisted of 10 sweeps at 100 mV/s to +0.8 V and back. Ferrocene oxidation peak current (first sweep) (left bottom graph) and ferrocene reduction peak current (first sweep)(right bottom graph) during the de-doping cycles.

pores within the polypyrrole coating change *in situ* during rapid cycling between reduced and oxidized states. The flow of anions into and out of the polymer clearly opens new channels and pores.

Moreover, by adjusting the conditions under which PPy-pTS and other control coatings were grown, substantially larger electrochemical areas could be obtained. Such coatings displayed clear hydrogen adsorption peaks in linear sweep voltammograms of the type displayed at points G in Figure 12.1, despite containing no ferrocene whatsoever. Thus, anodic shifts in the most positive potential for hydrogen generation could also be observed with PPy-pTS under optimum conditions.

Control coatings could, indeed, be obtained in which the maximum current per unit electrochemical area of the adsorption peaks, essentially matched or came very close to those obtained for PPy-**9** and PPy-**12**.

Thus, although the ferrocene groups assist in the acceleration of hydrogen generation in PPy-**9** and PPy-**12**, another effect clearly plays the more major role.

12.3 CATALYTIC EXPERIMENTS

As significant rate accelerations were observed after coating of the electrode with various forms of polypyrrole, the effect seemed to be associated with the polypyrrole support and, especially, with its electrochemical area.

Experiments showed, however, that the rate of hydrogen generation on a bare Pt electrode immersed in a 1 M acid solution is *not* enhanced when free pyrrole is added to the solution. Thus, individual pyrrole groups do not produce a catalytic effect. However, when polymerized into densely packed polypyrrole chains on the electrode surface, significant catalytic effects were produced.

12.3.6 Other Evidence for the Involvement of Polypyrrole in the Catalytic Cycle

All of the other polymers shown in Scheme 12.4 were also tested for hydrogen generation catalysis by coating on a Pt electrode in the manner described for PPy-**9** and PPy-**12**. After 12 h at -0.44 V, the currents plotted in Figure 12.7 were obtained.

What is immediately evident from Figure 12.7 is that a linear relationship exists between the rate of hydrogen generation and the electrochemical area of the coatings. This relationship is, in fact, the only relationship that could be identified between the different coatings.

Thus, for example, although Poly-**11** contains substantially more ferrocene than PPy-**11**, the two coatings display almost identical rates of hydrogen generation after 12 h at -0.44 V.

Moreover, the only coatings that do not fall on the line shown in Figure 12.7 are PPy-**14**, which contains thiophene, and PPy-pTS, which can be brought closer to the

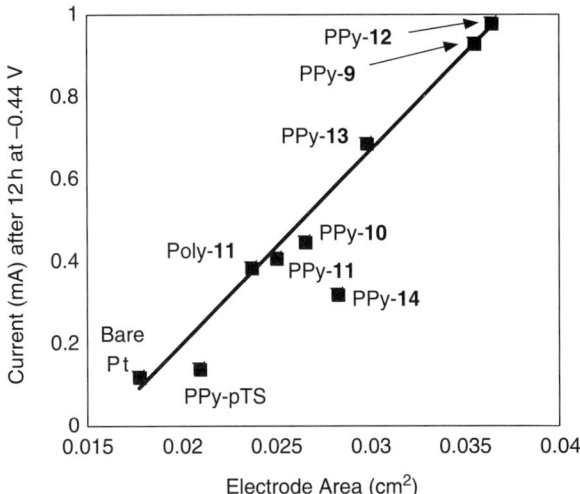

Figure 12.7. Plot of the comparable rate of hydrogen generation in 1 M H_2SO_4 after 12 hours at -0.44 V, of the polymer coatings depicted in Scheme 12.4. Each coating was deposited to 100 mC on a Pt electrode of 0.0177 cm^2.

line using special preparative conditions of the type described in the previous section. Poly-**15**, which contains only a terthiophene polymer, shows no activity whatsoever in hydrogen generation electrocatalysis.

Thus, pyrrole is clearly essential to the catalysis, whereas the relative amount of ferrocene present has, apparently, little overall influence on the catalytic rate (provided some ferrocene is present).

What is especially astonishing about Figure 12.7 is the fact that the rate of hydrogen generation does not increase in proportion to the increase in the electrochemical area. Thus, a doubling in the electrochemical area (from $0.0175 \, cm^2$ to $0.0350 \, cm^2$) results in a ca. 9-fold increase in the catalytic rate (0.10 mA to 0.90 mA).

This disproportionate rate increase indicates that the catalysis in these species is *not* of a typical heterogeneous character. A doubling of the surface area in a heterogeneous catalyst would normally result in a doubling of the overall catalytic rate, not an increase of 9-fold. This outcome occurs because doubling the surface area would double the number of catalytic sites present.

To put this into context for chemists who are unfamiliar with heterogeneous catalysis, imagine a beaker containing a certain quantity of a homogeneous catalyst and its reactants in a particular volume of solution. This beaker will generate products at a certain rate. If one now gets another beaker containing the identical mixture, then the two beakers *must* cumulatively generate product twice as fast as the one beaker alone. The same is true for the surface area of a heterogeneous catalyst since this is where all of the catalysis occurs. If the surface area of a heterogeneous catalyst is doubled, so must the overall rate of product generation.

However, in Figure 12.7, we see a ca. 9-fold increase in the rate of product generation when the surface area is doubled. And yet, the reaction rate is directly and linearly related to the electrochemical area.

The inherently amplifying nature of the system is further illustrated by the graph in Figure 12.8. This shows that the rate of hydrogen generation *per unit area* of each of the coatings in Figure 12.7 *also increases linearly with area*. In other words, the larger the area of the coating, the larger is the reaction rate *per unit area*. Figure 12.8 shows that an increase in the electrochemical area (from $0.0175 \, cm^2$ to $0.0350 \, cm^2$) results in an ca. 3.5-fold increase in the catalytic rate per unit area ($7.5 \, mA/cm^2$ to $25.0 \, mA/cm^2$).

To put this into context, consider again the earlier analogy of the beaker containing a homogeneous catalyst and its reactants. What Figure 12.8 is saying, in effect, is that the more identical beakers of solution one combines together, the faster is the rate of product formation *per beaker present*. This is clearly not a usual relationship.

The only reasonable explanation for these data is that the number of catalytic sites per unit area of the catalyst increases as the electrochemically active area is increased. That is, the number of catalytic sites per unit area is not fixed, but variable, and dependent on the physical conditions of the catalyst.

As noted in Sections 10.2.3 and 10.2.4 of Chapter 10, a feature of a combinatorial "statistical proximity" catalyst is that the number of catalytic sites is not fixed as it is in an intramolecular, time-dependent catalyst. Rather, it is dependent on the physical conditions present. As the concentration of the catalytic groups (or, in this case, the

12.3 CATALYTIC EXPERIMENTS

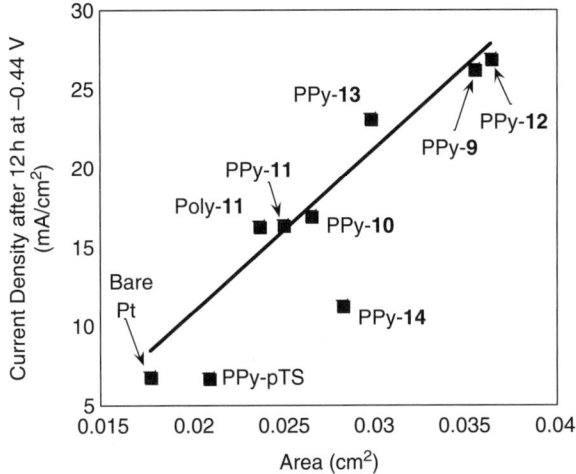

Figure 12.8. Plot of the comparable rate per unit electrochemical area of hydrogen generation in 1 M H_2SO_4 after 12 hours at -0.44 V, of the polymer coatings depicted in Scheme 12.4. Each coating was deposited to 100 mC on a Pt electrode of 0.0177 cm^2.

electrochemical area) is increased, the number of active sites per unit volume (or, in this case, area) will escalate. It is because the likelihood of two catalytic groups being optimally proximate and disposed to each other increases as the concentration (area) increases.

It seems to be precisely what is observed in this case, except that it is a function of the electrochemically accessible area of the polypyrrole, not its volume. In other words, it is a function of the electrochemically active portion of the polypyrrole that can be directly accessed by ions in solution.

12.3.7 The Pyrrole in Polypyrrole Is a Powerful, Time-Dependent, Combinatorial "Statistical Proximity" Catalyst

Pyrrole is known to be protonated in a rapid and reversible equilibrium in 1 M strong acids [16,17]. As polypyrrole, it also undergoes ready redox transformation between a reduced form **16** and an oxidized form **17** as depicted in Figure 12.9 [8]. At an applied potential of -0.44 V, the oxidized form is swiftly converted into the more favored reduced form.

Thus, it seems that the electrochemically active pyrrole units within surface polypyrrole undertake a catalytic cycle of the type shown in Figure 12.9. In such a cycle, pyrrole *N*-atoms are rapidly and reversibly protonated in a dynamic equilibrium. During the very brief period of protonation, the positive charge on the proton may, conceivably, be largely transferred to the *N*-atom, leaving the attached hydrogen as, effectively, highly reactive atomic hydrogen. If, in that brief period of its existence, another such atomic hydrogen species is present within van der Waals distance, the two activated protons will form dihydrogen, H_2.

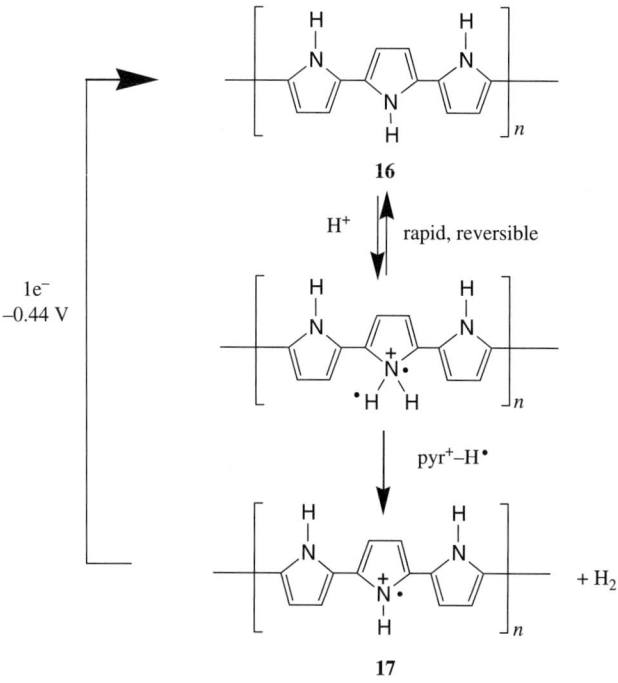

Figure 12.9. A likely mechanism for hydrogen generation catalysis by polypyrrole. The intermediate H–N cation is, presumably, formed only momentarily, with two atomic hydrogen species on separate, adventitiously proximate pyrroles combining to form H_2.

As the electrochemically active surface area of the polypyrrole is increased, more catalytic sites are created per unit area. How could this happen?

Given that the geometric area of the electrode remains fixed (at 0.0177 cm^2), an increase in the electrochemical area must involve more electrochemically active pyrrole units becoming accessible to solution ions within the same overall geometric area. In effect, there must be greater crowding of pyrrole units at the electrochemically active surface of the polypyrrole. This increased crowding clearly yields an increased proportion of pyrrole units that are adventitiously ideally proximate and disposed to each other for formation of an $H \cdots H$ transition state by attached protons.

Figure 12.10 provides a schematic illustration of how this may work. Figure 12.10(a) depicts a smooth electrochemically active surface of fixed, overall geometric area. If the electrochemical area is now progressively increased [as in Fig. 12.10(b)–(d)], without changing the geometric area, then the surface becomes rougher. In the process, more pyrroles are crowded into the same geometric area. This crowding can only be achieved by varying the orientations of the pyrroles. The more varied the orientations, the more catalytic sites that must be created. In other words, the rougher the surface, the more catalytic sites are likely to be present per unit area.

12.3 CATALYTIC EXPERIMENTS   315

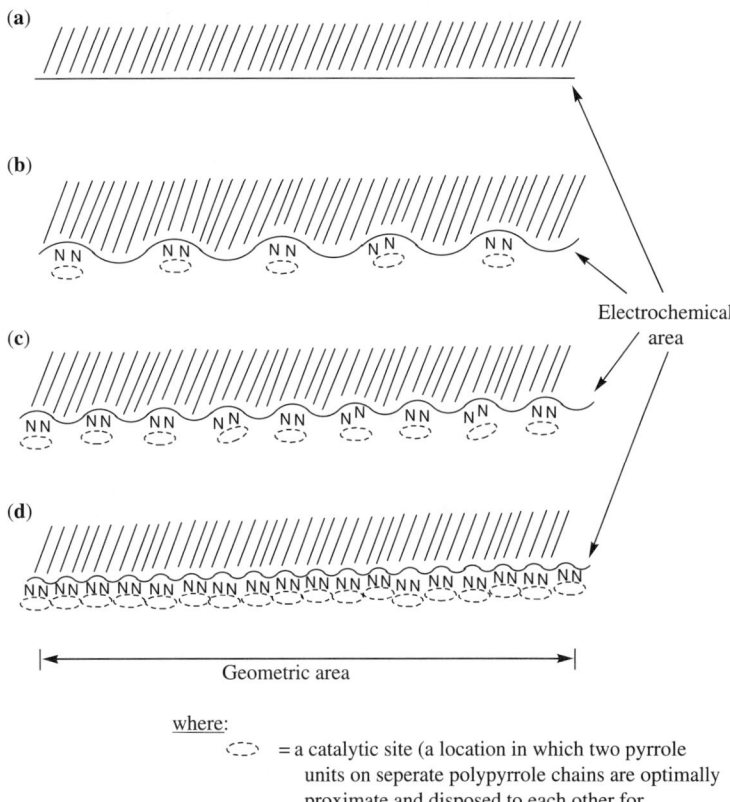

Figure 12.10. Schematic depiction illustrating how the number of time-dependent catalytic sites within a polypyrrole layer deposited on a platinum electrode may increase as the electrochemical area of the polypyrrole is increased within a fixed geometric area. Note that the number of catalytic sites increases per unit electrochemical area. Thus, both the overall catalytic rate and the catalytic rate per unit electrochemical area increases with the electrochemical area.

Thus, because of the sheer density of pyrrole units between separate chains of closely packed polypyrrole, a statistically relevant proportion of the pyrrole units must, necessarily, be adventitiously optimally proximate to each other for the facilitation of such a reaction. The more haphazardly the chains are packed at the surface, the greater the likelihood of optimum arrangement on the part of the pyrrole units, and the larger the number of catalytic sites per unit area. In effect, the polypyrrole seems to act as a bulk-phase, hybrid homogeneous–heterogeneous catalyst.

The ferrocenes undoubtedly also act catalytically. However, because so many more pyrrole units are present than ferrocenes (especially in **9**), the ferrocenes seem to play a decidedly more minor role. The ferrocenes are nevertheless important insofar as their presence seems to favor a more open polypyrrole morphology having a larger electrochemical surface area containing a greater density of pyrrole units.

12.4 CONCLUSIONS: A COMBINATORIAL "STATISTICAL PROXIMITY" CATALYST WAS OBTAINED AS A BULK, HYBRID HOMOGENEOUS-HETEROGENEOUS CATALYST

The mechanism depicted in Figure 12.9 conceptually mirrors the mechanism that operates for the ferrocene groups in [1.1]ferrocenophane as illustrated in Scheme 12.1. Thus, a statistical proximity catalyst was clearly obtained but just not the one that had been originally anticipated. Not only do the ferrocenes act as catalytic groups that momentarily stabilize $H^{\bullet}$ moieties in close proximity to each other as depicted in Scheme 12.6, it seems that the pyrrole groups also perform that function. These groups are sufficiently closely packed on separate polypyrrole chains to exploit the proximity effects present. Since there are many more pyrrole groups in the polypyrroles than there are ferrocenes present, and because pyrrole is more basic than ferrocene, this interaction provides the greater effect. The pyrrole groups are also more overwhelmingly represented at the interface between the coating and the bulk solution, which contains the proton reactants.

We should note here that it is also possible for the Pt electrode to play a role, although this must necessarily be minor from a catalytic point of view. Some $H^{\bullet}$ moieties that combine with each other on polypyrrole may, for example, originate from not only the ferrocene and pyrrole groups present, but also from exposed areas of the Pt surface, which is known to be capable of generating such species. "Spillover"-type mechanisms of this type are known. As discussed in Section 8.2 of Chapter 8, they constitute a form of synergy observed in heterogeneous catalysis. Even where this does occur, the catalytic effect still involves a combinatorial "statistical proximity" effect, since it derives from the closely packed nature of the pyrrole groups on adjacent chains.

The examples described in this chapter, therefore, provide a practical demonstration of the principles discussed in earlier chapters. Time-dependent homogeneous catalysts (or hybrid homogeneous–heterogeneous catalysts in this case) can be harnessed to generate remarkable catalytic effects that are not available using energy-dependent catalysts. Moreover, this may be done in systems that are easily prepared and readily applied. As such, time-dependent catalysis provides important new vistas and opportunities for catalysis science.

ACKNOWLEDGMENTS

The authors thank for their comments and assistance: Chuck Dismukes (Princeton University), David Trimm (CSIRO), and Tony Hollenkamp (CSIRO).

REFERENCES

1. Frey, M. *Chem. Bio. Chem.* **2002**, *3*, 153.
2. Collman, J. P.; Wagenknecht, P. S.; Hutchison, J. E. *Angew. Chem. Int. Ed. Engl.* **1994**, *33*, 1537, and references therein.
3. Mueller-Westerhoff, U. T. *Angew. Chem. Int. Ed. Engl.* **1986**, *25*, 702, and references therein.
4. Mueller-Westerhoff, U. T.; Haas, T. J.; Swiegers, G. F.; Leipert, T. K. *J. Organomet. Chem.* **1994**, *472*, 229, and references therein.
5. Waleh, A., Loew, G. H., and Mueller-Westerhoff, U.T. *Inorg. Chem.* **1984**, *23*, 2859.
6. (a) Karlsson, A.; Hilmersson, G.; Ahlberg, P. *J. Phys. Org. Chem.* **1997**, *10*, 590; (b) Karlsson, A.; Broo, A.; Ahlberg, P. *Can. J. Chem.* **1999**, *77*, 628.
7. Bitterwolf, T. E.; Ling, A. C. *J. Organomet. Chem.* **1973**, *57*, C15.
8. See for example: (a) Dong, S.; Che, G. *J. Electroanal. Chem.* **1991**, *309*, 103; (b) Gulce, H.; Ozyoruk, H.; Yildiz, A. *Ber. Bunsenges, Phys. Chem.* **1994**, *98*, 228.
9. Chen, J.; Swiegers, G. F.; Too, C. O.; Wallace, G. G. *Chem. Commun.* **2004**, 308.
10. (a) Chen, J.; Swiegers, G. F.; Too, C. O.; Wallace, G. G. *Electrochim. Acta.* **2004**, *49*, 691; (b) Chen, J.; Swiegers, G. F.; Too, C. O.; Wallace, G. G. *Electrochim. Acta.* **2002**, *47*, 4227; (c) Chen, J.; Burrell, A. K.; Collis, G. E.; Officer, D. L.; Swiegers, G. F.; Too, C. O.; Wallace, G. G. *Electrochimica Acta* **2002**, *47*, 2715; (d) Weinmayr, V. *J. Am. Chem. Soc.* **1955**, *77*, 3009.
11. Mueller-Westerhoff, U. T.; Nazzal, A. *J. Am. Chem. Soc.* **1984**, *106*, 5381.
12. Nicholson, R. S.; Shane, I. *Anal. Chem.* **1964**, *36*, 706.
13. Woods, R. *Electroanal. Chem.* **1976**, *9*, 1.
14. Gilman, S. *Electroanal. Chem.* **1967**, *1*, 111.
15. Bard, A. J.; Faulkner, L. R. Chap 6 in *Electrochemical Methods*, John Wiley and Sons Inc., 1980.
16. Jackson, A. H. Chap 3.2 in *Pyrroles. Part One. The Synthesis and the Physical and Chemical Aspects of the Pyrrole Ring*, Ed. Jones, R. A., John Wiley and Sons, **1990**, p. 305.
17. Inganas, O.; Erlandsson, R.; Nylander, C.; Lundstrom, I. *J. Phys. Chem. Solids* **1984**, *45*, 427.

13

TIME-DEPENDENT ("MECHANICAL"), NONBIOLOGICAL CATALYSIS. 3. A READILY PREPARED, CONVERGENT, OXYGEN-REDUCTION ELECTROCATALYST

Jun Chen, Gerhard F. Swiegers, Gordon G. Wallace, and Weimin Zhang

13.1 INTRODUCTION

The development of oxygen reduction catalysts for application in H_2/O_2 fuel cells has attracted considerable recent interest [1]. This interest can be attributed to a growing recognition that such cells offer a potential alternative, nonpolluting power source, especially in transportation applications, where their high-power densities allow them to compete with the internal combustion engine. H_2/O_2 fuel cells are currently used in buses and various other public transportation vehicles in several cities around the world [2]. Efforts are being made to extend their utility to automobiles. Finding efficient and, more importantly, *economical* dioxygen reduction catalysts is a significant part of this effort. Much work has focused on the development of new, inexpensive, non-noble metal electrocatalysts.

From a conceptual point of view, two key problems must be dealt with in catalytic dioxygen reduction. The first problem is that cleavage of the O–O bond in O_2 during dioxygen reduction results in the formation of two separate products, which are,

Mechanical Catalysis: Methods of Enzymatic, Homogeneous, and Heterogeneous Catalysis,
Edited by Gerhard F. Swiegers
Copyright © 2008 John Wiley & Sons, Inc.

ideally, water (H_2O) molecules according to the half-reaction:

$$O_2 + 4\,H^+ + 4\,e^- \rightarrow 2\,H_2O \tag{13.1}$$

For catalysis to be truly efficient, the two water molecules that are formed must be stabilized in separate locations during the course of the process. Thus, one typically and ideally requires a catalyst containing two catalytic groups to carry out the transformation. In other words, the optimum molecular catalyst for dioxygen reduction will comprise a dicentered species, in principle at least.

The second problem is that, being a neutral, nonpolar molecule, dioxygen has few handles with which to be bound. Collman considers small, gaseous, nonpolar molecules like O_2 to be "kinetically inert," which means that they typically bind prospective catalysts weakly and transiently [3]. This is, of course, a broad generalization; dioxygen binds certain metal species strongly.

If one puts the above issues together, it becomes clear that the "ideal" molecular catalyst for dioxygen reduction will, in principle, be a dicentered, time-dependent species. Such catalysts are specifically enabled to deal with substrates that bind dynamically to multiple catalytic groups.

A critical feature of such a catalyst would typically be that the two catalytic groups act in a *concerted* (or *coordinated*) fashion. In other words, the two O functionalities should be bound and mechanically pulled apart, in a very specific way that employs the lowest possible energy requirement. This could only occur if the two catalytic groups were disposed to act in a synchronized way. That is, it will only occur if their movements *converge* to create a concerted action.

As noted in Chapters 5 and 6, the best examples of multicentered catalysts that employ convergent catalytic actions are the enzymes of biology. Enzymes typically depend on very particular spatial arrangements and conformational movements in their active site for their catalytic effect [4]. Their need for, and use of, a coordinated action creates the great catalytic specificities for which they are famous.

Manmade, multicentered homogeneous catalysts do not normally act in a convergent manner, because the required synchronization is typically not readily achieved in a simple, molecular species. There is, nevertheless, great interest in developing catalysts that are capable of coordinated actions [5]. The rare few examples that exist in this respect rely on carefully designed and complicated covalent structures to bring about the necessary convergence [6]. However, the complexity of synthesizing such catalysts severely restricts their practical utility and economic viability. Discovering simple and inexpensive means with which to create convergence in multicentered homogeneous catalysis is therefore of particular interest.

In Section 10.2.4 of Chapter 10, we discussed the concept of a combinatorial, multicentered, time-dependent catalyst. We noted that it may be possible to create convergent actions by drastically concentrating dynamically binding, monomeric catalytic groups within a fixed volume. The resulting mixture would then contain a library of all possible structural arrangements, including those that are catalytically active. If such arrangements exist in any measurable proportions, they could conceivably act as highly active catalysts. In such a case, the necessary proximity in the

13.2 COFACIAL DIPORPHYRIN OXYGEN-REDUCTION CATALYSTS

catalytic groups would be created by the statistical fiat of local concentration. We consequently termed such combinatorial experiments "statistical proximity" catalysts.

In Chapter 12, we showed that the combinatorial approach could be employed to prepare highly efficient, time-dependent, hydrogen reduction catalysts. Unfortunately these catalysts were largely formed by the polypyrrole support rather than by the desired ferrocene catalytic groups. The catalysts therefore took the form of bulk, hybrid homogeneous–heterogeneous species.

In the present chapter, we will again examine the concept of a combinatorial time-dependent catalyst with the intention of seeing whether the use of a different monomeric catalytic group for a different reaction will result in a readily prepared, dicentered, time-dependent catalyst. Our choice of reaction and catalytic groups in this respect will be based on Collman's convergent, cofacial diporphyrin oxygen reduction catalysts that were discussed in Section 10.4.1.1 of Chapter 10. Out of this work we will describe an efficient oxygen reduction catalyst in which coordinated, convergent actions by commercially available, monomeric catalytic groups may be made statistically favored by the use of high concentrations. This catalyst is shown to be suitable for use in H_2/O_2 fuel cells.

13.2 COFACIAL DIPORPHYRIN OXYGEN-REDUCTION CATALYSTS

An example of a convergent nonbiological homogeneous catalyst is the cofacial dicobalt porphyrin **1** (Scheme 13.1) [7]. Compound **1** catalyzes the four-electron reduction of O_2 to H_2O at potentials negative of 0.71 V (vs. NHE) at

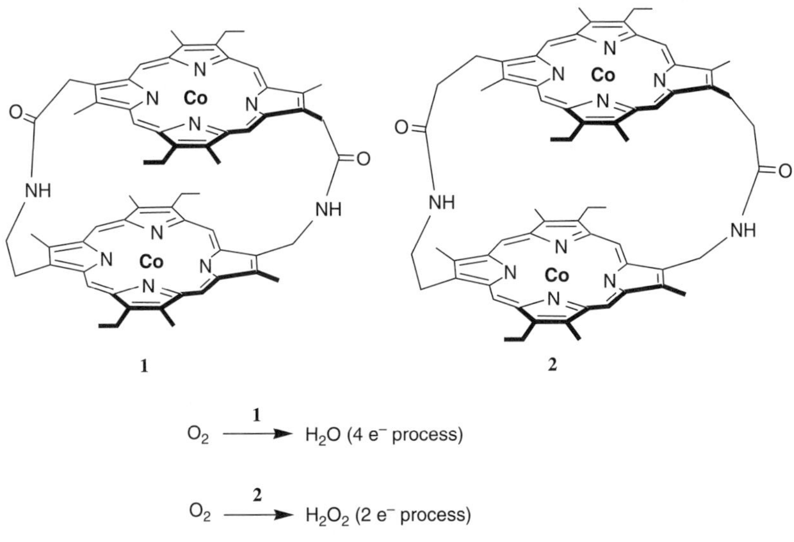

$$O_2 \xrightarrow{\mathbf{1}} H_2O \text{ (4 e}^-\text{ process)}$$

$$O_2 \xrightarrow{\mathbf{2}} H_2O_2 \text{ (2 e}^-\text{ process)}$$

Scheme 13.1.

pH < 3.5 when adsorbed on a graphite electrode [6]. Under identical conditions, the corresponding Co-porphyrin monomer and the equivalent diporphyrin **2** catalyze the two-electron formation of H_2O_2 [7,8].

Extensive studies have revealed that, in this class of catalyst, the O=O molecule must be *simultaneously* bound to both Co centers at the instant of reduction for O–O bond cleavage to be successful [7]. One of the Co centers needs, in fact, only to act as a Lewis acid [9].

Because of the brevity of Co–O_2 binding, the likelihood of this sort of coordinated, synchronized interplay between the Co catalytic groups in the corresponding monomers or in **2** is too low to yield substantial four-electron reduction. Instead, a slower two-electron reduction, involving O_2 bound to a single Co center at the instant of reduction, is favored [7].

Although **1** has potential utility as an oxygen reduction catalyst in H_2/O_2 fuel cells, this is limited, in practice, by the cost and complexity of its synthesis. How, then, can one use the effect present in **1** in an uncomplicated and practical manner?

Several approaches have been proposed in this regard. For example, when adsorbed on graphite, some mononuclear metallo-porphyrins and phthalocynanines that normally catalyze two-electron reduction of O_2 facilitate the four-electron reduction to H_2O in the same way as the cofacial metallodiporphyrins. This has elicited suggestions—as yet unconfirmed—that these species may adsorb in a side-on, pairwise arrangement that makes a coordinated interaction inevitable [10]. In Section 10.4.2.2, we discussed "statistical" approaches of this type to the problem of achieving convergence.

Another proposal involves covalently binding monomers in high densities to the backbone of a polymer, thereby increasing the likelihood of concerted action (see Section 10.4.1.8 in Chapter 10). For example, a cobalt phthalocyanine [10,11] and an iron(III) tetra(o-aminophenyl) porphyrin [10,12] that catalyze two-electron reduction of O_2 in open solution are reported to catalyze four-electron reduction when bound in high densities within polymers. These cases cannot, unfortunately, be unequivocally assigned as concerted, multicentered processes because the covalent connections with the support may alter the electronic structure of the monomer metal, allowing it to facilitate O–O bond unilaterally cleavage without participation by a second monomer. An electron-rich, monomeric, Ru-appended Co-porphyrin has, for example, been shown to catalyze O–O bond cleavage without involvement by a second monomer (see Section 10.4.3 in Chapter 10) [13]. Polymer supports may influence the catalytic process in other ways as well [10].

To eliminate these complications, it is also possible to concentrate the equivalent, free, neutral monomers within a restricted volume in the hope that coordinated action becomes statistically inevitable. A combinatorial, "statistical proximity" effect of this type has been demonstrated in the previous chapter, albeit in a bulk hybrid heterogeneous–homogeneous form [14]. However, it has not been observed for four-electron dioxygen reduction. This may possibly be because no means has existed to immobilize sufficiently high concentrations of neutral Co-porphyrin monomers within polymer layers.

13.3 VAPOR-PHASE POLYMERIZATION OF PYRROLE AS A MEANS OF IMMOBILIZING HIGH CONCENTRATIONS OF MONOMERIC CATALYTIC GROUPS AT AN ELECTRODE SURFACE

The traditional method of immobilizing metalloporphyrins on electrodes is to preload them onto a mesoporous material, such as carbon black particles, and then to coat these onto the electrode [15]. Unfortunately, composite electrodes of this type are insufficiently durable for use in a commercial device.

An attractive alternative is to incorporate the metalloporphyrins into conducting polymers, which have been applied in various electrochemical devices [16–20]. Among all conducting polymers, polypyrrole is the most popular, mainly because of its high electrochemical stability, conductivity, and mechanical robustness in different aqueous environments [21].

In recent work we have discovered a new technique for trapping small molecules within a thin-layer conducting polymer. The process employs a polypyrrole (PPy) film grown using so-called vapor-phase polymerization (VPP) [22]. During VPP, polypyrrole is deposited as a swollen and porous layer on the electrode. When this layer is subsequently washed with a suitable solvent, it undergoes dramatic shrinkage, becoming exceedingly dense. In so doing, it may be used to immobilize and concentrate molecular species within a thin layer of polypyrrole upon an electrode surface [22].

Given the catalytic properties of **1**, the question was whether we could induce a neutral Co-porphyrin monomer to undertake concerted, time-dependent catalysis of four-electron O_2 reduction to H_2O when concentrated within vapor-phase polymerized polypyrrole.

A convenient test species in this respect is commercially available, neutral Co-tetraphenylporphyrin **3** (Scheme 13.2). In monomeric form in open solution, **3**

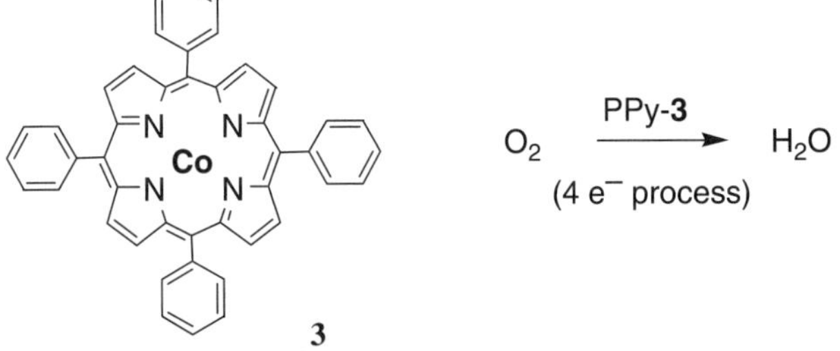

$$O_2 \xrightarrow{\text{PPy-3}} H_2O$$
(4 e⁻ process)

Scheme 13.2.

is known to be exclusively a catalyst of two-electron O_2 reduction to H_2O_2; when adsorbed on graphite, it generates a product mixture containing $>60\%$ H_2O_2 [10].

13.4 PREPARATION AND CATALYTIC PROPERTIES OF PPy-3

13.4.1 Vapor-Phase Preparation of Polypyrrolle-Co Tetraphenylporphyrin, PPy-3

Polypyrrole containing high local concentrations of Co tetraphenyl porphyrin **3** was prepared according to the procedures shown in Figure 13.1. A thin film of Fe(III) tosylate containing **3** was spin-coated onto indium tin oxide (ITO) glass at a speed of 1000 rpm from a tetrahydrofuran solution containing 20% Fe(III) tosylate and 0.5 mM **3**. The resulting glass slide was then exposed to pyrrole vapor, as described previously [22]. A thin film of polypyrrole (PPy) containing Co tetraphenyl-porphyrin **3** was thereby deposited on an ITO glass slide. Subsequent washing with acetonitrile removed the Fe(II) generated during polymerization and caused the polymer layer to shrink dramatically.

13.4.2 Electrochemistry of, and Oxygen Reduction by, Polypyrrolle-Co Tetraphenylporphyrin, PPy-3

Electrochemical characterization of the resulting electrode was performed in a standard three-compartment cell that was maintained under an O_2 or N_2 atmosphere (Fig. 13.2). The counter electrode was a Pt mesh separated from the working electrode by a porous glass frit. Ag/AgCl served as the reference electrode.

Linear sweep voltammograms (LSVs) of the resulting PPy-**3**/ITO electrode were carried out in an aqueous 0.5 M H_2SO_4 electrolyte solution. As shown in Figure 13.3,

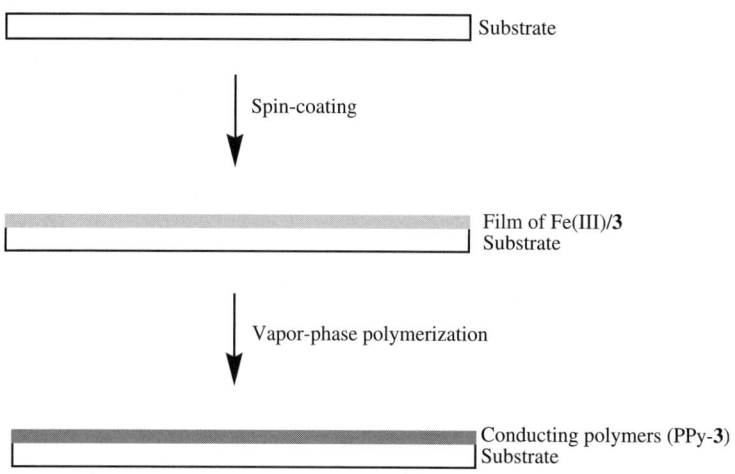

Figure 13.1. Schematic depiction of the steps used in the preparation of PPy-**3**.

13.4 PREPARATION AND CATALYTIC PROPERTIES OF PPy-3

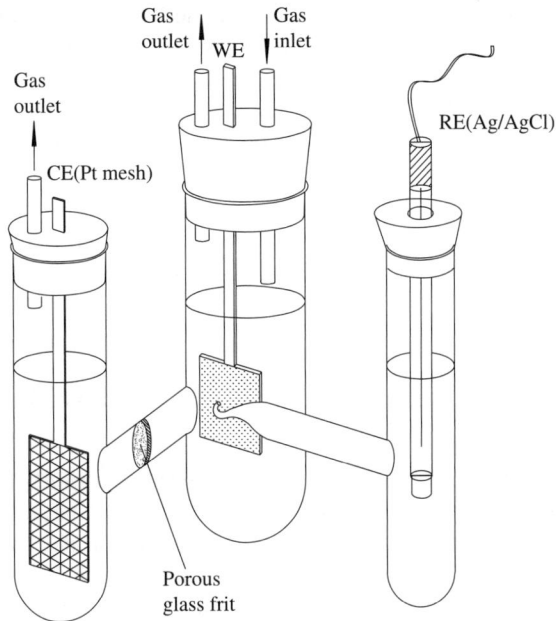

Figure 13.2. Illustration of the cell used to characterize electrochemically and study PPy-3.

no substantial electrochemical response was observed in sweeping from 0.5 V to 0.0 V under an N_2 atmosphere. However, under an O_2 atmosphere, a substantial reduction process negative of ca. 0.41 V was detected. This current flattens out from approximately 0.3 V. As the current is observed only when an O_2 atmosphere is employed, it is unequivocally caused by dioxygen reduction.

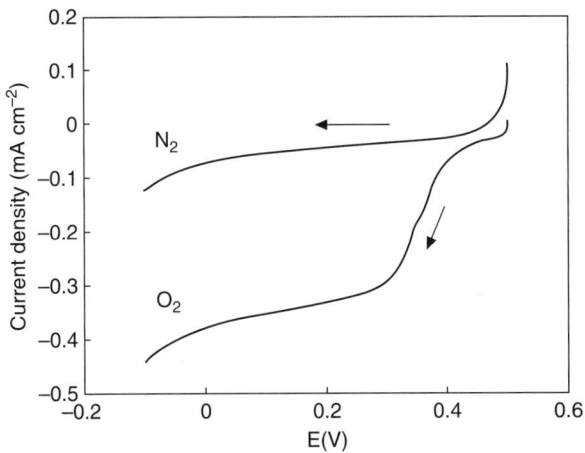

Figure 13.3. Linear sweep voltammograms (vs. Ag/AgCl) for O_2 reduction at a stationery PPy-3/ITO glass electrode in 0.5 M H_2SO_4. Scan rate 5 mV/s.

The profile of the electrochemical response under O_2 in Figure 13.3 is somewhat unusual. First-order reactions under diffusion control typically generate a clear and well-defined *peak* in their electrochemical response, not a *plateau* of the type shown in Figure 13.3 under O_2. Plateau-type currents under linear sweep conditions are more typical of reactions at steady state on, for example, a rotating disk electrode, a single microelectrode, or a microelectrode array.

This result seems to be consistent with a combinatorial "statistical proximity" catalyst, which would be expected to form a microarray of a few highly active H_2O-generating electrocatalytic sites surrounded by a sea of less active H_2O_2-generating sites.

13.4.3 Rotating Disk Electrochemistry (RDE) of Polypyrrolle-Co Tetraphenylporphyrin, PPy-3

To determine the number of electrons transferred during this O_2 reduction, linear sweep voltammograms were collected at different rotation speeds for vapor-phase polymerized PPy-3 similarly deposited on a glassy carbon electrode. Figure 13.4 shows the linear sweep voltammograms of dioxygen reduction at the PPy-3 modified glassy carbon electrode at different rotation speeds from 250 rpm to 1500 rpm.

The number of electrons involved in the dioxygen reduction was calculated using the Koutecky–Levich equation. This equation is valid for a first-order process with respect to the diffusion species, and the current i is related to the rotation rate of the electrode ω according to

$$\frac{1}{i} = \frac{1}{i_k} + \frac{1}{B\omega^{0.5}} \tag{13.2}$$

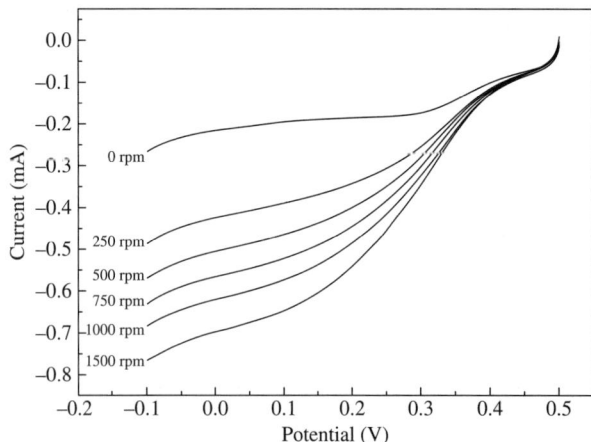

Figure 13.4. Linear sweep voltammograms for oxygen reduction at PPy-3/glassy carbon in O_2 saturated 0.5 M H_2SO_4/H_2O at different rotation speed. Scan rate 10 mV/s.

13.4 PREPARATION AND CATALYTIC PROPERTIES OF PPy-3

where i_k is the kinetic current and B is the Levich slope:

$$B = 0.2nFA(D_{O_2})^{2/3}v^{-1/6}C_{O_2} \tag{13.3}$$

In the above equations, n is the number of electrons transferred in the reduction of one O_2 molecule, F is the Faraday constant ($F = 96485$ C/mol), A is the geometric area of the electrode ($A = 0.24$ cm^2), D_{O_2} is the diffusion coefficient of O_2 ($D_{O_2} = 2.1 \times 10^{-5}$ cm^2 s^{-1}), v is the kinematic viscosity for sulfuric acid ($v = 1.07 \times 10^{-2}$ cm^2 s^{-1}), and C_{O_2} is the concentration of O_2 in the solution ($C_{O_2} = 1.03 \times 10^{-3}$ mol dm^{-3}). The constant 0.2 is adopted when the rotation speed is expressed in rpm.

The kinetic current i_k may be obtained by extrapolation of the Koutecky–Levich plots for $\omega^{-1/2} \to 0$. These plots are depicted in Figure 13.5. The excellent fit for a Koutecky–Levich rather than a Levich plot indicates that the currents are not controlled by O_2 diffusion within the polymer.

Table 13.1 depicts representative data, including the number of electrons transferred during O_2 reduction in the potential range -50 mV to 100 mV. As shown, between 3.3 and 4.0 electrons are involved in the reaction. This rate corresponds to a product distribution of 65–100% H_2O and 35–0% H_2O_2, respectively.

The fact that four-electrons are involved indicates that the reduction comprises direct conversion of O_2 to H_2O at a single catalytic site by, effectively, a concerted

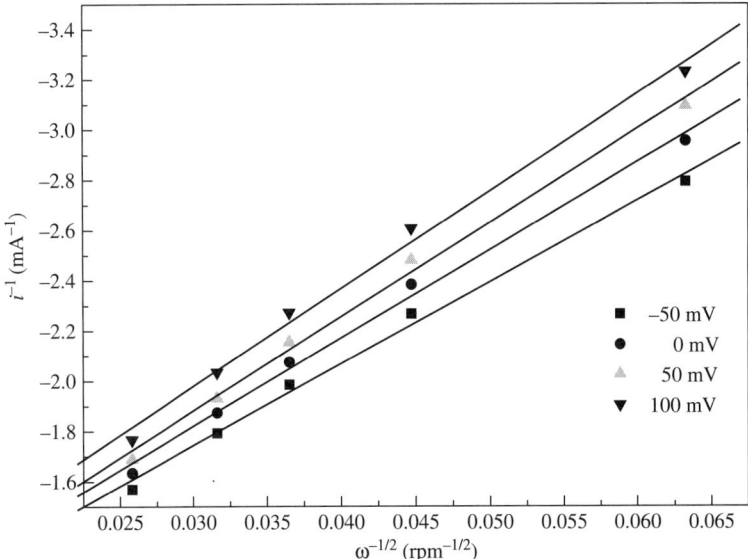

Figure 13.5. Koutecky–Levich plots for oxygen reduction at PPy-3/glassy carbon in 0.5 M H_2SO_4. Data taken from Figure 13.4.

TABLE 13.1. Slopes k_i of the Koutecky–Levich Plots and Corresponding Number of Electrons Transferred for the Catalytic Reduction of Oxygen at Polypyrrole-3 in O_2-Saturated 0.5 M H_2SO_4 Solution

i	E_i/mV vs. Ag/AgCl	k_i	$1/i_k$ (mA^{-1})	n_i
1	−50	32.4	−0.771	4.0
2	0	35.0	−0.769	3.7
3	50	37.4	−0.759	3.5
4	100	38.8	−0.815	3.3

process; this may possibly involve a short-lived, bound H_2O_2 intermediate. It cannot, however, involve two, two-electron reductions at separate sites, in a sequence of $O_2 \rightarrow H_2O_2$ followed by $H_2O_2 \rightarrow H_2O$.

The variation in product distribution as a function of applied voltage may be from differences in the rate of electron transfer, which have previously been proposed to influence product distribution [7,13].

13.4.4 Rotating Ring Disk Electrochemistry (RRDE) of Polypyrrolle-Co Tetraphenylporphyrin, PPy-3

These results were separately confirmed by a controlled rotating ring disk electrochemical (RRDE) study that employed Pt ring and disk electrodes rotating at 1500 rpm.

In the RRDE technique, any product formed at the inner disk electrode necessarily moves by convection during rotation over the outer ring electrode and is therefore separately detectable and quantifiable. As such, RRDE offers a powerful method to confirm the number of electrons and the percentage of peroxide involved in the dioxygen reduction.

In our studies, the ring electrode was poised at 1.0 V (vs. Ag/AgCl) in order to oxidize any H_2O_2 released during O_2 reduction at the disk electrode [7]. As H_2O is not oxidized at this potential, a comparison of the ring and the disk currents allows one to distinguish the proportion of H_2O_2 that is generated during the reduction. The released H_2O_2 travels by convection from the disk to the ring electrode during rotation [7].

The collection efficiency of the ring disk was measured in an independent experiment with the $Fe(CN)_6^{3+/4+}$ couple and was found to be $N_0 = 0.22$. Thus, 22% of the products released by the disk are detected at the ring.

Using RRDE, a linear sweep voltammogram was applied at the disk (Fig. 13.6). At 200 mV, the ring current (oxidation of H_2O_2) was 0.00837 mA, whereas the disk current was 0.275 mA (overall O_2 reduction). At 50 mV, a current of 0.011 mA was observed on the ring electrode, whereas the disk electrode produced a current

13.4 PREPARATION AND CATALYTIC PROPERTIES OF PPy-3

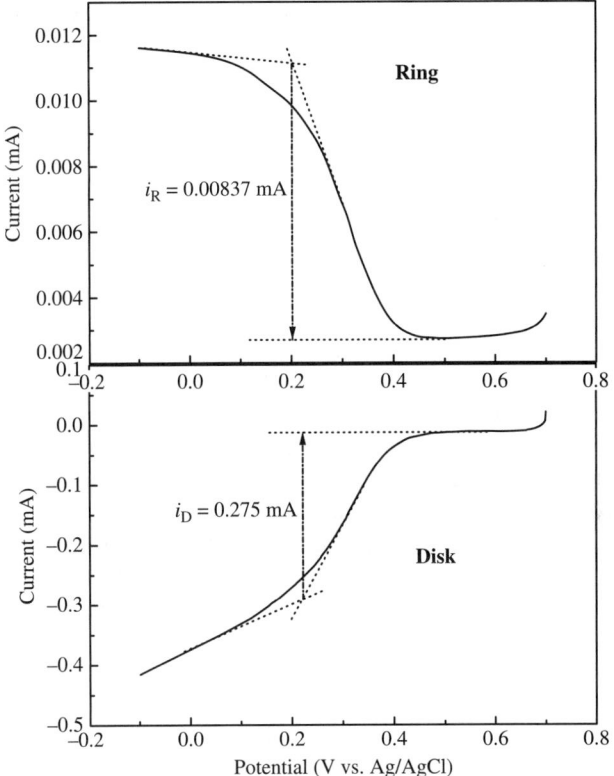

Figure 13.6. RRDE votammetry of oxygen reduction at a platinum electrode modified with PPy-3 in O_2 saturated 0.5 M H_2SO_4. Scan rate 10 mV/s.

of 0.352 mA. The low current observed at the ring confirms the efficiency of the four-electron reduction of O_2 to H_2O [7], which corresponds to at least 86% of the charge consumed at the disk electrode being involved in the conversion of O_2 to H_2O. The remaining 14% must be involved in the transformation of O_2 to H_2O_2.

13.4.5 The Product Distribution Relative to the Proportion of 3 in the Polypyrrolle-Co Tetraphenylporphyrin, PPy-3

Studies also examined the influence on the H_2O_2:H_2O product distribution of the relative proportion of **3** in the PPy-**3** layer. The data in Table 13.1 were obtained using PPy-**3** that had been deposited from an initial solution containing 0.5 mM **3**. However, PPy-**3** prepared from a five-fold more dilute solution (0.1 mM **3**) yielded substantially larger H_2O_2:H_2O ratios, of around 1:3 at -50 mV. Under these circumstances, nearly 26% of charge consumed is involved in the conversion of

O_2 to H_2O_2. By contrast, Table 13.1 shows that PPy-**3** deposited from an initial solution of 0.5 mM **3** yielded 100% formation of H_2O at -50 mV.

13.5 PPy-3 AS A FUEL CELL CATALYST

13.5.1 PPy-3 on Carbon Fiber Paper

As a very preliminary test of the durability and utility of PPy-**3** in a fuel cell role, we used vapor-phase polymerization to deposit PPy-**3** onto mesoporous carbon fiber paper (CFP) (Ballard Material Products Inc., Lowell, MA). Carbon fiber paper is a useful cathode substrate material because of its electrochemical and physical stability.

A solution containing Fe(III) tosylate and **3** was prepared by mixing a 20% Fe(III) tosylate/THF solution with a 5 mM solution of **3** (1:1 v:v). A piece of carbon fiber paper (2 cm × 5 cm) was then immersed in the above solution. When it was subsequently withdrawn, a thin film of Fe(III) tosylate/PPy-**3** was coated on the paper. The coated paper was dried in an oven (80–100 °C) for several minutes, after which it was placed for 15 min in a chamber filled with pyrrole monomer vapor. After the vapor-phase polymerization, the PPy-**3**-modified carbon fiber paper was washed with ethanol to remove the Fe(II) and Fe(III) from the film and to obtain a high density of PPy-**3**, which seemed to adhere strongly to the carbon fiber paper. The extent of adherence was found to affect strongly the long-term electrocatalytic performance of the coated paper during oxygen reduction.

13.5.2 Electrochemical Characterization of PPy-3 on Carbon Fiber Paper

The resulting PPy-**3**-modified carbon fiber electrode was characterized by cyclic voltammetry to study its electrochemical activity. Figure 13.7 shows the cylic voltammogram of PPy-**3**/Carbon Fiber Paper in 1.0 M $NaNO_3/H_2O$ solution. The typical polypyrrole response is observed. Polypyrrole was therefore successfully deposited onto carbon fiber paper with acceptable electroactivity. The Raman spectra of the modified carbon fiber paper electrode also indicated the presence of polypyrrole (insert in Fig. 13.7).

13.5.3 Morphology of the PPy-3 Carbon Fiber Composite Film

A scanning electron microscopy image of the modified carbon fiber paper electrode is shown in Figure 13.8. A thin coating of PPy-**3** covers the individual carbon fibers. The overall porous microstructure of the carbon fiber paper was retained. The deposited PPy-**3** film on each carbon fiber is therefore accessible to the dioxygen reductant. Such accessibility is normally a key parameter for electrocatalytic performance during oxygen reduction in fuel cells.

The distribution of **3** within the polypyrrole film deposited on the carbon fiber paper electrode was studied using SEM EDX-ray mapping. Figure 13.8 shows a typical SEM EDX-ray mapping image, where the bright spots correspond to Co

13.5 PPY-3 AS A FUEL CELL CATALYST

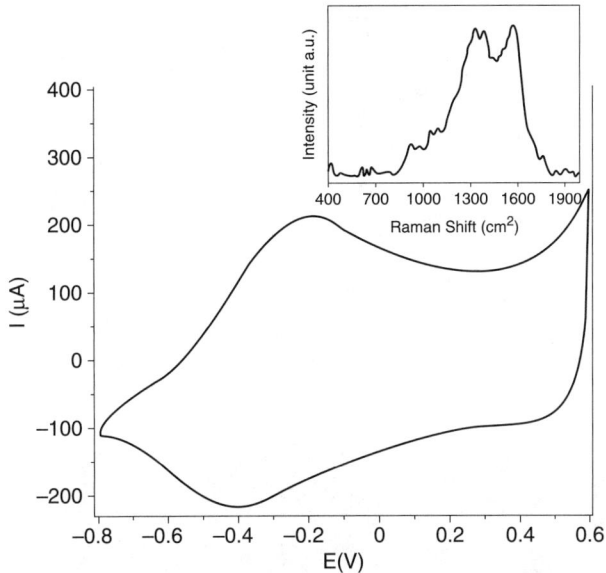

Figure 13.7. Cyclic voltammogram of PPy-3 modified carbon fiber paper electrode in 1.0 M NaNO$_3$/H$_2$O solution at the scan rate of 50 mV s^{-1}. The insert is the Raman spectra of PPy-3 with 632.8-nm diode laser excitation on a 900 lines/mm grating at room temperature.

sites incorporated within the porphyrin units. The Co tetraphenylporphyrin **3** was therefore uniformly immobilized in the coating without significant clustering.

13.5.4 Oxygen-Reduction Catalysis by the PPy-3 Carbon Fiber Composite Film in Simple Fuel Cell Test Apparatus

Oxygen reduction catalysis of the PPy-3-modified carbon fiber paper electrode immersed (and stationary) in an aqueous 0.5 M H$_2$SO$_4$ solution was studied using linear sweep voltammetry. As shown in Figure 13.9, a substantial reduction process was observed under an O$_2$ atmosphere. No such response occurred under an N$_2$ atmosphere.

Additionally, the consistent increase in the reduction current as the voltage is scanned toward more negative potentials is consistent with the previous finding (Fig. 13.3) that the reaction displays the character of a steady-state system, similar to that which may be generated by a rotating disk, a microelectrode, or a microelectrode array. It does this even though it is maintained stationary in the testing apparatus.

The PPy-3-modified carbon fiber paper was then tested in the simple fuel cell-type test apparatus depicted in Figure 13.10. A zero bias was applied to the carbon fiber electrode, which was pressurized in the right-hand side half of the cell for 2–5 min with a gas diffusion layer at 50 kg cm^{-2}. The gas diffusion layer was produced by

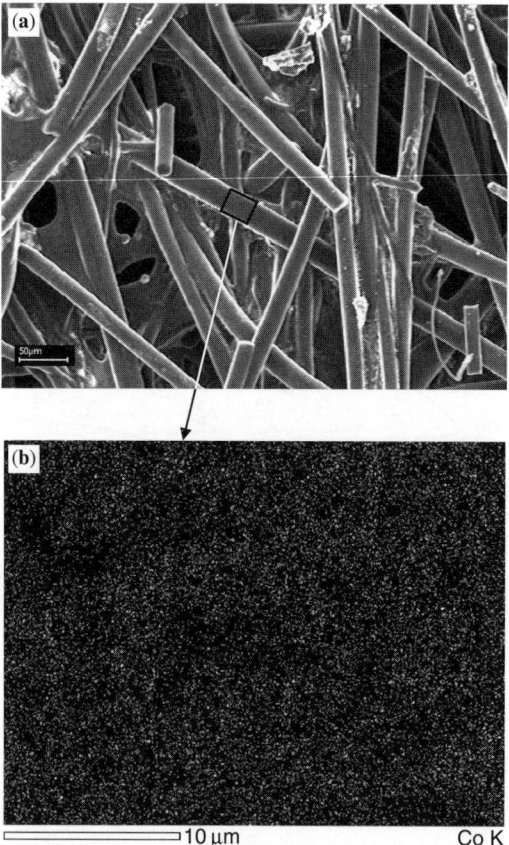

Figure 13.8. (a) SEM image of PPy-**3** modified carbon fiber paper electrode; (b) the EDX-ray mapping image of the Co distribution in the polypyrrole matrix.

a rolling method from a well-dispersed composite gel containing 62.5 wt% carbon black (Vulcan XC 72, Cabot) and 37.5 wt% PTFE.

The left half of the cell was filled with 0.5 M H_2SO_4 aqueous solution under an air atmosphere. A constant oxygen flow of 200 ml min^{-1} was maintained in the right-hand half of the cell (without any liquid solution) throughout the experiment.

The current density versus time pilot for the modified electrode is shown in Figure 13.11. After the initial decline, the oxygen reduction current at PPy-**3**/carbon fiber paper electrode leveled off, yielding a current density of 1.083 mA cm^{-2} after 15 h. The generation of a current of this size and persistence under a zero applied bias can only be caused by a sustained catalytic process.

This was confirmed by a control experiment, in which unmodified carbon fiber paper without PPy-**3** was investigated under identical conditions. After 15 h, it generated a current density of only 0.008 mA cm^{-2}.

13.5 PPY-3 AS A FUEL CELL CATALYST

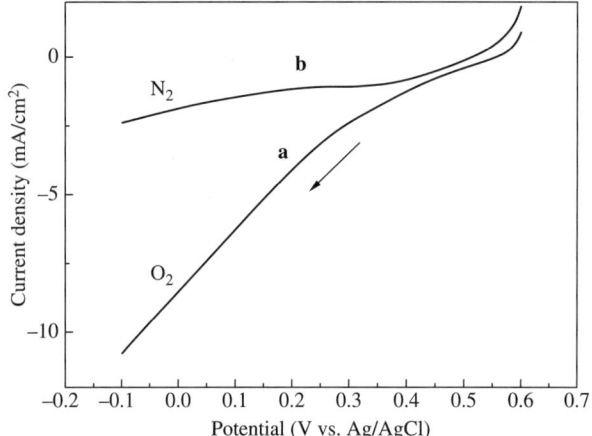

Figure 13.9. Linear sweep voltammograms (vs. Ag/AgCl) for oxygen reduction at PPy/Co-TPP modified CFP electrode in 0.5 M H_2SO_4 aqueous solution. Scan rate: $10\,\mathrm{mV\,s^{-1}}$.

The PPy-**3**-modified carbon fiber paper electrode, therefore, demonstrated a stable and sustained electrocatalytic activity for oxygen reduction. No obvious degradation to the polypyrrole film on the carbon fiber paper and no visible color change in the electrolyte was observed over 15 h. The polypyrrole film clearly provides both mechanical and electrochemical stability.

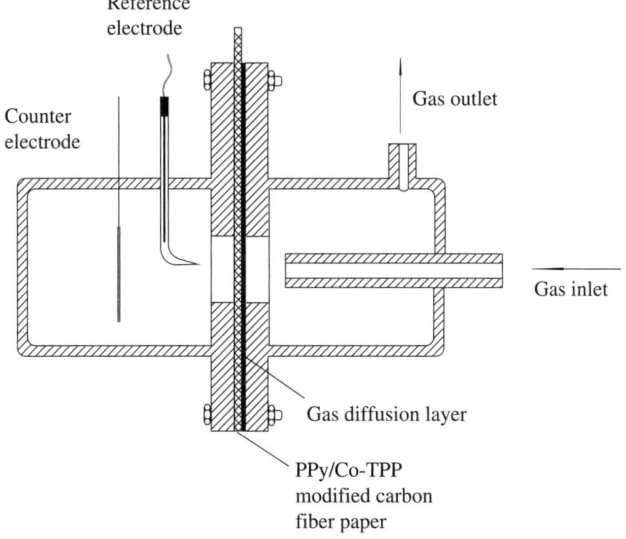

Figure 13.10. Schematic structure of the half-fuel cell system (PPy/Co-TPP = PPy-**3**).

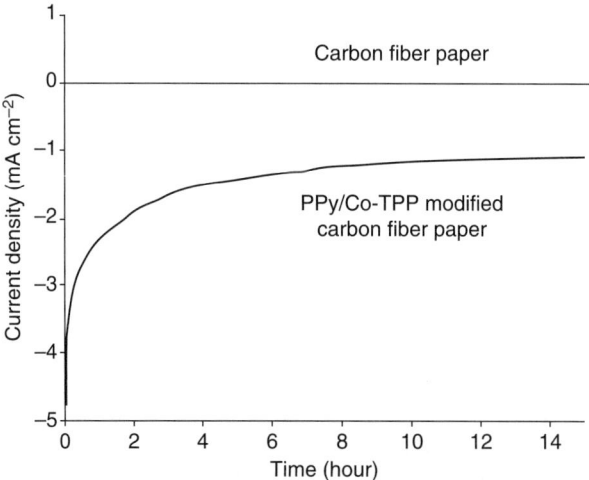

Figure 13.11. Current density of PPy-3 (PPy/Co-TPP) modified CFP electrode and pure carbon fiber paper electrode at zero bias for catalytic oxygen reduction in 0.5 M H_2SO_4 aqueous solution under O_2 gas flow, 200 ml/min.

13.6 CONCLUSIONS

These data indicate that, whereas monomeric **3** mainly facilitates two-electron reduction of O_2 to H_2O_2 in open solution or when adsorbed on graphite, it generates predominantly H_2O by four-electron reduction when trapped in a concentrated form within polypyrrole.

Moreover, the proportion of H_2O_2 produced declines as the concentration of **3** in the polypyrrole is increased. This finding is consistent with the neutral Co-tetraphenylporphyrin becoming statistically more disposed to a coordinated bimolecular interaction that favors four-electron H_2O formation upon concentration within polypyrrole.

Additionally, the apparent steady-state response observed in Figure 13.3 is consistent with, effectively, a microarray of a few highly active H_2O-generating electrocatalytic sites surrounded by a sea of less active H_2O_2-generating sites. This precise scenario is predicted to exist in a combinatorial, "statistical proximity" catalyst.

At this stage we cannot definitively say whether polypyrrole is involved in the catalytic process (as a possible Lewis acid?). Co-polypyrrole composites have recently been shown to offer useful catalysts in H_2–O_2 fuel cells [23].

Nor can we say whether H_2O_2 is a short-lived intermediate in the catalysis. The reduction potential of O_2 in Figure 13.3 (0.41 V vs. Ag/AgCl) is similar to two-electron reduction of O_2 to H_2O_2 ($E° = 0.68$ V vs. NHE), suggesting the presence of such an intermediate.

Detailed investigations are ongoing and will be described elsewhere.

REFERENCES

1. Brents, D. J.; Scullin, V. J.; Chang, B.-J.; Johnson, W.; Garcia, C. P. *NASA Report TM-2005-213381*.
2. Eudy, L.; Parish, R.; Leonard, J. *Proc. 2001 DOE Hydrogen Program Review, Paper NREL/CP-570-30535*.
3. Collman, J. P.; Wagenknecht, P. S.; Hutchison, J. E. *Angew. Chem. Int. Ed. Engl.* **1994**, *33*, 1537, and references therein.
4. See for example: (a) Hammes-Schiffer, S.; Benkovic, S. J. *Annu. Rev. Biochem.* **2006**, *75*, 519; (b) Benkovic, S. J.; Hammes-Schiffer, S. *Science* **2003**, *301*, 1196.
5. Breslow, R. *Acc. Chem. Res.* **1995**, *28*, 146, and references therein.
6. See for example: Breslow, R.; Dong, S. D. *Chem. Rev.* **1998**, *98*, 1997.
7. Collman, J. P.; Wagenknecht, P. S.; Hutchison, J. E. *Angew. Chem. Int. Ed. Engl.* **1994**, *33*, 1537, and references therein.
8. (a) Collman, J. P.; Denisevich, P.; Konai, Y.; Marrocco, M.; Koval, C.; Anson, F. C. *J. Am. Chem. Soc.* **1980**, *102*, 6027; (b) Collman, J. P.; Marrocco, M.; Denisevich, P.; Koval, C.; Anson, F. C. *J. Electroanal. Chem. Interfac. Electrochem.* **1979**, *101*, 117.
9. (a) Collman, J. P.; Hendricks, N. H.; Kim, K.; Bencosme, C. S. *J. Chem. Soc., Chem. Commun.* **1987**, 1537; (b) Liu, H. Y.; Abdalmuhdi, I.; Chang, K. C.; Anson, F. C. *J. Phys. Chem.* **1985**, *89*, 665; (c) Collman, J. P.; Kim, K. *J. Am. Chem. Soc.* **1986**, *108*, 7847.
10. (a) Vasudevan, P.; Santosh; Mann, N.; Tyagi, S. *Transition Met. Chem.* **1990**, *15*, 81, and references therein; (b) Abe, T.; Kaneko, M. *Prog. Polym. Sci.* **2003**, *28*, 1441, and references therein; (c) Chang, C. K.; Liu, H. Y.; Abdalmuhdi, I. *J. Am. Chem. Soc.* **1984**, *106*, 2725.
11. Behret, H.; Binder, H.; Sandstede, G.; Scherer, G. G. *J. Electroanal. Chem.* **1981**, *117*, 29.
12. Bettelheim, A.; Chan, R. J. H.; Kuwana, T. *J. Electroanal. Chem.* **1979**, *99*, 391.
13. Anson, F. C.; Shi, C.; Steiger, B. *Acc. Chem. Res.* **1997**, *30*, 437, and references therein.
14. Chen, J.; Huang, J.; Swiegers, G. F.; Too, C. O.; Wallace, G. G. *Chem. Commun.* **2004**, 308.
15. Deng, Z.; Wang, Z.; Li, C. M.; Cha, C. S. *Acta Chim. Sinica.* **1987**, *47*, 260.
16. Skotheim, T. A.; Elsenbaumer, R. L.; Reynolds, J. R. *Handbook of Conducting Polymers*, Dekker, 1998.
17. Dong, H.; Li, C. M.; Chen, W.; Zhou, Q.; Zeng, Z.; Luong, J. H. T. *Anal. Chem.* **2006**, *8*, 7424.
18. Park, W.; Yeo, I. H.; Suh, H.; Kim, Y.; Song, E. *Electrochim. Acta* **2000**, *45*, 3833.
19. Kvarnstrom, C.; Neugebauer, H.; Blomquist, S.; Ahonen, H. J.; Kankare, J.; Ivaska, A. *Electrochim Acta* **1999**, *44*, 2739.
20. Cosnier, S.; Gondran, C.; Wesesl, R.; Montforts, F. P.; Wedel, M. *J. Electroanal. Chem.* **2000**, *488*, 83.
21. Wallace, G. G.; Spinks, G. M.; Teasdale, P. R. *Conducting Electroactive Polymers*, Technomic Publishing Company, 1997.
22. (a) Winther-Jensen, B; Chen, J.; West, K.; Wallace, G. G. *Macromol.* **2004**, *37*, 5930; (b) Chen, J.; Winther-Jensen, B.; Lynam, C.; Nagmna, O.; Moulton, S.; Wallace, G. G. *Electrochemical and Solid-State Letters* **2006**, *9(7)*, H68.
23. Bashyam, R.; Zelenay, P. *Nature* **2006**, *443*, 63.

APPENDIX A

WHY IS SATURATION NOT OBSERVED IN CATALYSTS THAT DISPLAY CONVENTIONAL KINETICS?

In this appendix we provide a brief explanation for the rarity of saturation in conventional kinetics. For those readers interested in a more detailed discussion, a graphical illustration of the phenomenon of saturation is offered in Appendix B.

Consider reactions involving a single reactant and having *conventional kinetics* of the type depicted in Equation (5.3) of Chapter 5 (and reprinted here), with k_2 as the rate-determining step (C = catalyst, R = reactant, P = product):

$$C + R \xrightarrow{k_1} [CR] \xrightarrow{k_2} C + P \qquad (5.3)$$

In such a reaction, the catalyst effectively exists in only two forms:

(A) A *bound, reacting form*, where an attached reactant is being converted into the product, which will then be released.
(B) A *free form*, where after releasing the product, the catalyst waits for the next reactant to bind.

The bound, reacting form (A) has a fixed, average lifetime after each occasion of reactant binding. This average lifetime is essentially invariant and unaffected by the concentration of the reactant in the surrounding solution. That is, once the reactant binds the catalyst, there is a fixed average time before it is converted into product; this is unaffected by the concentration of reactant in the surrounding solution.

An increase in the reactant concentration about such a catalyst, therefore, only has the effect of shortening the lifetime of the free form (B) after each occasion of product

---

Mechanical Catalysis: Methods of Enzymatic, Homogeneous, and Heterogeneous Catalysis,
Edited by Gerhard F. Swiegers
Copyright © 2008 John Wiley & Sons, Inc.

release. In other words, only one variable in the catalytic process is affected by changes in the reactant concentration. Such catalysts start becoming saturated when the average lifetime of form (B) approaches or is smaller than the average lifetime of form (A). At this stage, incremental increases in reactant concentration [with parallel decreases in the lifetime of form (B)] have an ever-decreasing effect on the overall rate. For most catalysts, this will occur only at a high *absolute concentration* of reactant.

Now consider a catalyst that displays *Michaelis–Menten kinetics* (also known as "saturation kinetics"). This characteristic originates from the rapidly equilibrating nature of the catalyst – reactant complex, as depicted in Equation (5.2) of Chapter 5 and reprinted here in a form that employs C = catalyst, R = reactant, P = product:

$$C + R \xrightleftharpoons{K} [CR] \xrightarrow{k_2} C + P \qquad 5.2(a)$$

In this scheme, the catalyst is constantly binding and dissociating reactant (substrate) before product formation. This binding and dissociation process forms part of the rate-limiting step.

In addition to forms (A) and (B), a third form of the catalyst exists:

(C) A *bound, inert form*, in which the catalyst binds the reactant only to then release it before reaction.

Catalysts displaying Michaelis–Menten kinetics will undergo a certain number of bind-release steps, on average, before the step of product formation is undertaken. Under these circumstances, an increase in the substrate concentration will have the following two effects:

(i) It will increase the frequency of these bind-release steps, thereby decreasing the time needed to get from the first occasion that a new reactant binds the catalyst to the point of product formation.
(ii) It will decrease the time between a product leaving the catalyst and the next reactant attaching to it.

Two processes are therefore affected by reactant concentration, not *one*. Because of the cumulative time needed for the bind-release steps, the first of these processes will, necessarily, have a more pronounced effect on the overall rate than the second, whose lifetime is much shorter. When the first process is maximized, a change in the order of the reaction will consequently be observed.

Thus, saturation occurs when process (i) above can go no faster. This will occur at a certain excess of the reactant over the catalyst; that is, at a certain *relative concentration* of the substrate in respect of the catalyst. As the time between a product leaving the catalyst and the next reactant attaching to it is small relative to the cumulative time needed to form the product, process (ii) will then have an ever-decreasing influence on the overall rate. Catalysts of this type therefore become saturated at a relatively low reactant concentration, especially when small concentrations of catalysts are employed.

This is precisely why Michaelis and Menten, and later Haldane and Briggs, had to postulate an equilibrating intermediate in enzyme kinetics.

The difference between the Michaelis–Menten scheme in Equation (5.2) and the conventional kinetics in Equation (5.3) is therefore that the former involves rapid reactant binding and release steps, whereas the latter does not.

APPENDIX B

GRAPHICAL ILLUSTRATION OF THE PROCESSES INVOLVED IN THE SATURATION OF MOLECULAR CATALYSTS

It is not necessary to the arguments presented in Chapter 5 or Appendix A to understand why saturation kinetics is more common in molecular systems involving rapidly equilibrating intermediates. The fact that enzymes are known to involve such intermediates tells us that they are time-dependent catalysts. However, for those readers wishing to understand the connection between saturation and rapidly equilibrating intermediates, we present below an attempt to illustrate the processes involved. To do this we employ comparative graphical representations of the so-called "transit times" of the events that take place during catalysis. Each figure below is drawn to the same scale, so that the represented transit times in the different figures can be compared in a simple and direct manner.

B1 CONVENTIONAL KINETICS

In reactions of the following type from Equation (5.3) of Chapter 5 (C = catalyst, R = reactant, P = product):

$$C + R \xrightarrow{k_1} [CR] \xrightarrow{k_2} C + P \qquad (5.3)$$

Mechanical Catalysis: Methods of Enzymatic, Homogeneous, and Heterogeneous Catalysis,
Edited by Gerhard F. Swiegers
Copyright © 2008 John Wiley & Sons, Inc.

APPENDIX B

Two forms of the catalyst exist:

(A) A *bound, reacting form*, where an attached reactant is being converted into product, which will be released (depicted as a thick black line in the figures below).
(B) A *free form*, where after releasing the product, the catalyst waits for the next reactant to bind (depicted as a broken, dashed line in the figures below).

To illustrate saturation in such a system, consider Figure B1. Schemes (a) and (b) depict, from left to right, the typical (average) timeline of events that occur for one particular catalyst in solution.

The dashed lines depict periods during which the catalyst is unbound and free [that is, periods during which form (B) above exists].

The thick black lines depict periods during which the reactant is bound to the catalyst and undergoing reaction (that is, form (A) exists).

At the end of each thick black line, an asterisk indicates the point of product release.

Thus, starting on the far left in Figure B1(a), the catalyst is free. At point (i), it binds the reactant. At point (ii), it releases the product. Between point (ii) and the next point (i), it is free. At that next point (i), it binds another reactant. At the following point (ii), it releases another product.

The average lifetime of the bound, reacting form (A) is the time t_A between points (i) and (ii). This is invariant and the same as the average time between all other points (i) and (ii).

The average lifetime of the free form after each occasion of product release is the time between the first point (ii) and the second point (i), t_B. In Figure B1, this time is

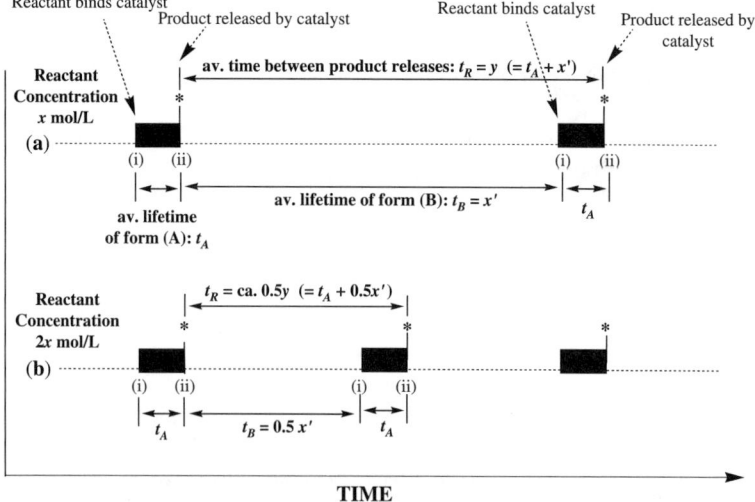

Figure B1. Transit times in a catalyst system displaying conventional kinetics within a moderately dilute solution of reactant.

much longer than t_A. The lifetime of form (B), t_B, is inversely related to the concentration of the reactant in the solution. The more concentrated the reactant, the shorter, on average, will be t_B after each occasion of product release. In (a), the average lifetime of form (B) has been labeled as $t_B = x'$. This lifetime is related to the reactant concentration, which is x mol/L.

The time between product releases is labeled t_R. This time is inversely related to the observed rate of the reaction. The shorter t_R, the greater will be the observed reaction rate. In scheme (a), t_R is labeled as $t_R = y = t_A + x'$.

Scheme (b) depicts the same sequence of events as in (a), but at double the reactant concentration, which is now $2x$ mol/L. The doubling in the reactant concentration means that the average lifetime of form (B) in scheme (b) must, necessarily, be half that in scheme (a). That is, $t_B = 0.5\ x'$.

This change causes t_R to also become shorter. In scheme (b), $t_R = t_A + 0.5x'$. If t_A is small relative to x', then t_R is equal to approximately $0.5y$. Thus, the rate of the reaction doubles.

In other words, a doubling of the reactant concentration results in a doubling of the reaction rate. This occurs because $t_B > t_A$.

However, a very different situation exists if $t_B \leq t_A$. In that case, an increase in the concentration will not result in an equivalent increase in the reaction rate. The reason for this is illustrated in Figure B2, which is the same as Figure B1, but with $t_B \leq t_A$.

In scheme (c) in Figure B2, the reactant concentration is much higher, which means that t_B is short. A direct and simple comparison of t_B with Figure B1 suggests it is about one tenth as short as the t_B in scheme (a) in Figure B1. Thus, the reactant

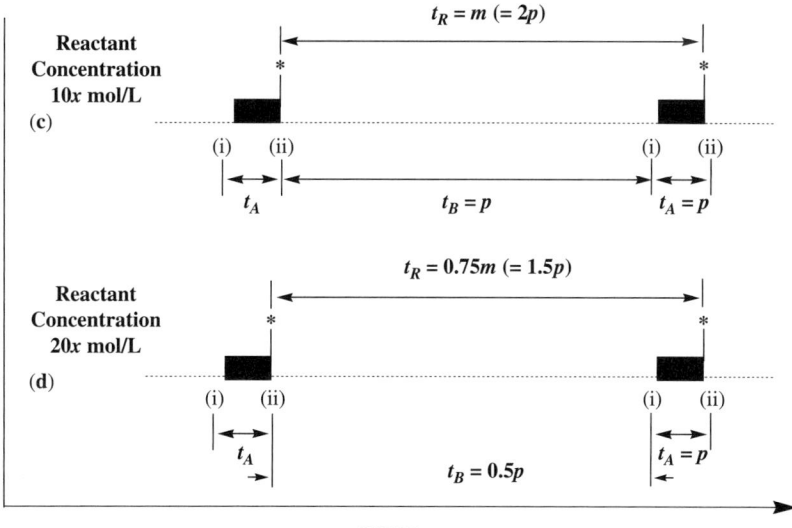

Figure B2. Transit times in a catalyst system displaying conventional kinetics within a concentrated solution of reactant.

concentration in (c) is about $10x$ mol/L. As shown, t_B in scheme (c) is approximately equal to t_A. For convenience, we will label $t_B = t_A = p$.

The high concentration also means that t_R is short. That is, the reaction rate is very rapid. We will label it $t_R = m = 2p$.

When the reactant concentration is doubled from (c) to (d) above, it increases to $20x$ mol/L, which means that t_B must necessarily halve. Thus, $t_B = 0.5p$ in (d). However, t_R does not halve. It decreases only to $t_R = t_A + t_B = p + 0.5p = 1.5p = 0.75m$. Thus, the system is starting to saturate. That is, a doubling in the reactant concentration does not result in an equivalent doubling in the reaction rate. Instead, the reaction rate increases by only 1.5-fold.

Additional increases in reactant concentration will elicit ever-smaller increases in reaction rate. Complete saturation will occur when t_B approaches zero. That is, when all of the active sites are occupied all of the time. This will occur only at a *high absolute concentration* of the reactants.

In effect, these figures show that catalysts displaying *conventional kinetics* start becoming saturated only when the average lifetime of form (B) approaches or is smaller than the average lifetime of form (A). At this stage, incremental increases in reactant concentration [with parallel decreases in the lifetime of form (B)] have an ever-decreasing effect on the overall rate. This will occur only at a high absolute concentration of reactant for most catalysts.

B2 MICHAELIS–MENTEN KINETICS

Now consider Michaelis–Menten kinetics. This property originates from the rapidly equilibrating nature of the catalyst–reactant complex (C = catalyst, R = reactant, P = product):

$$C + R \xrightleftharpoons{K} [CR] \xrightleftharpoons{k_2} C + P \qquad 5.2\text{(a)}$$

In the Michaelis–Menten scheme, the catalyst undergoes binding and dissociating of the substrate before product formation. Thus, a third form of the catalyst exists in addition to forms (A) and (B):

(C) A *bound, inert form*, in which the catalyst binds the reactant only to then release it before reaction (depicted as a thick greyline in the figures below).

Each catalyst will undergo a certain number of bind-release steps, on average, before the step of product formation is undertaken.

To illustrate saturation in such a system, consider Figure B3. The existence of form (C) above is indicated by the thick gray lines between each of the points (iii) and (iv). For convenience, we will say that the average lifetime of form (C) is the same as that of form (A) on each occasion of binding; that is, it is t_A.

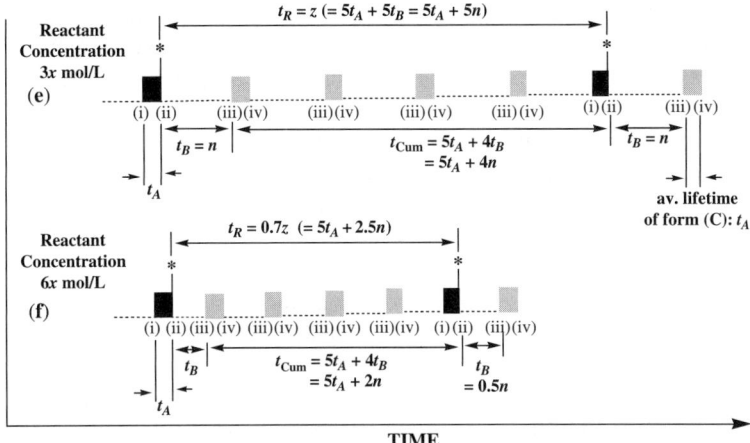

Figure B3. Transit times in a catalyst system displaying Michaelis–Menten kinetics within a concentrated solution of reactant.

A totally different sequence of events occurs in this case, compared with the earlier case.

Starting on the left of (e) in Figure B3, a reactant binds the catalyst at (i) and is converted to product at (ii) [form (A) is indicated by the thick black lines].

At point (iii), another reactant binds the catalyst. It is released without reaction at point (iv). Thus, between points (iii) and (iv), form (C) above exists (thick gray lines). Four more bind-release steps (iii) to (iv) then occur; these are indicated by the thick gray lines. For this particular catalyst, we will decide that four bind-release steps need to occur, on average, before product formation. Finally, productive binding occurs at the second point (i), with release of the next product at the second point (ii).

The first thing to be noticed is that t_A in scheme (e) is much shorter than t_A in the previous figures. This is because binding is more transient in this type of kinetics than it is in conventional kinetics. The second thing to notice is that $t_B > t_A$. In scheme (e), t_B is the time between points (ii) and (iii) or between points (iv) and (iii). In (e), we will set $t_B = n$.

As in the previous examples, t_B is inversely related to the reactant concentration. A simple and a direct comparison with t_B in scheme (a) of Figure B1 suggests that it is about one third of the size. That puts the reactant concentration at ca. $3x$ mol/L in (e).

The time between the first point of reactant binding in each reaction sequence, point (iii), and the next event of product release, point (ii), is cumulatively t_{Cum}. In scheme (e), $t_{Cum} = 5t_A + 4t_B = 5t_A + 4n$.

The time between product releases in scheme (e) is, consequently, $t_R = z = 5t_A + 5t_B = 5t_A + 5n$. The catalyst will display a rate consistent with such a t_R.

Now, we will double the reactant concentration to $6x$ mol/L in scheme (f), which means that t_B must necessarily halve to $t_B = 0.5n$. The time between product releases in scheme (f) therefore becomes $t_R = 5t_A + 5t_B = 5t_A + 2.5n$.

If $t_B = 2\, t_A$, this means that $t_R = 0.66z$, not $0.5z$. That is, a doubling of the reactant concentration leads to only a *1.5-fold* increase in rate. Thus, the system is already starting to saturate, even though t_B is much larger than t_A and the concentration is very much lower than that in scheme (d) above.

Systems involving equilibrating intermediates, therefore, tend to saturate at lower reactant concentration than systems displaying conventional kinetics.

The analysis above does not take into account several factors that may be present in enzymes. For example, the times between points (iv) and a following point (iii) in (e) and (f) may be substantially shorter than t_B if the same substrate is released and then bound again. Nor does it take account of the fact that t_A may be longer in enzymes than in simpler catalysts because of the influence of the many different individual binding interactions that are possible. These factors may be expected to amplify the effects noted above and to accelerate the onset on saturation.

In enzymes, saturation is observed when the frequency of the bind-release steps prior to product release cannot go any faster. Because the cumulative time needed for these bind-release steps is so much longer than the time period between product release and the binding of the next reactant to the catalyst, the system then saturates.

Thus, enzymes become saturated at a relatively low reactant concentration, especially if a small concentration of catalysts is used.

INDEX

Activation
  of a reactant, 69, 71, 74, 84, 89, 107, 110, 111, 113, 116, 117, 118, 119, 121, 123, 124, 125, 128, 139, 141, 142, 143, 149, 150, 158, 172, 176, 183, 184, 202, 212, 213, 214, 222, 226, 229, 230, 234, 238, 239, 252
  (deactivation), 254, 260, 263, 270, 271
Activation energy
  concept of, (see Glossary), 17, 18, 19, 24, 26, 29, 30, 48, 57, 58, 59, 60, 61, 65, 66, 67, 68, 69, 70, 72, 73, 75, 77, 79, 80, 82, 86, 88, 89, 94, 95, 98, 99, 100, 105, 107, 113, 127, 131, 132, 133, 140, 148, 149, 165, 167, 170, 171, 172, 173, 174, 176, 182, 187, 211, 215, 270, 299
Adsorption theory
  of enzymatic catalysis, 163
Arrhenius plots
  curvature of, 65, 66, 67
Aryl couplings
  Pd-catalyzed, 46, 112

Benkovic, Stephen, 24, 100
Bi-directional catalysis, 113, 114, 117, 129, 132, 133, 155, 169, 195
Biomimetic chemistry (Breslow), 49, 100, 155, 156, 197, 198, 219, 220, 221
Biomimicry, 221
Bis(cinchona)
  alkaloids, as time-dependent catalysts, 188, 189, 195, 196, 197, 205
Boltzmann
  distribution in high-/low-pressure limit of Hinshelwood-RRK theory, 62, 63

Carbenoid reactions
  catalysts of, 189, 249
Carbonic anhydrase, 41, 73, 114, 210
Cascade (see Glossary), 15, 19, 20
Catalytic groups, 25, 27, 31, 41, 42, 43, 46, 48, 49, 101, 111, 112, 113, 116, 117, 118, 119, 121, 122, 123, 124, 127, 128, 129, 131, 132, 141, 143, 144, 146, 148, 149, 150, 151, 153, 157, 158, 190, 191, 192, 193, 194, 195

Catalytic groups (*Continued*)
    196, 197, 198, 200, 202, 203, 220,
    221, 222, 223, 224, 225, 226, 227,
    228, 229, 230, 231, 232, 233, 234,
    235, 237, 245, 248, 250, 254, 255,
    259, 261, 270, 271, 298, 300, 301,
    313, 316, 320, 321, 322, 323
Causality, 5, 34
Chaos (see Glossary), 9, 11, 13, 35
Cinchona, catalysts (Corey)
    as examples of time-dependent catalysts,
        188, 189, 195, 196, 197, 205
Co tetraphenylporphyrin, catalysts, 323,
    324, 326, 328, 329, 331, 334
Cofacial, porphyrin catalysts
    (Collman), 43, 235–240, 321
    others, 43, 240, 260, 262
Combinatorial
    catalysts, 31, 34, 216, 226, 227, 228, 229,
        233, 235, 269, 300, 301, 308, 312,
        313, 316, 321, 322, 326, 334
    experiments, 31, 214, 215, 216, 226,
        227, 228
Complementarity/Complementary,
    (see Glossary)
    functional (see Glossary), 115, 116, 190,
        192, 194, 195, 202, 205, 218, 220,
        229, 230, 234, 252, 257, 259
    structural, for the transition state, 27, 31,
        42, 48, 49, 51, 99, 111, 112, 113,
        117, 119, 121, 122, 125, 127, 128,
        129, 131, 132, 144, 146, 148, 149,
        154, 155, 158, 162, 164, 165, 167,
        169, 174, 175, 176, 194, 195, 196,
        197, 198, 200, 213, 214, 220, 221,
        222, 223, 225, 226, 227, 270, 298
    synergy, 30, 115, 116, 190, 192, 194,
        195, 201, 202, 248
Complex Systems Science (see Glossary),
    14, 115, 187, 190
Complexity
    feedback loops, 14, 16
    hierarchies, 14
    memory, 14
    nonequilibrium, 7, 8, 14, 62, 68, 75, 90,
        94, 154
    nonlinear, 14, 27, 127, 154, 195, 228
Convergence/Convergent, (see Glossary)
    actions, 193

coincidental, 193, 194, 198, 199
functional (see Glossary), 115, 116, 129,
    132, 192, 194, 195, 201, 202, 203,
    204, 205, 218, 220, 230, 233, 235,
    237, 241, 243, 244, 248, 249, 252,
    254, 258, 260, 261
"pseudo" (see Glossary), 197, 198, 199,
    201, 203, 204, 205, 245
Coupled protein motions, 24, 30, 56, 100,
    106, 168, 170, 171, 173
Cu–Fe, catalysts, 43
Cubane, catalysts of water-oxidation
    (Dismukes), 272–278, 282–294
Cyclodextrin, catalysts containing
    (Breslow), 49–52, 198–200,
    245–248
Cyclophilin A, 168

Determinism/deterministic, 5, 7, 11, 13, 15,
    138, 153
Deterministic chaos, 11, 13
Dicobalt, catalysts, 43, 235–240,
    321–334
Dicopper, catalysts, 43
Diels–Alder
    time-dependent catalyst of, 188, 189
Diffusion-control/led, 19, 67, 72, 73, 77,
    82, 210, 211
Dihydrofolate reductase, 24, 56, 100
Dihydroxylation catalysts
    of olefins, 189, 195–197, 240
Diiron, catalysts (μ-carboxylate
    bridged), 43
Dimanganese, catalysts, 43, 240
Dinuclear Schiff base, catalysts, 43
Dirhodium, catalysts, 189, 248, 249
Dye-sensitized, solar cells, 279–282,
    291, 292

Effective molarity, 166, 167, 169, 170, 172,
    173, 176, 258
Electronic activation
    mechanism of synergy in heterogeneous
        catalysis, 184
Eley–Rideal
    mechanism of catalysis, 71
Encounter (chemical term; see under
    "Collision Frequency" in
    Glossary), 56

INDEX

Energy hypersurface, 11, 61, 62
Energy-dependent catalysis
  Process of, 141–146
Enthalpy, 5, 6, 10, 90, 156, 268
Entropy
  concept of, 5, 6, 10
  "trap", 168, 169, 170, 172
Enzymatic catalysis
  brief description of, 41, 42, 99–100
Epoxidation, catalysts of, 49, 50, 51, 101, 102, 198, 199, 200, 247, 248, 252, 257, 258, 259, 260
Exceptions, to Sabatier's Principle, 93
Eyring, Henry, 23, 28, 61

Facial selectivity, 184, 186, 189, 203
Free Energy, 5, 10, 85, 114, 172

Glucose dehydrogenase, 24
Glycerol kinase, 115
α-Glycerolphosphate, 115
Goodness-of-fit, 51
Gravitational potential energy/field, 1, 3, 8, 10, 137
Grignard, catalysts, 112

Heterogeneous, catalysis/catalysts, 28, 30, 31, 32, 37–42, 48, 69, 71, 77–96, 104, 105, 106, 107, 117, 124, 129, 163, 181–184, 197, 209, 210, 211, 212, 214, 215, 216, 217, 218, 227, 253, 254, 257, 261, 279, 297, 312, 315, 316, 321, 322
High-dilution
  as example of time-dependent reaction, 63, 64, 65
High-pressure limit
  of Hinshelwood–RRK theory, 23, 28, 59, 61
Hinshelwood–RRK, theory, 23, 28, 59, 61, 62, 68
Horseradish peroxidase, 24
Hybrid Homogeneous–Heterogeneous, catalyst, 40–41, 197, 227, 315, 316, 321, 322
Hydride, abstraction of, 275
Hydroformylation, catalysts of, 248, 249
Hydrogen bonds/-ing, 49, 101, 110, 111, 118, 214, 234

Hydrogenation, catalysts of, 90, 252–254
Hydrolysis
  time-dependent catalyst of, 188

"Ideal" catalyst, 28, 78, 89, 107, 158, 181, 183, 203, 204, 205, 320
Induced-fit, theory (Koshland), 99, 165
Insertion, reaction, 44, 45, 185
Intangible, 7, 10, 11, 75, 137
Intermolecular, catalysts/reactions, 166, 167, 168, 169, 233, 235, 252, 259–261
Internal energy, 5, 6, 7, 10
Intramolecular, catalysts/reactions, 100, 103, 143, 146, 156, 165, 166–167, 168, 169, 170, 171, 172–173, 174, 176, 193, 231, 232, 233, 234, 235–259, 262, 300, 312
Intramolecularity, 165–167, 169, 171, 172–173, 174, 176, 193
Isotactic, polypropylene, 44, 185, 186, 189

Kinetic perfection
  in a catalyst, 73, 115, 210, 211
Koutecky–Levich, equation, 326, 327

Langmuir–Hinshelwood
  mechanism of catalysis, 69, 71
"Lock-and-key", theory, 163
Lorenz, Edward, 13
Low-Pressure limit
  of Hinshelwood–RRK theory, 28, 59, 62, 68

Mechanical
  abiological catalyst, 132, 240
  biological catalyst, 150, 151, 153, 154
  process, 2–5, 7–17
Mechanics, applications including
  acoustics, 8
  astrodynamics, 8
  celestial mechanics, 8
  fluid mechanics, 8
  hydraulics, 8
  statics, 8
Mechanics, laws of, 7
Mechanics, statistical, 6

350

Mechanics, theories including
  classical, 4
  quantum, 5
  relativistic, 5
Menger, Frederic, 24, 100, 169, 170, 176
Metal cluster, catalysts, 43, 183, 252–254, 261, 269, 270
Michaelis–Menten, kinetics, 26, 87, 94, 98, 101–109, 116, 129, 151–153, 182, 188, 189, 195, 228, 249, 287, 338, 339, 344
Mimicry, of enzymatic catalysis, 42, 43, 155–156, 175, 176, 197, 221, 223, 225
Mn, porphyrin catalysts, 49, 50, 240
Molecular recognition, 51, 161, 196, 197
Multicentered, catalysts
  class A type, 230, 231, 235–241
  class B type, 230, 231, 240, 241–245
  definition of and conventions with, 44, 46, 47, 48, 101, 103
  examples, intermolecular
    functionally complementary, 259
    functionally convergent, 260–261, 301–316, 321–334
  examples, intramolecular
    clusters, 253, 254, 269, 270, 282–291
    functionally complementary, 252
    functionally convergent, 235–245
    polymers, 255, 257, 302, 316, 322, 334
    possible convergent, 250
    probable convergent, 247
    pseudoconvergent, 245–248
  processes in, 111, 112, 115, 117–122, 123, 129, 130, 134, 139–155, 212–218
  rational design of, 230–235
  synergy in, 190–205
Multicentered activation
  mechanism (of synergy) in heterogeneous catalysis, 183
Multifunctional, catalysts, 231
Multisite, 39, 48, 184, 213, 216
Mutual enhancement/mutually enhancing (synergy) (see Glossary), 184, 187, 188, 189, 190, 201, 202, 218

Nafion, 34, 262, 272, 278, 279, 283–292, 292–293

Orbital steering, theory, 167, 170
Overpotential, 244, 281, 285, 286, 293, 297

Path-dependent, 9, 24, 26, 55, 138, 153, 182
Pauling, Linus, 28, 42, 99, 100, 107, 148, 155, 161, 162, 165, 169, 170, 174, 175, 197
Photocurrent, 284–290, 292
Photoelectrochemical, cells, 269, 279–282
Photosynthesis, 269
Photosystem II, 33, 226, 253, 267, 268, 269, 271, 293
Polarization curves, for water electrolysis, 306
Polymer, -supported catalysts, 41, 198, 235, 239, 254–259, 260, 279, 282–291, 292, 300, 302–316
Polypyrrole, 34, 300, 302–316, 321, 323–334
Potential energy, 1, 3, 10, 60, 61, 65, 68, 69, 88, 110, 150, 281
Precursor
  mechanism of catalysis, 69, 71
Propinquity effect, 168
Proximity effect, 168
Pseudo-convergent, type of synergy, 197–200, 202–205, 245–248

Rational design, of time-dependent homogeneous catalysts, 25, 31, 32, 157, 182, 183, 203, 219, 221–230, 301
Reaction-controlled, 73, 210, 211
Residence time τ, 79, 82, 83, 84, 85, 86, 87, 88, 89, 92, 93, 104, 109, 124, 129, 139, 146, 171, 211, 214
Rh
  Dirhodium catalysts, 189, 249
  phosphine hydroformylation catalysts, 247
Rotating disk electrochemistry, 328
Rotating ring disk electrochemistry, 328, 329

Ru
  dodecacarbonyl hydrogenation catalysts, 253
  metallocenophanes, 244
  water oxidation catalysts, 233, 250, 251, 279, 281

Sabatier's principle, 29, 77, 78, 79, 89, 90, 93, 94, 95, 204, 205
Sachtler–Fahrenfort, plot, 82, 90
Saturation, kinetics of, 87, 89, 102, 103, 104, 105, 108, 109, 110, 132, 133, 189, 196, 285, 286, 287, 308, 337–346
Selectivity, 39, 41, 42, 49, 92, 114, 128, 132, 156, 184, 185, 186, 188, 189, 190, 195, 196, 197, 198, 199, 200, 202, 203, 204, 205, 227, 228, 238, 239, 245, 246, 247, 248, 255, 259
Self-assembled, catalysts, 261
Single-centered, 44–49, 184–190, 202, 203, 212, 215, 217, 218, 247, 259, 262
Single-site, 39, 42, 46, 48, 213, 217
Spatiotemporal
  action, 9, 16, 24, 26, 55, 56, 138, 153, 182
  hypothesis (Menger), 24, 30, 100, 169, 170, 171, 176
Spillover
  mechanism (of synergy) in heterogeneous catalysis, 184
Statistical, mixture of products, 65
"Statistical proximity", catalyst, 224, 226–229, 233, 235, 300, 301, 308, 312, 313, 316, 321, 322, 326, 334
Stochastic (see Glossary), 11, 13, 23, 24, 26, 55, 56, 100, 138
Strain, theory (Haldane), 164, 165, 171
Stress, theory (Fersht), 165, 171
Structure-insensitive, catalysts, 111, 121, 146
Structure-sensitive, catalysts, 112, 122, 129, 132, 148, 195, 196
Supramolecular catalysis, 245
Syndiotactic, polypropylene, 44, 185, 186, 189
Synergy, in heterogeneous catalysts, 183, 184

Tanaka–Tamaru, plots, 90
Thermodynamic
  Equilibrium, 5–7
  laws of, 5–7
  Process, 1, 2, 5–7, 9–17
Thermosynechoccus sp. (cyanobacterium), 269
Time-dependent catalysis
  as a consequence of combinatorial experiments, 214, 215, 224, 226–230, 260–261, 297–316, 319–334
  process of, 116–128, 130–133, 146–155, 156–157, 194–197, 235–245
"Time required for transition state formation" t_{TS}, 71, 73, 74, 108
Transit times, 83, 122, 341
Transition state, theory (Eyring), 23, 29, 61–62, 68, 75, 90, 94, 153, 182
Triad, enzymatic catalyst, 112
Triosephosphate isomerase, 73, 168, 210
Turnover frequency, 247, 288, 289

Unification
  of heterogeneous, homogeneous, and enzymatic catalysis, 157, 159, 209–218
  of theories of enzymatic catalysis, 161–177
Unimolecular, reactions, 58, 62, 166

Vapor-phase, polymerisation, 323, 324, 326, 330
Volcano, plot, 79–91, 92, 93, 98, 204, 205

Water oxidation, catalysts of, 33, 233, 250, 251, 252, 260, 278, 279, 281, 282–291, 292–294
Water-oxidizing center/complex, in Photosystem II, 33, 226, 267, 269, 271, 272
Water-splitting, 280, 292
Williams, R. J. P. (Bob), 24, 100, 169
Wurtz, cyclization as example of time-dependent reaction, 63, 64, 65

Zirconocene, polymerisation catalysts, 45, 46, 185